Advanced Microsystems
for Automotive Applications 2012

Gereon Meyer (Ed.)

Advanced Microsystems for Automotive Applications 2012

Smart Systems for Safe, Sustainable and Networked Vehicles

Springer

Editor
Dr. Gereon Meyer
VDI/VDE Innovation + Technik GmbH
Berlin
Germany
E-mail: gereon.meyer@vdivde-it.de

ISBN 978-3-642-43146-3 ISBN 978-3-642-29673-4 (eBook)
DOI 10.1007/978-3-642-29673-4
Springer Heidelberg New York Dordrecht London

Softcover reprint of the hardcover 1st edition 2012

Printed on acid-free paper

Springer is part of Springer Science+Business Media (www.springer.com)

Preface

Road transport is facing a multitude of challenges, ranging from the need to increase fuel efficiency and to reduce greenhouse gas emissions to the goal of accident-free traffic. Intensive automotive research and development efforts of recent decades lead to major innovations such as efficient engines and electronic stability control, technologies which are widely deployed in today's cars. Nonetheless, further and even more revolutionary progress will be required for the automobile of the future to provide mobility solutions which are sustainable in view of the global issues of societal, economic and environmental development. It can be foreseen that this implies a paradigm shift in the concept of the car regarding its propulsion technology, materials, and architecture, and calls for an intelligent integration into the systems of transportation and power.

Information and communication technologies (ICT) like smart components and systems have been essential for a multitude of recent innovations. They will be key enabling technologies for the changes ahead, both inside the vehicle and at its interfaces for the exchange of data and power with the outside world. Data from sensors, cameras, transmitters and road maps can be combined to warn the driver of imminent threads like lane departures or collisions. Such advanced driver assistance systems, subject to a multitude of recent research projects, are now available to contribute to the safety and comfort of contemporary cars. They pave the way to more automated and maybe even autonomous driving. At the same time, electronic circuits, digital controls and wired or wireless communication can replace mechanical systems to enable greater precision and adaptivity of vehicle steering and power train control. Hybrid and electric vehicles are particularly qualified to benefit from the application of smart systems which may lead to substantial gains in energy efficiency and driving range. A complete redesign of architecture of the car may thus be needed to make electrified vehicles mature for mass deployment. Synergies from combining innovations in different technology fields, like e.g. driver assistance and electrification, will potentially accelerate such developments.

It has been the objective of the International Forum on Advanced Microsystems for Automotive Applications (AMAA) for almost two decades to detect such novel trends and to discuss their implications for the innovation of road transport and for securing the leadership of the involved industries from day one on. The topic of the 16th AMAA conference, held on 30 and 31 May 2012 in Berlin, is "Smart Systems for Safe, Sustainable and Networked Vehicles". Organizers of the AMAA are VDI/VDE Innovation + Technik GmbH and the European Technology Platform on Smart Systems Integration (EPoSS), supported by the two EU-funded Coordination Actions of the Public Private Partnership European Green Cars Initiative, ICT4FEV and CAPIRE.

As AMAA editor and conference chair I would like to express my gratitude to all authors for the time and efforts they spent on their contributions. The conference papers published worldwide in this book summarize the excellent results and findings of most recent research and development in the fields of ICT, components and systems and other key enabling technologies for the automobile and road transport of the future. The application fields considered include electrification, power train and vehicle efficiency, safety and driver assistance, networked vehicles, as well as components and systems.

I feel very fortunate to have had the great help of the members of the AMAA Steering Committee in selecting the contributions for the conference programme. Furthermore, funding by the European Commission and continuous support by a multitude of industrial partners is greatly acknowledged.

It should be pointed out that the AMAA 2012 conference and the book at hand result from the tireless efforts of a team of highly committed colleagues at VDI/VDE-IT. Great appreciations should be given to Iohanna Gonzalez for running the AMAA office in an efficient, reliable and proactive manner, as well as to Anita Theel and René Stein for the brilliant work of preparing the master copy of this book. Special thanks also go to Beate Müller and Jan-Henrik Fischer-Wolfarth for their big help with regards to content. Last but not least, I am particularly grateful to Wolfgang Gessner and Jürgen Valldorf for continuous encouragement and inspiring discussions.

Berlin, May 2012

Dr. Gereon Meyer

Funding Authority

European Commission

Supporting Organisations

European Council for Automotive R&D (EUCAR)

European Association of Automotive Suppliers (CLEPA)

Strategy Board on Electric Mobility (eNOVA)

Advanced Driver Assistance Systems in Europe (ADASE)

Zentralverband Elektrotechnik- und Elektronikindustrie e.V. (ZVEI)

Mikrosystemtechnik Baden-Württemberg e.V.

Hanser Automotive

Organisers

European Technology Platform on Smart Systems Integration (EPoSS)

Coordination Action "Information and Communication Technologies for the Fully Electric Vehicle"

Coordination Action "PPP Implementation for Road Transport Electrification"

VDI|VDE Innovation + Technik GmbH

Honorary Committee

Eugenio Razelli	President and CEO, Magneti Marelli S.P.A., Italy
Rémi Kaiser	Director Technology and Quality Delphi Automotive Systems Europe, France
Nevio di Giusto	President and CEO Fiat Research Center, Italy
Karl-Thomas Neumann	Executive Vice President E-Traction Volkswagen Group, Germany

Steering Committee

Mike Babala	TRW Automotive, Livonia MI, USA
Serge Boverie	Continental AG, Toulouse, France
Geoff Callow	Technical & Engineering Consulting, London, UK
Bernhard Fuchsbauer	Audi AG, Ingolstadt, Germany
Kay Fürstenberg	Sick AG, Hamburg, Germany
Wolfgang Gessner	VDI|VDE-IT, Berlin, Germany
Roger Grace	Roger Grace Associates, Naples FL, USA
Klaus Gresser	BMW Forschung und Technik GmbH, Munich, Germany
Riccardo Groppo	Fiat Research Center, Orbassano, Italy
Horst Kornemann	Continental AG, Frankfurt am Main, Germany
Hannu Laatikainen	VTI Technologies Oy, Vantaa, Finland
Jochen Langheim	ST Microelectronics, Paris, France
Günter Lugert	Siemens AG, Munich, Germany
Steffen Müller	NXP Semiconductors, Hamburg, Germany
Roland Müller-Fiedler	Robert Bosch GmbH, Stuttgart, Germany
Andy Noble	Ricardo Consulting Engineers Ltd., Shoreham-by-Sea, UK
Pietro Perlo	IFEVS, Sommariva del Bosco, Italy
Detlef E. Ricken	Delphi Delco Electronics Europe GmbH, Rüsselsheim, Germany
Christian Rousseau	Renault SA, Guyancourt, France
Jürgen Valldorf	VDI|VDE-IT, Berlin, Germany
Egon Vetter	Ceramet Technologies Ltd., Melbourne, Australia
David Ward	MIRA Ltd., Nuneaton, UK

Conference Chair:

Gereon Meyer	VDI|VDE-IT, Berlin, Germany

Table of Contents

Electrified Vehicles

Safety & Driver Assistance

Networked Vehicle

Components & Systems

Electrified Vehicles

Battery Management Network for Fully Electrical Vehicles Featuring Smart Systems at Cell and Pack Level

A. Otto, S. Rzepka, T. Mager, B. Michel, Fraunhofer ENAS
C. Lanciotti, Kemet Electronics Italia Srl
T. Günther, O. Kanoun, Chemnitz University of Technology

Abstract

Current limitations of battery systems for fully electric vehicles (FEV) are mainly related to performance, driving range, battery life, re-charging time and price per unit. New cell chemistries are able to mitigate these drawbacks, but are more prone to catastrophic failures due to a thermal runaway than current solutions. Therefore, new and more advanced management strategies are necessary to be able to safely prevent the energy storage system from ever coming into this critical situation. In this paper, a novel battery management system (BMS) architecture is introduced, which will be able to meet these high requirements by introducing a network that has smart satellite systems in each macro-cell or even in each individual cell. In addition, the issues of safety and reliability to be considered for the integration and packaging technologies of the smart satellite systems will be described as well. The work reported is part of the European project ‚Smart-LIC', which is supported by EPoSS, and the German project ‚HotPowCon'.

1 Introduction

It is widely accepted that electrified mobility offers much potential for the reduction in greenhouse gas emission, primary energy consumption, and traffic noise. However, current hurdles are mostly related to the electrochemical energy storage system itself. Thus, great effort is made to further improve energy and power density, efficiency, safety, lifetime and cost of the battery cells. Besides a proper selection of materials, improved production processes, and advanced mechanical designs, great importance is also attached to the electronics being responsible for managing the battery by dedicated monitoring and controlling activities, i.e., to the so called 'battery management system' (BMS). It needs to assure the safety of the battery and the user, shall extend the battery lifetime, and keep the battery in the desired operating status, i.e. all time ready to use. The concrete implementation concept depends on actual cell chemistry as well as on the application scenario. It can range from occasional manual readings of single parameters to complex schemes of fully automatic online monitoring and controlling. In case of FEVs, the main functionalities and features of a BMS can be summarized as follows:

- Charge management and balancing
- Demand and power management
- Monitoring of various cell & battery parameters
- Computation and determination of derived state variables, like state-of-charge (SoC), state-of-health (SoH) and state-of-function (SoF)
- Data storage for cell and battery models, lookup tables, log book functionalities etc.
- Thermal management
- Crisis management
- Communication to central computing unit of the vehicle

An advanced management of the power and energy flow within the battery system as well as within the vehicle is of utmost importance to finally ensure an efficient usage of the limited driving range. This is particularly noteworthy in case of FEV batteries, since the energy capacity and lifetime of such complex multi-cell systems strongly depend on how equally the charge is distributed among the cells.

Monitoring the relevant cell and battery parameters like cell voltages, in- and out-going currents, electrical impedance, and temperature is mandatory for the determination of SoC, SoH, and SoF. It allows predicting the future behaviour of the battery and thus enables a prognosis of the vehicle's remaining driving range, which is one of the most essential inputs for the FEV driver.

Furthermore, based on the battery state information, the thermal management will be able to keep the battery always in optimal temperature range, while the crisis management will be responsible for preventing critical situations early enough and for triggering fast and effective counteractions such as emergency warnings to the FEV driver or intensification of the cooling system. Especially high temperatures will reduce the battery lifetime and can even trigger a thermal runaway, which can lead to fire or explosion and needs to be avoided under all circumstances [1]. At the same time, it should be noted that the hazard potential also strongly depends on the battery size and energy density which can be expected to increase in the future, when new cell chemistries such as lithium sulphur (Li-S) will be introduced. Thus, safety related management activities and strategies will become more and more important for future BMS to be able to meet the high safety standards being demanded in the automotive sector.

2 Novel BMS Solution

2.1 System Architecture

Current battery systems are usually composed of large numbers of electrochemical cells, mostly based on lithium compounds, which are organized in different cluster levels and compositions, e.g. in parallel and/or in series, depending on the specific requirements for voltage and current ratings. A battery system can be equal to a battery pack, or in case of large systems, as could be found in FEVs, realized by several battery packs being connected together. Each battery pack contains battery modules, cooling and interconnection system and the BMS electronic. State-of-the-art in BMS is one central unit per battery pack with a clear trend to distribute some of the BMS functionalities, like dedicated monitoring and balancing activities, to the next lower level, i.e. to the module. A possible hierarchical constellation with its different packaging levels is depicted in Fig. 1, but many different variations exist especially at module level. Note that a mono-cell or elementary cell corresponds to the basic cell chemistry, whereas a common battery cell is built up by a stack of many mono-cells (e.g. 20).

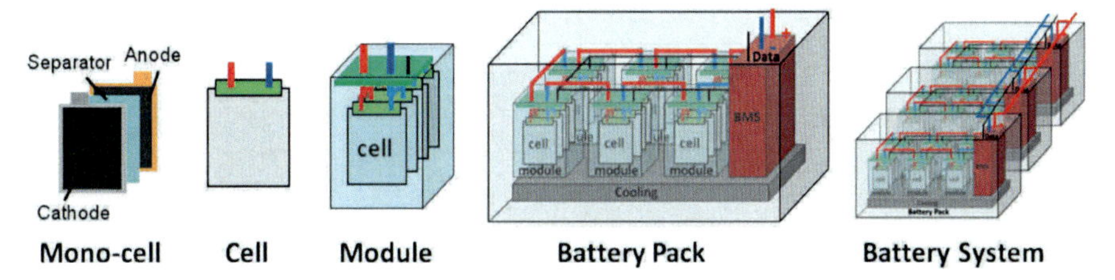

Fig. 1. Hierarchy of different packaging levels of a typical FEV battery system.

The approach presented in this paper intends to continue the trend of distributed functionality by adopting a higher partition of the BMS, i.e., by introducing smartness at cell level. Dedicated BMS features will directly be integrated into the lithium-ion cells in order to enable a more advanced and efficient cell and battery management. The principle idea of the envisaged smart cell as well as a state-of-the-art cell is depicted in Fig. 2. In addition to passive safety features like over-pressure valves and current interrupt devices (CID), which can already be found in common lithium-ion cells, a compact and smart system with advanced functionalities will now be directly integrated into the cell. This smart satellite system will basically consist of the following components:

- Sensing elements (for measurement of voltage, current, temperature, electrochemical impedance spectroscopy, and pressure)

- Actuators for balancing and switching purposes
- Signal conditioning circuitry
- Data (pre-) processing and storage
- Communication interface

As a consequence of this new system architecture, each cell can now be monitored and managed individually, which opens the way to more rigorous cell and battery management. Moreover, key parameters like temperature or pressure, which are susceptible to change in case of failure or accident, can directly be acquired inside the lithium-ion cell without any delay. As a result, the crisis management and thus the overall safety will be improved to a level potentially even allowing new cell chemistries such as Li-S to be introduced for the sake of higher energy density providing longer driving range and lower battery weight without any effective increase in the safety risks for the passengers.

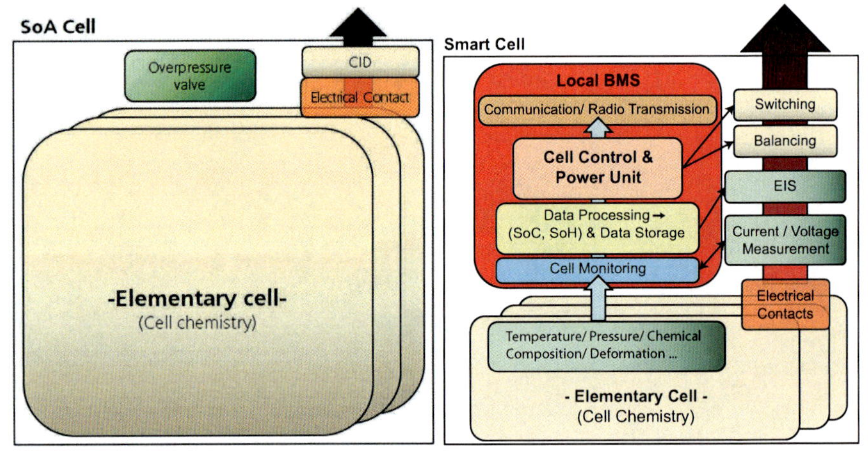

Fig. 2. Schematic drawing of a state-of-the-art lithium-ion cell with incorporated passive safety features (left) and of the proposed smart lithium-ion cell with incorporated smart system (right).

These ambitious goals are currently under development within the European project 'Smart-LIC'. The challenges to be overcome by this project are manifold and exist not only in technical, but also in economical terms. For this reason, one main objective of the project is to show the benefits of such a distributed BMS network, like improved safety and battery state determination or increased lifetime and flexibility, with respect to effort and initial costs. In addition, the reduction in total cost of ownership is a main project target.

In order to minimize the technical difficulties and to obtain a further system architecture variation for benchmarking purposes, an intermediate step in terms of a smart system being applied at macro-cell level will be included. In this project, the macro-cell consists of four pouch cells with 20Ah each that are connected in series. The maximum relative current is set to 5 C (100 A). A possible realization of the smart macro-cell is shown in Fig. 3. The local BMS electronic, consisting of a power module and a microcontroller board, will be mounted at the face side.

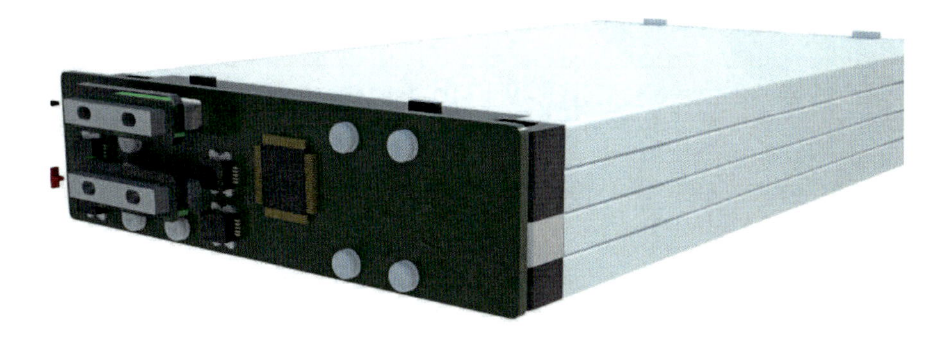

Fig. 3. Smart macro-cell.

The power module will host low-resistance semiconductor switches plus the driving stage for active balancing between the macro-cells and the power connectors. Apart from that, they will also be used as complementary active security devices to the already existing passive versions. This way, it will now become possible to disconnect individual (macro-) cells from the remaining battery system instead of needing to shut down whole modules and packs. Therefore, a high degree of safety for passengers and rescue teams is provided in case of failures or accidents. Simultaneously, a highly efficient 'limp home' functionality can be realized as well.

The electronic board in turn contains the microcontroller, sensing elements with signal condition circuitry and communication elements. In addition, switches plus resistors for passive balancing are implemented likewise. Thus, a hybrid active-passive balancing approach can be realized which provides an optimum between energy consumption and complexity.

2.2 Advanced Cell and Battery Monitoring Techniques

In addition to the measurement of cell parameters such as voltages, currents, and temperatures, the novel BMS solution will also bring in new measuring techniques and concepts like the monitoring of the safety relevant cell pressure and the implementation of the electrochemical impedance spectroscopy (EIS).

EIS is based on a small signal analysis. Different internal effects of the cell influence the system behavior in dependency on the perturbation signal frequency. Therefore, the complex impedance of the battery is measured over a wide frequency range to capture all the frequency dependent phenomena. The interpretation of the impedance spectra is fulfilled by matching a battery model to the measured data. The model parameters allow information extraction about physical and chemical states of the single cells [2, 3]. Furthermore, the derivation to high-level information like SoC and SoH is possible since changes in the spectra depend, e.g., on the state-of-charge (see Fig. 4) or on the cyclic age of the cell.

For determining the behavior of nonlinear systems, EIS is combined with other techniques like coulomb counting and open circuit voltage analysis, which enhance the voltage response prediction and support deriving high-level parameters as state-of-health.

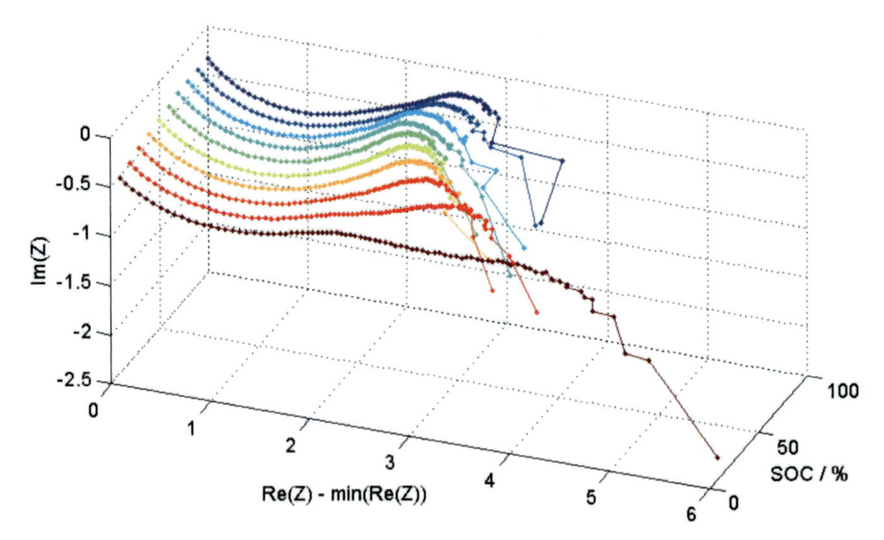

Fig. 4. Change of the EI-Spectra characteristics of a lithium-ion cell due to state-of-charge.

A main goal of the current work is to reduce the computational effort for measurement as well as for signal preprocessing and high level analysis in order to fit into a smart integrated system with its reduced computational capabilities.

2.3 Communication Strategies vs. EMC

The evaluation of different communication concepts for realizing a bi-directional communication link between the smart satellite cell systems and the central BMS unit is a further main objective within the 'Smart-LIC' project. Thereby, shielding and electromagnetic compatibility issues (EMC), caused by 'Signal and Power Integrity' (SI/PI), need to be taken into account carefully. Possible realizations are either wire-based (electrical or optical) or wireless, whereas different aspects should be considered in order to identify the optimum solution.

State-of-the art solutions are mainly based on wired CAN busses, but have the disadvantage of comparatively high costs and error-proneness. Another option would be the use of optical fibers, which are not affected by EMC issues, but lack of adequate reliability and durability. Power line communication techniques (PLC) via the battery poles would be a further alternative. However, they are quite complex in case of multi-cell and bi-directional communication and seem not to be commercial feasible from today's point of view.

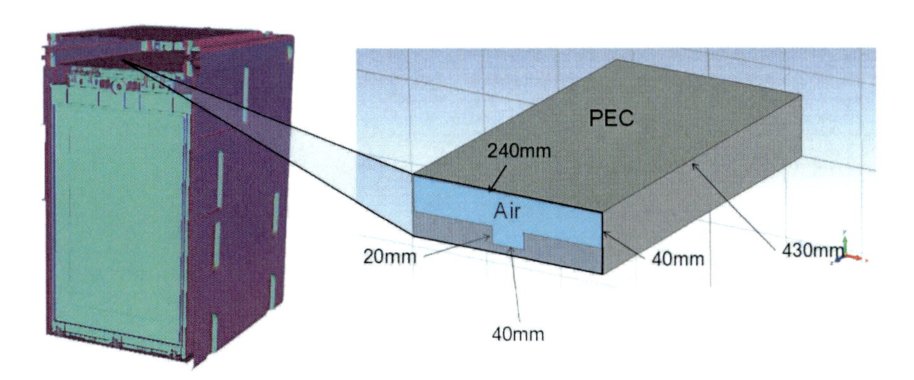

Fig. 5. Used geometry model for first investigations on different wireless communication approaches.

For this reason, wireless communication approaches will be investigated in particular, since they avoid excessive wiring, simplify maintenance and show no difficulties with different voltage levels. Thereby, three different options are conceivable:

- Transmission in existing electric cavities and channels
- Transmission in waveguide mode with open wedge-faces or absorber-faces
- Inductive near field communication

The model dimensions and results of the 3-D numerical simulations are shown in Fig. 6., respectively. The goal has been to determine the boundary conditions of different transmission options. This is crucial to ensure optimal communication coverage at all positions within the cell pack. Yet, further efforts are needed to fully comply with all EMC/EMI requirements for the reliable communication with signals of millivolt and milliampere next to spurious signals of several hundreds of volts and amps.

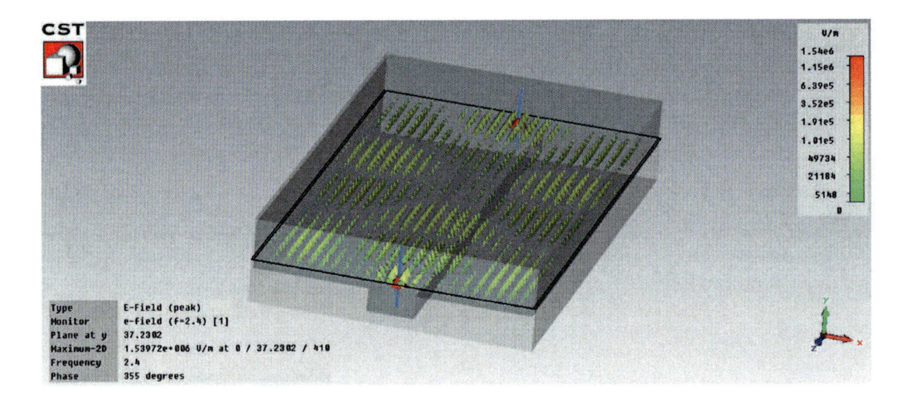

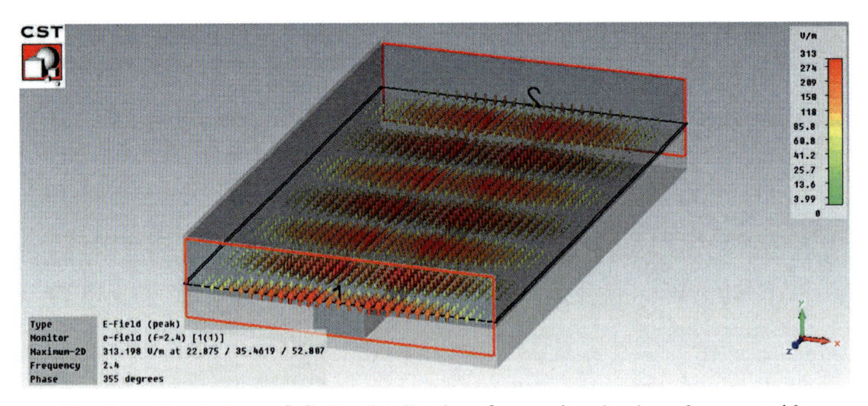

Fig. 6. Simulation of field distribution for cavity (top) and waveguide (bottom) approach, respectively.

2.4 System Integration

The system integration implies different aspects, and reaches from the packaging of the electronics to the integration of the smart (macro-) cells into the battery system. The former one is of particular importance here, since the packaging of the smart systems - especially in the 'smart cell' case - must be able to withstand harsh environmental conditions within the electrochemical cell including aggressive electrolytes, high temperatures in case of failure or mechanical stress. Still, reliability and thus the overall safety must not be compromised. For this reason, new overmolding techniques and materials are applied for a safe, cost-effective, and durable encapsulation [4].

Furthermore, integration concepts on how to bring the smart system directly in a lithium-ion cell during the fabrication process are of concern. Here, suitable solutions are mandatory to ensure that the production process will not get unduly difficult and that the functionality of the cell is not negatively affected by the assembly steps.

3 Addressing of Reliability Issues

Rather than defects in the electrochemical cells, failure mode analyses on existing battery systems have shown the electronic boards, which are supposed to manage them, as well as the wiring system to be the major weakness points [5]. Further introducing complexity to the battery system by adding the smart satellite systems, this issue becomes even more important, since reliability and lifetime of the BMS electronics is in fact tightly linked to the overall safety of FEVs, which needs to be guaranteed under all circumstances. To ensure this, the BMS shall be operational longer then the battery, whose lifetime is stretched till the end-of-life of the entire vehicle by current developments. Therefore, 15 ... 20 years are seen as realistic BMS lifetime target.

Reliability assessments usually involve accelerated tests by overstressing the test samples in a well defined way. This reduces the testing time compared to real life. However, the current tests mostly apply single loading factors like thermal cycles or mechanical vibration while complex environmental conditions exist in the real life. The new BMS satellites within the lithium-ion cell will even be exposed to additional factors like aggressive chemicals (electrolytes etc.). Testing all factors sequentially would really be time consuming, as accelerated thermal cycle tests typically take three full months already. In addition, the sequential procedure is not able to account for interactions between the failure factors and, hence, does not generate realistic failure modes fully comparable to the real life conditions. Therefore, new accelerated and combined reliability test methods are needed, which will be investigated within the 'Smart-LIC' project as well. This will be achieved by performing single loading factor tests independently, followed by various combined tests in order to finally obtain correspondence charts for the loading profiles with respect to the ageing mechanisms. Thus, the validity of superposition and the interaction occurring in complex loading situations can systematically be analyzed. As a consequence, the specification of new and more advanced tests to shorten the assessment periods, to trigger more realistic ageing behaviour leading to more accurate reliability predictions and, finally, faster time-to-marked for new BMS products can be achieved simultaneously.

Furthermore, there is a demand for BMS electronics being capable for still working at temperatures above 250°C, and hence, to be able to perform crisis management and safety switching activities when most needed, i.e. in case of serious malfunctions and even at the onset of thermal runaway, in which they must be able to stop the further spread of the crisis by shutting down and isolating the affected cells. State-of-the-art solder alloys used for electronic packaging do not provide the required thermal stability. Therefore, the German joint research project 'HotPowCon' has been set up to develop material systems, tools, and technologies for simultaneously joining sensor, control, and power devices to thermally stable substrates. Here, a new paste system containing tin and copper powders allows fabricating joints that finally consist of intermetallic compounds only with no eutectic regions left. However, the drawback of these temperature resistant materials is their high stiffness and brittleness. Therefore, design rules for electronic systems as well as joining processes must be further optimized in order to meet the required level of reliability.

4 Conclusions

The novel architecture proposed for BMS in FEV battery systems consequently continues the trend to more distributed functionality among the battery by bringing parts of the BMS directly into each lithium-ion cell or macro-cell. This will enable a more rigorous and sophisticated management of the overall battery system compared to current solutions. Hence, performance and efficiency of the battery will greatly be improved due to optimal usage of the given capacity, the reduced weight due to much less cabling, the increased lifetime of each cell and consequently of the whole pack. In addition, the safety is significantly increased, which opens the way to new cell chemistries with higher energy densities but inherently lower safety level. The added complexity and cell costs can be countered by highly integrated and cost efficient system in package

(SiP) technologies and the reduction in the overall cost of ownership for the FEV user due to increases in reliability and lifetime. A concrete demonstrator, to show the feasibility of this newly approach, is right now under development within the 'Smart-LIC' project. The new BMS architecture will in general be applicable to all cell types and chemistries.

Furthermore, joining technologies and materials with higher temperature stability as well as new reliability tests combining multiple loading factors are under development in order to better meet the high requirements concerning safety and reliability of the future FEVs to become well accepted by end users, i.e., the general public.

Acknowledgements

The authors would like to acknowledge the European Commission as well as the German ministry of education and research (BMBF) for supporting these activities within the project 'Smart-LIC' (project number: 284879) and 'HotPowCon' (project number: 13N11513), respectively.

References

[1] P.G. Balakrishnan, R. Ramesh, T. Prem Kumar, Safety mechanisms in lithium-ion batteries, Journal of Power Sources, Volume 155, 401–414, 2006.
[2] U. Tröltzsch, O. Kanoun, H.-R. Tränkler, Characterizing aging effects of lithium ion batteries by impedance spectroscopy, Electrochimica Acta, Volume 51, 1664–1672, 2006.
[3] P. Büschel, U. Tröltzsch, O. Kanoun, Use of stochastic methods for robust parameter extraction from impedance spectra, Electrochimica Acta, Volume 56, 8069–8077, 2011.
[4] M. Steinau, et al., New Concepts for System Integration in Smart Battery Management Systems and Transmission Control Units, 2011 EPoSS Annual Forum "Towards Smart Systems Products", Barcelona, Esp., Oct 6, 2011.
[5] V. Hennige, Real World testing: Reliability & Safety of Lithium Ion Battery Packs, At: Joint EC / EPoSS / ERTRAC Expert Workshop 2010 "Electric Vehicle Batteries Made in Europe", Nov 30, 2010.

Alexander Otto, Sven Rzepka, Bernd Michel
Fraunhofer Institute for Electronic Nano Systems ENAS
Department Micro Materials Center
Technologie-Campus 4
09126 Chemnitz
Germany
alexander.otto@enas.fraunhofer.de
sven.rzepka@enas.fraunhofer.de
bernd.michel@enas.fraunhofer.de

Thomas Mager
Fraunhofer Institute for Electronic Nano Systems ENAS
Department Advanced System Engineering
Warburger Str. 100
33098 Paderborn
Germany
thomas.mager@enas-pb.fraunhofer.de

Claudio Lanciotti
Kemet Electronics Italia Srl
Via San Lorenzo 19
40037 Sasso Marconi (BO)
Italy
claudiolanciotti@kemet.com

Thomas Günther, Olfa Kanoun
Chemnitz University of Technology
Reichenhainer Str. 70
09126 Chemnitz
Germany
thomas.guenther@etit.tu-chemnitz.de
olfa.kanoun@etit.tu-chemnitz-de

Keywords: fully electric vehicle, battery management system, lithium-ion battery, smart cell, safety, wireless communication, electrical impedance spectroscopy

Smart Battery Cell Monitoring with Contactless Data Transmission

V. Lorentz, M. Wenger, M. Giegerich, S. Zeltner, M. März, L. Frey,
Fraunhofer IISB

Abstract

The market breakthrough of electric vehicles is mainly delayed by the still too high costs of the battery system. The smart battery cell monitoring presented in this article enables further cost reduction. It consists of battery cells integrating the monitoring electronics together with a data transfer interface for communicating in a bidirectional way with the battery management system. The data transfer interface presented in this paper is based on a differential contactless data transmission bus using galvanically isolated, capacitively coupled links to each single smart battery cell. Since neither galvanic contacts nor connectors are needed, the proposed concept provides simultaneously a very high level of reliability and robustness, and a highly cost-efficient manufacturing process, thus allowing a significant reduction of the final battery pack costs. This paper describes a possible implementation of such a differential contactless data transmission for monitoring and managing battery cells in electric vehicles.

1 State of the Art in Battery Systems

Despite the great efforts that have been made during the last years to promote electric mobility, the electrical vehicle still has not had its final breakthrough. An important reason for the lack of broad acceptance is the cost of electric vehicles that is still too high. Especially the battery system is and will remain a major cost factor in the near future. This article describes a novel approach to the battery system, which is aimed to significantly reduce the component costs, the development time and the manufacturing costs of the battery system, while increasing its flexibility and reliability.

1.1 Centralized Battery Monitoring Architectures

A battery system for an electric vehicle or a hybrid electric vehicle typically provides voltage levels around 400 V. In case of Lithium-ion batteries, it is necessary to series connect up to 100 and more battery cells in order to reach this high voltage [1]. The series connected cells are arranged in modules, which typically consist of groups of 4 to 14 cells. The series connection of all modules forms the battery pack. A battery pack can consist of 1 to 25 modules or even more. Each module is equipped with, at least, a cell voltage monitoring and balancing circuit. The monitoring circuit acquires the voltage of every single cell and the temperature of each or only some of the cells. It passes the measurement data on to the higher level battery management system (BMS), commonly via an internal CAN bus (iCAN) [2][3]. The BMS processes the data to determine the current state of the battery and communicates with the vehicle control unit via an external CAN bus (eCAN) [4] [5]. Fig. 1 shows the architecture of a battery system like the one described [6].

The battery monitoring and balancing circuit board is typically mounted onto the battery module itself. The core part of its circuit is a dedicated battery monitoring integrated circuit (IC). It comprises a number of analog-to-digital converters to measure the cell voltages and the temperatures. It is also in charge of cell balancing. In most cases, a passive balancing method is used, thus allowing a simple balancing circuit by only using a transistor and a resistor. More sophisticated monitoring ICs feature an active cell balancing method, where the energy of the most charged cells is redistributed to weaker cells. The monitoring IC is controlled by a microcontroller, which pre-processes the data acquired by the monitoring IC and passes it on to the BMS via the iCAN bus. As the monitoring circuits operate at voltage levels up to 400V, the communication and power supply lines have to be galvanically isolated from the rest of the vehicle [7].

A state-of-the-art battery system as described above was designed, tested and demonstrated in the European ENIAC JU Project E3CAR [6]. The module size was chosen to be 4 cells and the pack consisted of 25 modules. However, one of the major issues with this kind of battery systems are the wiring and contacting of cell terminals for voltage monitoring and cell balancing, as well as the temperature measurement, since each cell terminal and each temperature sensor has to be contacted separately. Also, the galvanic isolation of the supply and signal lines require a lot of attention during the conception and a huge effort in terms of component choice and integration space. Fig. 2 shows a photograph of the developed state-of-the-art module with 4 prismatic cells, the monitoring circuit and the corresponding wiring. It becomes obvious that the number of cells per module has a huge impact on the overall system cost, due to the same parts that have to be used for every module.

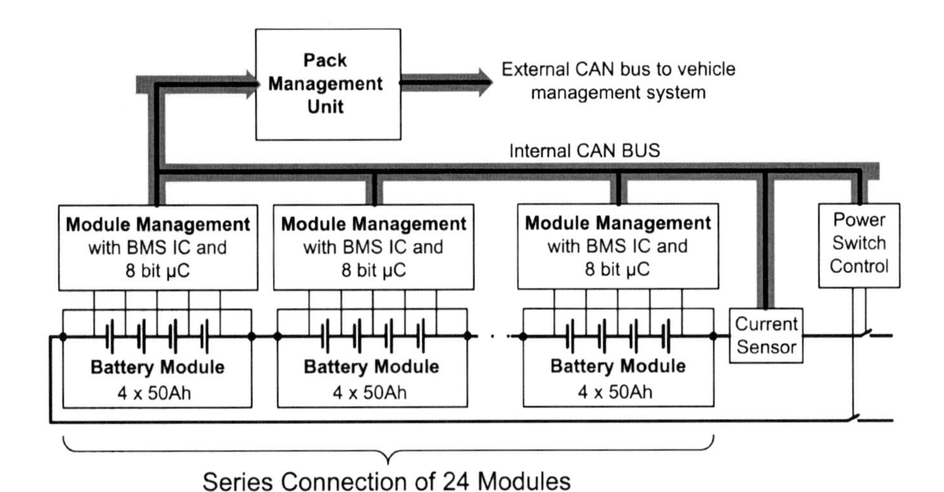

Fig. 1. Example of a state-of-the-art battery system architecture [6]

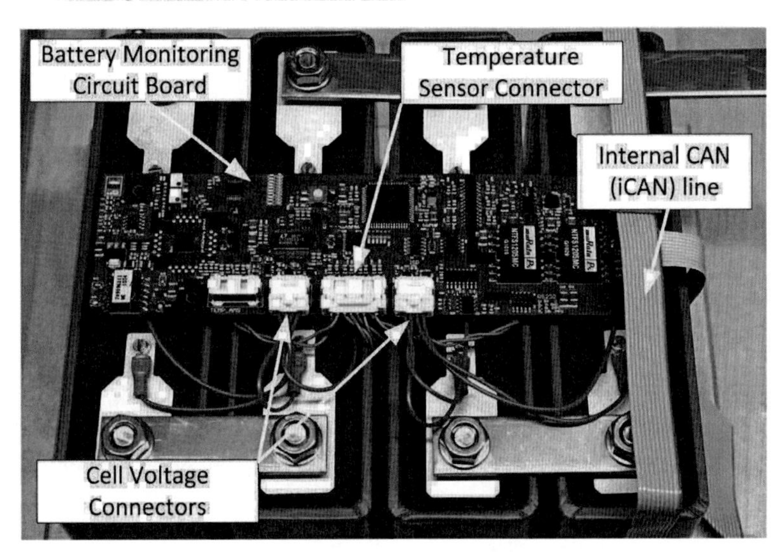

Fig. 2. Photograph of a 4 cell module with the monitoring and balancing circuit board [6]

1.2 Distributed Battery Monitoring Architectures

A different approach is based on distributing the monitoring circuit to every battery cell, thus making the battery monitoring board at module level needless. This concept is highly modular and costs are independent of the module size, but it requires even more wiring and galvanic isolation for communication purposes. This can be solved by transferring only very little data through a one wire communication line [8]. This is not ideal for safe and reliable automotive battery systems. Another possible option is to use a wireless RF transmission solution integrated in every battery cell, thus no more wiring for data transmission is required. However, this transmission method is rather expensive and interference-prone, since the transmitter and receiver are located beside high switched currents.

In the following section of this paper, the proposed novel concept will be described in detail. The functions of the distributed battery cell monitoring implemented in a smart battery cell will be shown [9]. Finally, an implementation strategy of the core of the data transmission interface will be presented.

2 Novel Smart Battery Cell Monitoring Concept

The state-of-the-art centralized battery monitoring architecture has major drawbacks, which obstruct the cost-efficient development and production of battery modules and battery packs:

- The assembly and contacting of the battery cells for monitoring their voltage and temperature is very expensive, even if fabricated in high volumes.

- The high amount of connectors causes reliability issues and is a source of failures.

- Different applications, such as different types or models of electric vehicles, have different requirements and specifications. In most cases, this leads to a full redevelopment of the battery module and the battery monitoring circuit boards for each new type and model.

- The redevelopment of battery monitoring boards increases the time-to-market, the development and the production costs, which makes the centralized battery monitoring architectures economically not viable for the mass market.

These drawbacks were the reason for designing a battery monitoring and management system requiring less manufacturing and assembling effort.

The novel smart battery cell monitoring concept consists of single battery cells with integrated electronics allowing bidirectional contactless data transfer to the battery management system. The proposed distributed battery cell monitoring uses a differential capacitive galvanic isolated data transmission bus to communicate with the battery management system, thus providing following advantages:

- ▶ The smart battery cells provide all the necessary electronics to be fully monitored and balanced.

- ▶ The monitoring and balancing functions are integrated into each smart battery cell, thus improving the reliability, simplifying the assembling and reducing the contacting effort.

- ▶ The effort required for contacting the smart battery cells is reduced to two power contacts (battery poles) and two galvanically isolated capacitive differential contacts to the data bus used for communicating between the integrated battery cell monitoring electronics and the battery management system.

- ▶ The cell temperature measurement, cell voltage measurement, cell pressure measurement, cell current measurement, cell balancing current measurement can be provided by the monitoring electronics implemented in the smart battery cell without any additional contacting effort needed when constructing the battery modules and the battery pack.

- ▶ Smart battery cells enable much shorter time-to-market when developing novel battery packs for electric vehicles, since no more complex application specific battery monitoring circuits and boards need to be developed.

- ▶ Smart battery cells dramatically improve the usability and the modularity during battery modules and battery pack design.

- ▶ The hardware of the battery management system can be standardized and reused in new and different developments simply by maintaining and improving the firmware (i.e. embedded software).

- Costly galvanically isolated CAN transceivers are no longer needed for the data transmission bus between the battery modules and the battery management system.

- The electronics integrated into the smart battery cells allows protection against counterfeit.

The new smart battery cell monitoring concept not only solves the issues of state-of-the-art centralized battery monitoring architectures, but also adds modularity and flexibility, and alters the traditional role of the battery manufactures, automotive manufacturers and automotive suppliers. The different parts needed for designing the proposed smart battery cell monitoring concept based on distributed battery monitoring are described in the following subsections.

2.1 Central Battery Management System (Master)

The battery management system acts as the master in the proposed concept. It controls the energy storage at system level and determines the current state of the battery. It collects status data from all the devices of the battery system (e.g. battery cells, safety switches), uses this input to compute the overall state of the battery pack and sends this information to the vehicle management system. Besides that it fulfils functions like cell balancing control, thermal conditioning and drives different kinds of actuators like high voltage switches and safety functions (see Fig. 3).

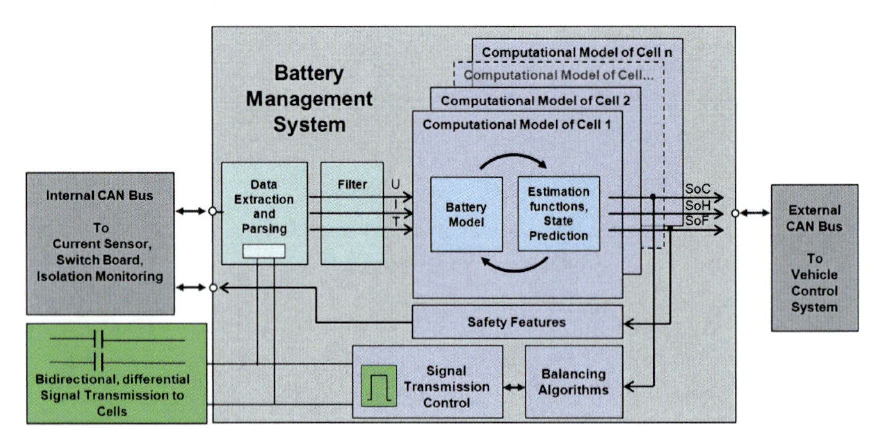

Fig. 3. Architectural overview of battery management system (BMS)

To calculate the state of charge, state of health and state of function of the battery pack, the BMS requests the sensor data from each single smart battery cell monitoring circuit. The differential bidirectional data transmission bus is used for direct communication with every single smart battery cells, in addition to the common automotive communication bus systems like the iCAN bus, which communicates with the internal devices of the battery pack, like the current sensor and the safety switch boards. With the information gathered from the sensors on each battery cell (e.g. voltage, temperature, balancing current, pressure), the internal state of each battery cell is computed, based on battery cell models and using prediction strategies like Kalman filter and variations, fuzzy logic and neural networks [10]. To meet the requirements concerning safety and computational power, an electronic control unit has been developed using the 32-bit TriCore microcontroller from Infineon (i.e. TC1797) in combination with a safety watchdog IC (i.e. CIC61508). This automotive microcontroller offers a clock frequency of 180 MHz, DSP functionalities and short interrupt latency. The safety watchdog allows an automotive safety integrity level up to the level ASIL-D.

2.2 Differential Bidirectional Data Transmission Interface

Different data transmission technologies can be used for data transmission interfaces: ohmic contact, radio frequency, optical (e.g. opto-coupler), inductive (e.g. transmission over a transformer), or capacitive [11]. The ohmic contacting requires a high contacting effort; therefore, it is economically not an option. The RF solution is interesting but requires complex transmitters and receivers. This technology is also not best suited to harsh environments where high currents are switched, causing electromagnetic interferences and other resulting perturbations. The optical solution is promising in theory, but the reliability issues will not make it a practicable way currently. The inductive solution can be very cheap, but for a proper magnetic coupling, a ferromagnetic core is required, which makes it not as simple as the finally preferred capacitive solution used in this paper, which is the method of choice for transmitting data since it does not require a high amount of efforts for contacting [11]. Different topologies for capacitive data transmission can be used: daisy-chained with single-ended signals, daisy-chained with differential signals, or finally the preferred parallel connection with differential signals (see Fig. 4).

Using the parasitic capacitors provided by simple copper strips directly put onto the battery cells makes this solution extremely cost efficient and allows a bus bandwidth between 500 kbps and 1 Mbps. Large tolerances are acceptable for the capacitors, since they are transmitting data signals differentially. Experimentally, capacitance values of approximately 25 pF were obtained

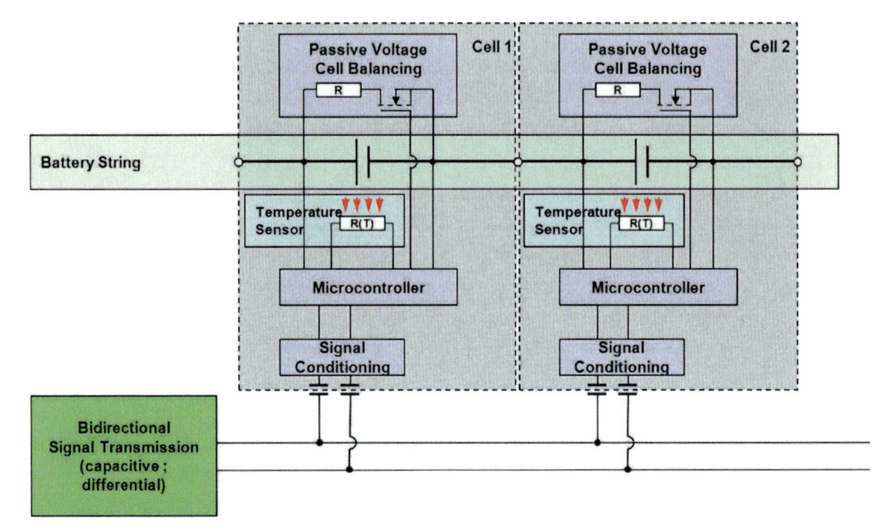

Fig. 4. Schematic of the distributed battery cell monitoring concept with the differential data transmission interface

for 1 cm^2 copper surfaces and 50 µm thick polyimide film used as dielectric. Matching between the capacitors can be optimized by using 90° strips structures as shown in Fig. 5.

The BMS (master) controls every single battery cell (slaves). The bidirectional data transmission interface between the master and the slaves uses 2-wire 5V differential signals. Only data are transmitted on the differential data wires: there is no clock. The length of a typical data frame is 13 bytes and the length of a pure command frame is only 7 bytes. Table 1 shows the structure of the frames. The data includes the voltage, the temperature and the timestamp, all coded on 16 bit. The cyclic redundancy check is used for error detection.

START	Cell ID	CMD	DATA			CRC	END
			Voltage	Temperature	Timestamp		
[1 Byte]	[1 Byte]	[2 Bytes]	[2 Bytes]	[2 Bytes]	[2 Bytes]	[2 Bytes]	[1 Byte]

Tab. 1. Structure of a typical data frame

The necessary bandwidth when all the battery cells are connected over one single bus is shown in table 2. For 1 full measurement set (i.e. the voltage and the temperature of all the 100 battery cells), 18 kb (i.e. 19256 bit) of data are exchanged between the BMS and the battery cells. By performing a full

Broadcast	Perform Measurement (Master → Slaves)	56 bit [7 Bytes]
100 times (100 cells)	Read cell data from Cell ID I (Master → Slave)	56 bit [7 Bytes]
100 times (100 cells)	Get data from cell with Cell ID I (Master ← Slave)	104 bit [13 Bytes]
1 full measurement set	Total without resend	16.056 bit
1 full measurement set	Total with additionally 20% resend	19.256 bit

Tab. 2. Needed bandwidth during full battery cell connection using one single bus

measurement set 10 times in a second, 180 kbps of data are exchanged, thus providing a bus load of 36% for a 500 kbps bus.

2.3 Smart Battery Cells (Slaves)

The proposed distributed battery monitoring concept is perfectly suited for prismatic battery cells, but can also be adapted to pouch cells (i.e. coffee-bag) or to cylindrical battery cells. Applications in other types of energy storage components are also possible. The battery cell electronics consists of an integrated voltage and temperature sensor, a differential capacitive transmission, a signal conditioning electronic and a monitoring and driving circuit, which can be a low-cost automotive microcontroller, or a system-in-package produced in high volumes and using a cryptoprocessor providing additionally advanced anti-counterfeit procedures for identifying the smart battery cells. This microcontroller (slave) also manages the data conversion from analog (e.g. cell voltage, cell temperature, cell balancing current, cell pressure) to digital and sends the values to the BMS (master) via the capacitive transmission. It is supplied directly by the battery cell voltage. Its role is also to activate the passive cell balancing when the battery management has sent the instruction to balance the cell. Each battery cell has its own identification address stored in the microcontroller, so that the BMS can request information independently from each cell sequentially. Further, the voltage measurement is triggered by the BMS via a single command addressing simultaneously all the connected battery cells (broadcast). Without any specific synchronization effort, this method provides the possibility of measuring the voltages (or other physical quantities) of all battery cells within a maximum timing interval of 1 clock period (e.g., maximum of 100 ns delay between the first and the last measurement at a 10 MHz clock frequency).

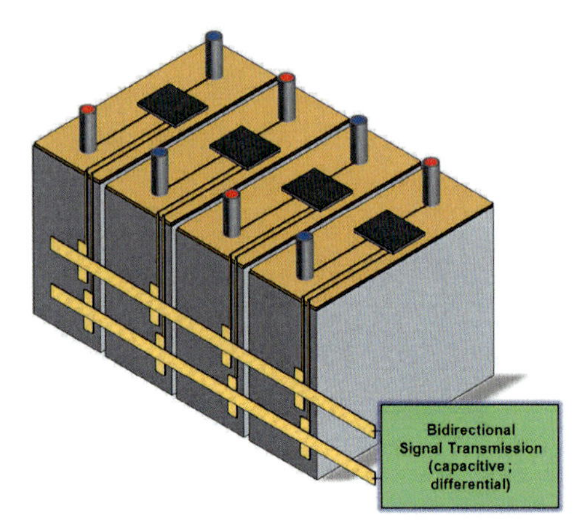

Fig. 5. Battery cells with distributed battery monitoring connected to the differential data transmission bus

3 Conclusion

In this paper, a novel data transfer interface based on a differential bidirectional data transmission bus using galvanically isolated capacitively coupled links to each single smart battery cell in the battery pack was presented and analysed. Since neither galvanic contacts nor connectors are needed, the proposed concept provides simultaneously a very high level of reliability and robustness, and a highly cost-efficient manufacturing process, thus allowing a reduction of the final battery pack costs. Further, also the time-to-market when developing a new battery pack can be dramatically reduced, since the development of a battery monitoring circuit for each single battery module becomes needless.

Acknowledgments

The research leading to these results has received funding from the European Union Seventh Framework Program (FP7/2007-2013) under grant agreement n°285224 ("SuperLIB") and grant agreement n°285739 ("ESTRELIA"). The authors thank Justin Salminen from European Batteries Oy (Finland) for his support by providing battery cell components.

References

[1] T. Stuart, W. Zhu, "Modularized battery management for large lithium ion cells," Journal of Power Sources, vol. 196, pp. 458-464, 2011.
[2] H. Kim, K. Shin, "DESA: Dependable, efficient, scalable architecture for management of large-scale batteries," IEEE Transactions on Industrial Informatics, no. 99, pp. 1-12, 2011.
[3] J. Douglass, "Battery management architectures for hybrid/electric vehicles," Electronic Product Design, March 2009.
[4] K. Cheng, B. Divakar, W. Hongjie, D. Kai, F. Ho, "Battery-management system (BMS) and SOC development for electrical vehicles," IEEE Transactions on Vehicular Technology, vol. 60, no. 1, pp. 76-88, 2011.
[5] D. Andrea, "Battery management systems for large lithium-ion battery packs", Norwood, MA, Artech House, 2010.
[6] M. Brandl, M. Wenger, V. Lorentz, M. Giegerich et al., "Batteries and battery management systems for electric vehicles" Design, Automation & Test in Europe Conference & Exhibition (DATE), 2012, accepted.
[7] D. Xu, L. Wang, J. Yang, "Research on li-ion battery management system," International Conference on Electrical and Control Engineering (ICECE), pp. 4106-4109, 2010.
[8] EV Power Australia Pty Ltd., EV Power Battery Management System, http://www.ev-power.com.au/-Thundersky-Battery-Balancing-System-.html, December 2011.
[9] Patent Pending: Fraunhofer-Gesellschaft zur Förderung der angewandten Forschung e.V., „Energiespeicherzelle, Energiespeicherzellenmodul und Trägersubstrat", Deutsches Patent, Amtliches Aktenzeichen 102011088440.8, 2011.
[10] J. Zhang, J. Lee, "A review on prognostics and health monitoring in li-ion battery", Journal of Power Sources, vol. 196, pp. 6007–6014, 2011.
[11] S. Zeltner, "Research into Insulating Low-Loss Compact Driver Circuits with an Integrated Control Unit for Regulating the Load Current", Ph.D. Thesis, 2011.

Vincent Lorentz, Martin Wenger, Martin Giegerich, Stefan Zeltner, Martin März, Lothar Frey
Fraunhofer IISB
Schottkystr. 10
91058 Erlangen
Germany
vincent.lorentz@iisb.fraunhofer.de
martin.wenger@iisb.fraunhofer.de
martin.giegerich@iisb.fraunhofer.de
stefan.zeltner@iisb.fraunhofer.de
martin.maerz@iisb.fraunhofer.de
lothar.frey@iisb.fraunhofer.de

Keywords: smart battery cells, distributed battery monitoring, battery management system, differential bidirectional data transmission bus, capacitive coupling

New Concepts of High Current Sensing by Using Active Semiconductors for the Energy Management in Automotive Applications

K. Rink, W. Jöckel, Continental

Abstract

Monitoring the state of charge of batteries in start-stop, hybrid- and electrical vehicles makes precise current sensing one of the key functions of today's and future car architectures. Huge required measurement ranges of up to ±600A conflict with fine resolution and high precision requirements – which currently could only be solved by using high performance shunts in connection with specialized ICs offering highest resolution ADCs. The invention follows the approach of a variable, percentage resolution, offering required fine resolution at low currents and the minimum required resolution at high currents. A realization approach using state-of-the-art MOSFET technology and innovative control loop design with a standard µC is presented.

1 Introduction

In today's vehicles, current measurement is applied for a lot of applications. These sensors are e.g. part of control loops for monitoring limits or for the measurement of battery currents e.g. for start-stop systems. The search for new powertrain concepts by using regenerative energy approaches concentrates on electric- and hybrid-electric vehicles. Within these new electrical architectures, the monitoring of the state of charge and the overall state of the drive battery gains far higher importance. With the precise measurement of discharge and charge currents, the state of charge of the battery is determined. Besides this, conclusions are drawn concerning state of health and state of function by monitoring of the behaviour of the internal resistance. To enable these functions, the precise measurement of voltage and current is required; for battery voltages, a measurement span up to 1000V in mV steps is possible while discharge currents up to 600A occur. In particular, the large dynamic range of these currents from 10mA up to 1000A (discharge and charge range) – a factor of 10E5 – while keeping accuracies in the range of <1% is a big challenge.

Actually, the most common approach is the measurement of the voltage drop over an ohmic shunt resistor within the electrical circuit. To avoid high power dissipation, the maximal ohmic value of the shunt resistor is limited to 100μΩ. Besides this approach, magnetic transducers are used – they work contact-free and are based on Hall- or magneto-resistive technology. Also known are optical transducers, especially used in industrial applications for very high currents. The common challenges are to cover the large dynamic span of the current measurement while keeping the required accuracy and the very small voltage drop over the shunt, which causes EMV problems due to disturbing external electromagnetic fields, which are known for the automotive environment. Therefore, costly high resolution AD converters and EMV protection measures are required. This pushes the costs while the society asks for electrical powertrain concepts at reasonable costs. The presented solution approach is based on the idea to enable a constant resolution over the whole measurement span. To achieve this, actively controllable components are required, which enable the realization of a non-linear transfer function, e.g. with power MOSFETs. They are used within a control loop where the drain-source voltage is kept constantly at a low voltage level. In this architecture the control voltage is a measure for the source-drain current of the MOSFET. The characteristic of the transfer curve allows a constant resolution. This results in a significantly more simple signal conditioning, which could be realized with standard approaches using e.g. microcontrollers with standard 12 bit AD converters (instead of the >16bit AD converter of the shunt solution). In addition, the modern state of the art MOSFET technology enables the availability of such actively controllable components, especially concerning the key parameters performance, power dissipation and costs. Furthermore, with this holistic approach, not only the desired performance but also the improvement of electro-magnetic robustness can be achieved at reasonable costs. As additional outlook, multiple options for the internal failure surveillance of the sensor and its components exist.

2 Motivation and System Solutions

As CO_2 emission is seen as one of the global warming drivers, its reduction is recognized as an important aim future car powertrain architectures can contribute to. This is manifested by CO_2 emission and fuel consumption targets, which may be different by regions, but which are enforcing in general more complex powertrain and energy management approaches to cope with. Besides measures to further increase the efficiency of the exploitation of fuel energy with superchargers, high injection pressures, downsizing and decreased number of cylinders to reduce internal friction losses, intelligent strategies to manage the energy consumption of electrical consumers of the EE architecture, the electrification of the powertrain is seen as the major approach to reduce emission and consumption. This electrification takes place in different steps of approaches: Start-Stop systems have a long history, but only when realized with intelligent battery management compromising a low level of recuperation using the existing generator and smooth quick starting capabilities e.g. within the Efficient Dynamics package of BMW, this system started it triumphal procession around the world, which is still in full swing. Hybrid systems add electrical engines to support acceleration of the car (micro-, mild-, full- & plug-in hybrids) or even constant driving at short distances (mild hybrids) or longer distances (full- & plug-in hybrids), enabling the use of smaller, more efficient combustion engines – and a higher degree of recuperation of braking energy. Full electrification of electric vehicles allows the full exploitation of all advantages of the high powertrain efficiency factor, independence from parallel combustion engines and potentially high recuperation rates – but suffers from high battery costs and weight, a significant disadvantage of today's solutions, resulting in intense research activities for new battery technologies – and bridging solutions like the range extender. With increasing degree of electrification of these systems, performance and safety functions are more and more relying on energy management sensors, under which the current sensor is the most important one to evaluate the energy flows.

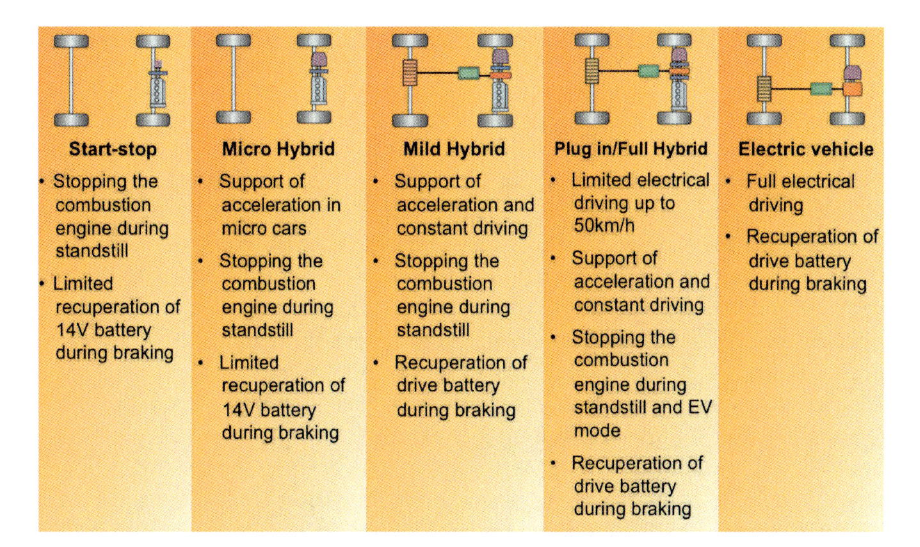

Fig. 1. Overview of different hybrid systems

3 System Requirements

In the above mentioned systems, the main functions of the energy management are the monitoring of the battery state of charge (SOC), state of health (SOH) and state of function (SOF) for the following purposes:

- to enable the disposal of energy meeting the demands not only for the engine but for all other consumers of the car
- to enable precise prediction of remaining driving distance and efficient management of remaining energy
- to keep the charge level of the battery within the desired limits (e.g. from 30% to 80%)
- to determine the possible degree of permitted recuperation
- to determine the electrical load capacity
- to ensure health of the battery and enable a maximum lifetime
- to detect failures like defect battery cells, short circuits etc.

So far no direct state of charge sensing principle for automotive applications is known. These basic and fundamental functions could be realized by knowledge of the battery cell voltages, temperatures and especially by sensing the charge and discharge current of the battery pack. This means that for detection of state of charge, the current value has to be integrated over time and checked for plausibility via voltage measurement or e.g. at charge-end state. The derived requirements for current sensing span a huge range: for the different electrified powertrain systems, the battery systems are mainly staggered by different voltage levels (e.g. 14V systems, 48V systems and systems with battery voltages >>60V), while the measuring range for the current stays in the same dimension – e.g. from ≈150A up to ≈600A for the discharge current, and in most cases the same values for the charge currents (even though quick charge capabilities even with 400V connection are limited). This results in max. measuring ranges from -600A to +600A.

To enable the determination of the state of charge of the battery, high precision is required for the current measurement; as every error falsifies the correct result of integration – resulting in continuously faulty state of charge values. Additionally, also quantization errors have a contribution, so than a chosen high resolution especially within the most frequent current ranges is essential. Typical values for the needed accuracy are below ±1 to some % (containing all influences of temperature, lifetime drift etc.), an offset accuracy of some 100mA, while resolution steps tend in the direction of around 10mA – which directly leads to one of the biggest challenges for current sensing: e.g. 10mA resolution within a measurement range of 1200A in total leads to huge 17 bit quantization range. Furthermore, the current sensor also has to contribute to the operating safety of these new electrified vehicles; the detection of overcurrents (some factors of nominal current) is essential and safety relevant – and besides that, additional self diagnosis functions are in discussion to realize an optimal decomposition of safety requirements within the whole battery management system.

4 Approach

Different approaches are used for current sensor, the main difference is the technological approach for the measurement transducer:

- Hall cells and –cell configurations are used for low cost measurements, more complex transducers containing flux guiding circuits and state-of-the-art signal conditioning enable also applications in some high performance applications (industry measurement applications, but also automotive applications)
- Anisotropic magnetoresistive concepts address the same application areas, without the need for flux guiding circuits
- Shunt resistors enable high precision measurement as material configuration and –properties are well known. Coming from industry applications, this approach is the most commonly used one for first generation electrified vehicles- and especially start-stop systems. In comparison to above mentioned magnetic operating transducers, which measure current by means of resulting magnet field characteristics, a shunt resistor is directly part of the electric circuit from battery to consumers – which results in strong restrictions for the resistance of the shunt (typical values: 100µΩ) to limit power dissipation.

Fig. 2. Actual Continental current sensor for hybrid and electric vehicles

The common challenge of all these approaches is the very small output voltage – and on the other side, the high resolution required. For example, at a current of 10mA at 100μΩ resistance, the voltage drop is 1μV, which must be measured to within 1%. At 1000A , the voltage drop is 100 mV, which also must be measured very accurately. Small output voltages are susceptible to disturbances and require intense signal protection and – conditioning, the high resolution span results in specialized components with >16bit analogue-digital converters, leading to high system costs (or alternatively slow analogue-digital conversion rates), which worsen the already bad cost situation of electrified vehicles.

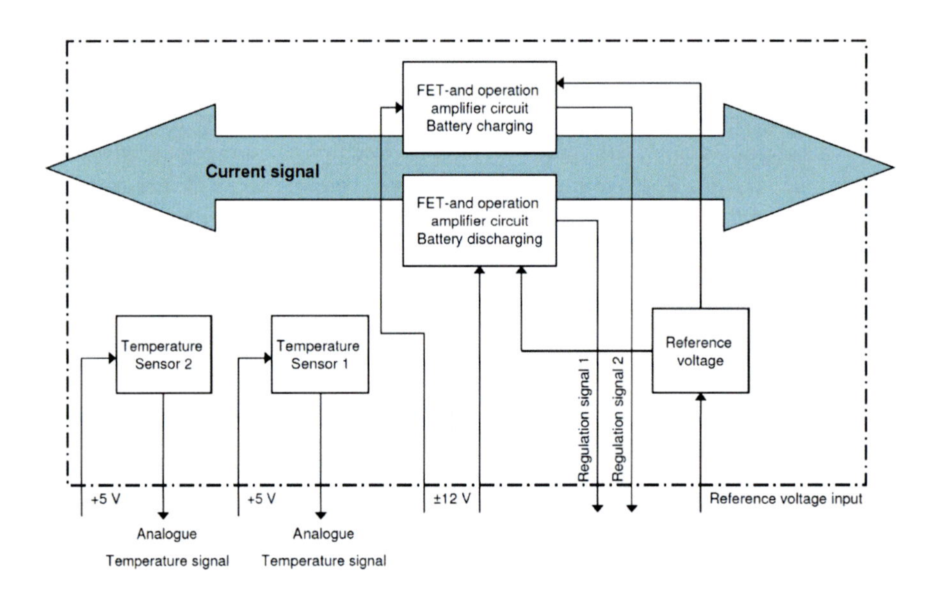

Fig. 3. Basic control loop circuit block diagram of the "active shunt" approach

The new solution approach is based on the idea to enable a constant resolution over the whole measurement span – as fine resolution steps are not necessarily needed at higher currents, because e.g. signal noise exceeds resolution steps in these areas significantly. To achieve this, actively controllable components are required, which enable the realization of a non-linear transfer function. This could be achieved already by state-of-the-art power MOSFETs, which have been "bred" for the last years in direction of lowest resistance for high current applications like IGBTs.

These power-MOSFETs are used as part of a control loop as "active shunt" - instead of a passive resistive shunt. The drain-source voltage of the transistors –one for each current direction- is kept at a constant level by two control loops, defined by the two reference sources, regardless of the current flowing through the transistor. With this approach, the transistors operate as a controlled resistor. The voltage to be regulated is far below the forward voltage of the parasitic diodes, at a few mV. One control loop operates during the discharging of the battery, the other during charging of the battery. The controlled variables *Regulation signal 1* and *Regulation signal 2* follow proportionally the current through the transistors and are a measure for the current to be determined. Both values contain offsets. The temperature dependence of transistor characteristics could be determined by measuring the transistor temperature and canceled by temperature compensation of the raw data.

The characteristic of the transfer curve allows a constant, nearly percentage resolution – from fine resolution at small currents to sufficient resolution at high currents. With this, standard microcontrollers with integrated ADCs could be used, which greatly reduces the costs. The output voltages of the control loop are within the robust mV range. Both *Regulation signal 1* and *Regulation signal 2* voltages are generally positive down to zero – this enables immediate detection of the current direction - and also the recognition of failures, as the operation of the control loop components could be monitored.

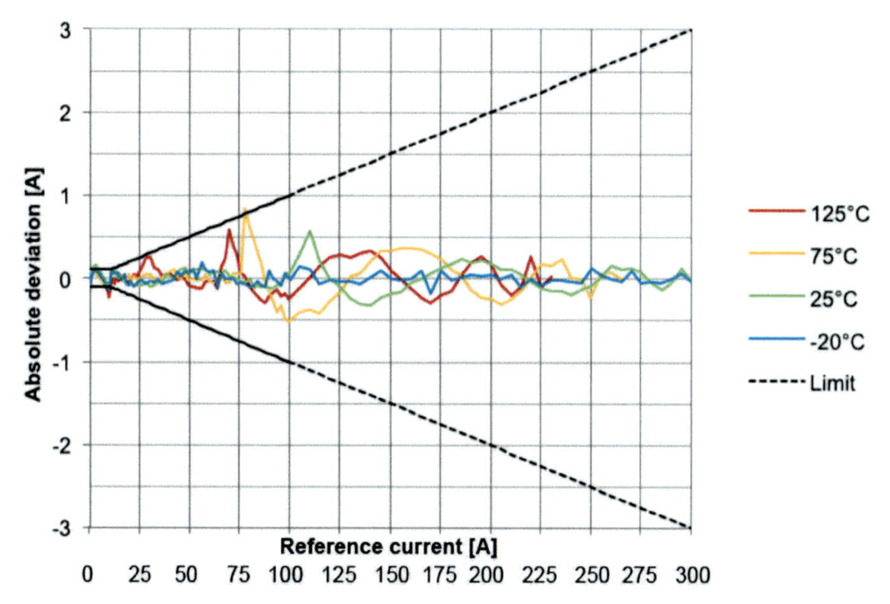

Fig. 4. First measurement results

5 Results

Simulations and experimental samples were set up using state-of-the-art power MOSFETs and standard microcontrollers. Offset cancelation, temperature effect compensation and generation of a linear transfer curve are challenging, but solvable tasks, as first measurements show (Fig. 4).

Results are within exemplary limits over a wide range and over temperature. An experimental linearization algorithm was used for this sample; long-term and great volume characterization experience has to be extended. Yet the power dissipation is not in the desired range, but will be significantly improved by future MOSFET technology steps and bare die connection technologies.

6 Conclusions

A completely new transducer approach to measure currents was invented; first designs were dimensioned and experimentally characterized. This approach offers the potential for:

- Voltages in mV-, not in µV-range – compared to shunt approach
- Measuring range scalable for charge and discharge channel
- Accuracy behavior which can be optimized in regards to most frequently used current regions
- Use of standard components: no special shunt or ASIC needed (µC sufficient)
- No bi-directional ADC which is needed, ADC with lower resolution possible (17bit vs. 12bit)
- New potential for packaging flexibility, lower weight (Al- busbars...)
- Multiple safety options and features: Surveillance of internal short circuit, etc.
- Intrinsic cost reduction potential (technology steps)

These potential advantages are the motivation for a further examination of this new alternative approach. It could be an enrichment in the field of current sensing technologies, of which each one has its specific set of advantages – and disadvantages.

Klaus Rink, Wolfgang Jöckel
Continental
Guerickestr. 7
60488 Frankfurt
Germany
klaus.rink@continental-corporation.com
wolfgang.joeckel@continental-corporation.com

Keywords: battery management, state of charge, current sensor, shunt, MOSFET, variable resolution

Comparison of Gapped and Gapless Designs for an Automotive DC-DC Converter Inductor

R. Demersseman, Z. Makni, Valeo Group

Abstract

This paper is aimed at comparing two designs, one comprising gaps in its magnetic circuit, the other one without, for a power inductor operating in the medium frequency range (10 – 20 kHz) and crossed by a triangular current superimposed with a DC bias. The goal is to find the design that satisfies the target inductance value using the minimum volume of materials. Two kinds of material are considered for the magnetic core: soft amorphous alloy to be used with the gapped configuration and powder to be used with the gapless one. The designs are calculated through a process that includes an analytical preliminary calculation and a numerical optimization. In the case of soft magnetic alloy, the magnetic non-linearity of the material is neglected in the analytical model. On the contrary, non-linearity of powder material can not be neglected. It requires the use of an iterative calculation. The numerical optimization is based on 2D Finite Element Analysis associated to a Simplex-type algorithm. It completes the analytical calculation to reach an appropriate solution for which the target inductance and minimum volume are reached simultaneously.

1 Introduction

For the emerging hybrid and full electric automotives, applications of power electronics are considerably increasing and becoming more and more critical. Actually, the viability of electrified power trains requires the use of components which ensure the best compromises between efficiency, weight and cost. Thus, the design of energy storage inductors, which are typically the heaviest and most expensive components, must be optimized. In the literature several works dealing with this optimization can be found [1] [2] [3]. Here we present the methodology developed at Valeo to dimension a foil-coil inductor for a new DC-DC converter [4]. Gapped and gapless configurations are considered for the magnetic circuit. Analytical and numerical calculation steps are respectively presented in sections 3 and 4. Results obtained for 2605SA1 soft amorphous alloy from Metglas and High Flux 26μ powder [5] from Magnetics, to be used with the two configurations respectively, are reported and compared in section 5.

2 Inductor Design Specifications

The gapped and gapless configurations of the inductor and associated geometric parametrizations are shown in figure 1. The coil is made of n turns of a wound insulated foil of copper. The considered K_b copper to foil volume ratio is 0.9. The magnetic core is assembled of four U-shape parts. Two kinds of soft material are considered for these parts: powder, which is press-molded and sintered, and insulated tape of amorphous alloy, which is wound into rounded rectangle forms that are cut in halves to yield the U-shapes. With the former, the parts are glued in direct contact while for the latter thin resin plates are inserted between them (for powder core no macroscopic gap is required since it is "replaced" by the numerous residual microscopic gaps between the grains). The inner surface of the coil is insulated from the cores by a plastic socket. The thickness c of the latter and lengths d_1 and d_2 of the air gaps between the outer surfaces of the coil and the inner surfaces of the cores are considered to be respectively 2, 3 and 5 mm. In the gapped configuration, the length e of the gaps comprised in the magnetic circuit is considered to be 1.5 mm. Because of the mechanical design of the DC-DC converter into which it will be integrated, the length l_1 and width l_2 of the inductor are constrained to be respectively 71 mm and 54 mm. In operation, the waveform of the current through the inductor is expected to be a triangle ripple $\delta i(t)$ superimposed with a DC bias i_0. Considered values for the latter and for the peak-to-peak amplitude Δi of the ripple will be respectively 37 A and 20 A. Frequency and duty cycle α will be typically adjusted within the 1 – 20 kHz and 0.5 – 0.7 ranges. The target inductance value L^* and maximum current density inside the coil turns J_{max} are respectively 450 μH and 10 A*mm-2.

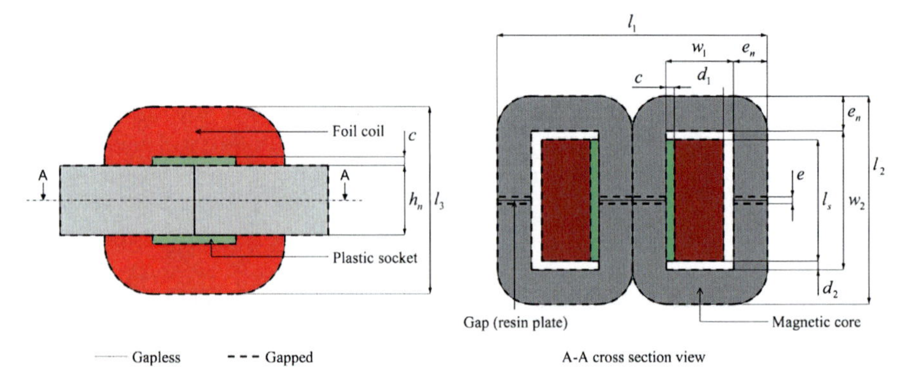

Fig. 1. Inductor Structure

3 Preliminary Analytical Calculation

Given the design specifications detailed in the previous section, the unknowns to be calculated for both configurations of the inductor are five: the dimensions e_n and h_n of the core cross section, the thickness e_s and width l_s of the coil foil and the number of coil turns n. Analysis of the geometry and imposition of the maximal current density J_{max} inside the coil turns yields three over the five equations that are required to calculate these unknowns:

$$\frac{l_1}{2} - 2e_n = c + d_1 + ne_s \tag{1}$$

$$l_2 - 2e_n = l_s + 2d_2 \tag{2}$$

$$K_b l_s e_s J_{max} = i_{max} \tag{3}$$

where $i_{max} = i_0 + \Delta i/2$ is the peak current. The analytical expression of the inductance provides a fourth equation. Noting φ_0 and $\delta\varphi(t)$ the DC and ripple components of the flux inside the magnetic circuit, which result respectively from the DC i_0 and ripple $\delta i(t)$ components of the current, neglecting the hysteresis of the core and assuming that 1) there is no flux leakage out of the magnetic circuit (in particular, in the gapped configuration, there is no flux fringing out of the gaps) 2) the fields H_c and B_c inside the cores are uniform 3) in the *gapped configuration*, the fields H_e and B_e inside the gaps are uniform and H_e is very high compared to H_c, which enables to neglect the magnetic non-linearity of the core 4) in the *gapless configuration*, the relationship between the ripples $\delta B(t)$ and $\delta H(t)$ is linear, in other words, the error in evaluating B from H through a local linearization of the magnetization curve at the point of $H = H_0$ abscise is small, the inductance can be calculated as:

$$L_{\text{gapped}} = n\frac{\delta\varphi(t)}{\delta i(t)} = \frac{n^2\mu_0 K_t e_n h_n}{e} \tag{4}$$

$$L_{\text{gapless}} = n\frac{\delta\varphi(t)}{\delta i(t)} = \frac{2n^2\mu_{rev}(H_0)e_n h_n}{l_1 + 2l_2 + (\pi - 8)e_n} \tag{5}$$

In the first expression, K_t is the notation for the magnetic alloy to tape volume ratio and μ_0 is the vacuum permeability. In the second expression $\mu_{rev}(H_0)$, commonly named *reversible permeability*, is the proportionality coefficient between $\delta H(t)$ and $\delta B(t)$ or the slope of the locally linearized magnetization curve or the derivative of the relationship between H and B at the point of $H = H_0$ abscise. The field H_0 can be linked to the DC bias current i_0 through the Ampere's law applied on the mean path of a half of the magnetic circuit:

$$H_0 = \frac{ni_0}{l_1 + 2l_2 + (\pi - 8)e_n} \tag{6}$$

The fifth equation is derived from the minimization of the component volume. In the case of the gapped configuration, the latter is achieved by imposing the core to work at its *magnetic saturation* limit:

$$B_{c,\max} = B_{sat} \tag{7}$$

where $B_{c,max}$ and B_{sat} are respectively the maximum flux density reached in the core and the saturation flux density of the core material. On the basis of the hypothesis mentioned above, (7) can be written under the form:

$$\mu_0 n i_{\max} = 2eB_{sat} \tag{8}$$

The latter equation ensures that the core cross section is as small as possible or, in other words, that the magnetic flux is as concentrated as possible. In the case of the gapped configuration a linear system of five equations can thus be constituted enabling an analytical calculation of the five unknowns. Contrariwise, such calculation can not be carried out in the case of the gapless configuration because of the non-linearity of the relationship between μ_{rev} and H_0. The following semi-numeric alternative can be used: from (1), (2) and (3) express n, e_s and l_s as functions of the single variable e_n:

$$n(e_n) = \frac{K_b J_{\max}}{4i_{\max}} \times \left[\left(c + d_1 + 2d_2 - \frac{l_1}{2} - l_2 + 4e_n \right)^2 - \left(c + d_1 - 2d_2 - \frac{l_1}{2} + l_2 \right)^2 \right] \tag{9}$$

$$e_s(e_n) = \frac{1}{n(e_n)} \left(\frac{l_1}{2} - 2e_n - c - d_1 \right) \tag{10}$$

$$l_s(e_n) = \frac{i_{\max}}{K_b J_{\max}} \frac{1}{e_s(e_n)} \tag{11}$$

and then express the inductance L and volume V of the component as functions of the remaining unknowns e_n and h_n. Mathematically the problem to be solved can then be formulated as: given a domain $D_0 = [e_{n,min}; e_{n,max}] \times [h_{n,min}; h_{n,max}] \subset IR^2$ of admissible values for the couple (e_n, h_n), search the subdomain $D_1 \subset D_0$ such as

$$\forall (e_n, h_n) \in D_1, \quad L(e_n, h_n) = L^* \tag{12}$$

or, in an equivalent way,

$$\forall (e_n, h_n) \in D_1, \quad F(e_n, h_n) = \left| \frac{L^* - L(e_n, h_n)}{L^*} \right| = 0 \tag{13}$$

i.e. search the points in D_0 in which the target value for the inductance is met, and then search the point $(e_{n,s}, h_{n,s})$ in D_1 in which V reaches its global minimum. Practically, the problem above can be solved using a computer program which "sweeps" a grid discretization of the domain D_0 of steps Δe_n and Δh_n, detects the points in which F is below a certain small predefined ε, evaluates the volume in that points and stores the point $(e_{n,s}, h_{n,s})$ in which the volume reaches its minimum.

4 Finite Elements Optimization

Finite Element (FE) analysis is performed to refine the preliminary designs calculated with the analytical methodology previously described. Actually, the non-linearity of both materials is taken into account through the definition of an apparent inductance which is optimized with respect to the target inductance value. A FE-based optimization is employed to find the inductor dimensions which enable to satisfy the calculation objective i.e. reach the target inductance value and minimize the volume. Two freeware tools, Scilab and FEMM, were used to carry out this optimization. They were coupled using the *scifemm.sci* script included in FEMM distribution. It is based on the open source *Lua* scripting language.

A dedicated Scilab script was developed to automate the execution of FEMM instructions that run FE calculations and pre\post processing required by the optimization loop. After loading the results of the preliminary analytical calculation, to use them as a "starting point", the program enters a loop, the first step of which is to call the 2D FE analysis software FEMM and to pass on the dimensions, current wave parameters and magnetic properties of the cores. FEMM builds and meshes the geometry shown in the A-A cross section view of Fig. 1. and performs N analysis of the inductor, one per value i_k of the sampling of the current ON-interval shown in Fig. 2. Samples $\delta\varphi_k$ of the magnetic flux ripple are then loaded in the Scilab environment and used to calculate the apparent inductances as:

$$L_k = n\frac{\delta\varphi_k}{\delta i_k} \quad \forall k \in [1, N], k \neq q \tag{14}$$

where δi_k are the current ripple samples and q is the index of the sample taken in the middle of interval, i.e. such as $i_q = i_0$. For this sample the apparent inductance can not be defined because of nullity of the current ripple. $\{L_k\}1 \leq k \leq N,\ k \neq q$, which are all different because of the non-linear magnetization of the magnetic materials (Fig. 3), are then used to evaluate the objective function f_{obj} driving the optimization.
The latter is expressed as:

$$f_{obj}(X) = C_L \frac{\sum_{k=1}^{N} \left| L_k(X) - L^* \right|}{N} + C_V V(X) + p_n(X) \tag{15}$$

where $X = (e_n, h_n)$ is the array of optimization variables, $V(X)$ is the calculated inductor volume, p_n is a penalty function and C_L and C_V are weighting coefficients which enable to balance the contributions of the two optimization criteria to the objective function. Thus, no criterion is hidden by the other while the optimization algorithm is minimizing the objective function. The above f_{obj} expression enables to reach the target inductance value and to minimize the volume. If the minimization criteria of f_{obj} are met then the program stops else it defines new values for the optimization variables (e_n, h_n) and iterates.

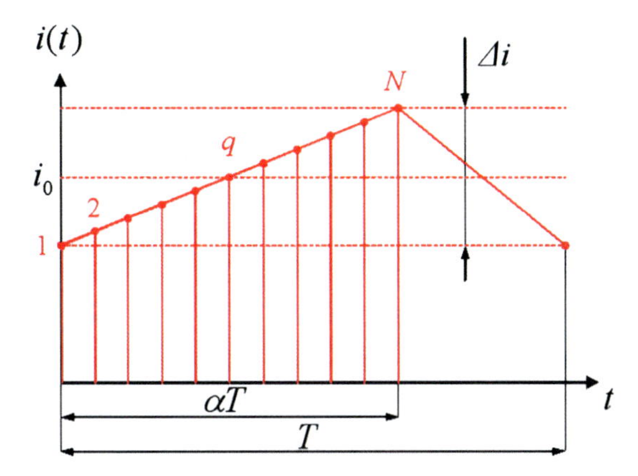

Fig. 2. Current ON-interval sampling

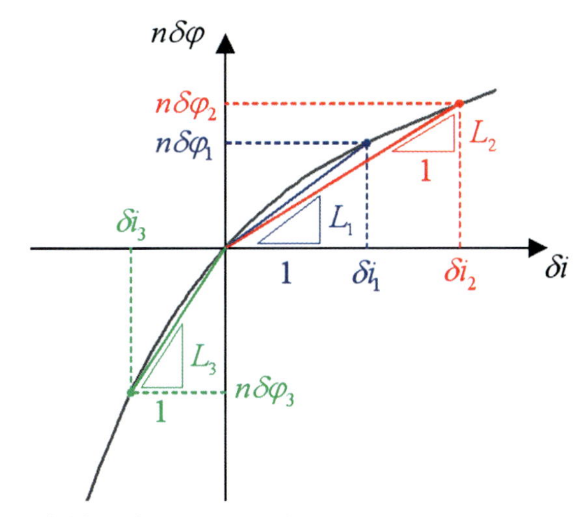

Fig. 3. Definition of the apparent inductance

The Nelder-Mead algorithm, which is a pattern search method for use on unconstrained non-linear models, was selected to carry out the optimization. It is available via the *fminsearch* Scilab function. The method is based on the concept of Simplex, which calculates $m+1$ testing points to solve m dimensions problem. Since it does not require a calculation of the objective function gradient, it suits well for the FE discrete calculation. However the Nelder-Mead algorithm is unconstrained, hence a penalty function is used. The latter is the sum of two terms that increase f_{obj} if one of the two optimization variables approaches its limits. Each term corresponds to a bathtub curve, the associated function of which is almost equal to zero as long as the variable remains in its tolerated interval.

5 Comparison of the Two Designs

As mentioned in the previous section, FE optimization is performed in order to ensure that both compared designs satisfy the target value of inductance and make the least use of material volume. The same optimization procedure is carried out for the gapped geometry with 2605SA1 soft amorphous alloy and the gapless one with High Flux 26µ powder. During optimization, the material volume and the average deviation relative to the target inductance value (ΔL) evolve progressively until the minimum of f_{obj} is reached. Evolutions of these criteria are shown in Fig. 4.

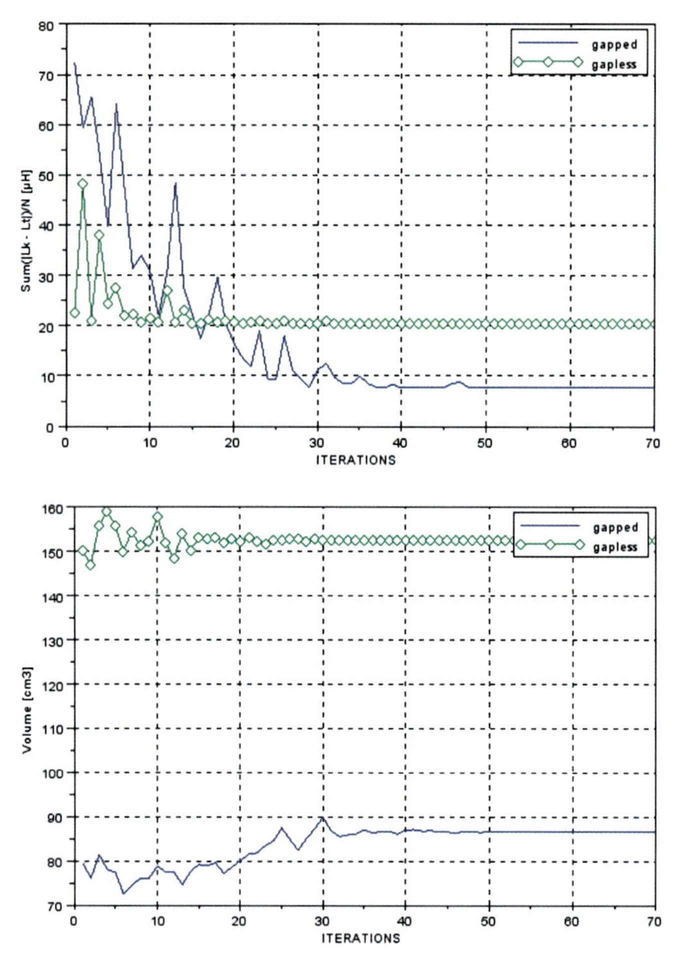

Fig. 4. Evolution of average deviation relative to target inductance (top), volume (bottom)

	Gapped					Gapless				
	ΔL[μH]	V [cm³]	e_n [mm]	h_n [mm]	n	ΔL[μH]	V [cm³]	e_n [mm]	h_n [mm]	n
Analytical	72.5	79.5	7.9	12.2	80	22.5	150.2	8.2	35.3	75
Numerical	7.7	86.6	9.7	18.9	53	20.4	152.6	8.18	36	75

Tab. 1. Comparison of gapped and gapless designs of the inductor

These curves show similarities of optimization behaviour in the two cases. Starting from the analytical solution (first iteration calculation), the optimization converges to a numerical solution (e_n, h_n) which slightly modifies the inductor volume and reduces the average deviation of L. Main analytical and numerical calculation results for both designs are given in table 1. Variations of L versus the current ripple magnitude for the gapped and the gapless structures are plotted in Fig. 5 and Fig. 6 respectively. This variation is estimated by FE analysis carried out for both analytical and numerical output designs.

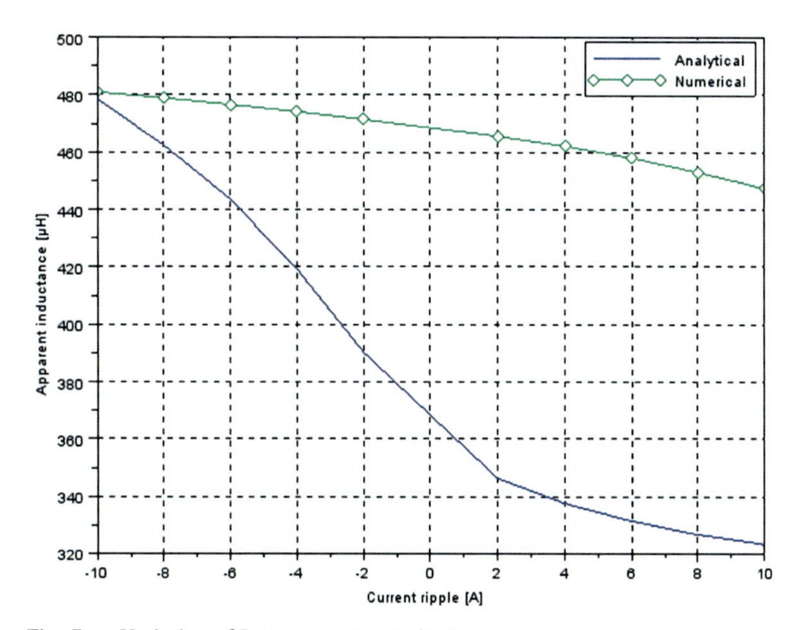

Fig. 5. Variation of L versus current ripples: gapped

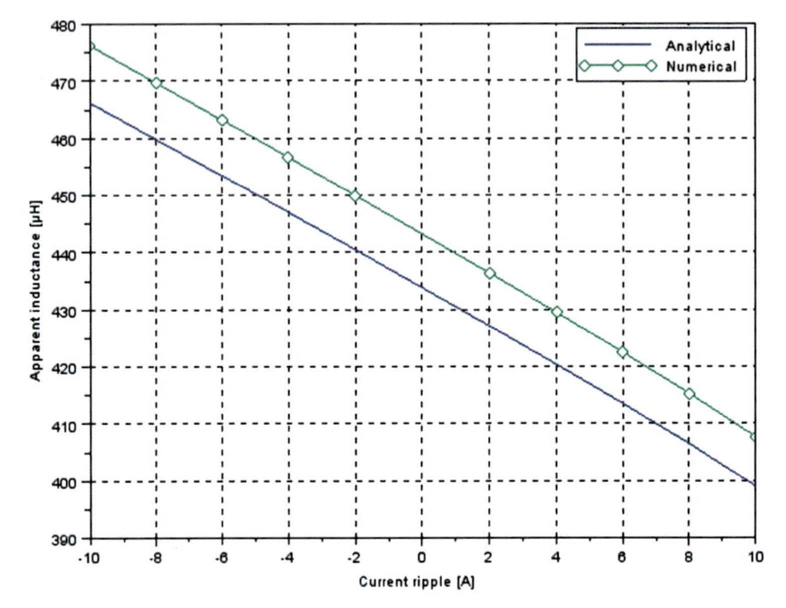

Fig. 6. Variation of L versus current ripples: gapless

FE optimization improves to 89% and 9.33% the deviation relative to the target inductance by correcting to 55% and 1.6% the core cross section area for the gapped and the gapless structures respectively.

As illustrated above, the preliminary analytical design is close in performance to the numerical one for the gapless structure. In contrast the preliminary design for the gapped structure is significantly improved after FE optimization. The analytical imprecision may be due to the assumed linearity of the magnetic circuit in the analytical model of the gapped configuration.

In the light of these results, it can be concluded that the gapped design better serves the need of target inductance using minimum volume of material. In fact, the gapped inductor is almost twice smaller than the gapless one and has a reduced average deviation relative to the target inductance (gapped: 7.7 µH / gapless: 20.4 µH).

6 Conclusions

Gapped and gapless designs of a power inductor for a new reversible Buck/Boost converter were compared. The comparison is aimed at identifying, during the early design stages, the couple structure-material that satisfies the target inductance using the minimum volume of material. The calculation methodology comprises a preliminary analytical calculation and a numerical optimization. The latter completes the analytical calculation to reach an appropriate solution that better matches the requirements. Considering the two criteria of target inductance and minimum volume, the gapped structure of 2605SA1 soft amorphous alloy was identified as the best of both. The advantage of the gapped design has to be confirmed taking into account the effects of the gaps in terms of radiated electromagnetic interference and induced losses in the coil due to the fringing flux and considering materials costs.

References

[1] Y. Bong-Gi and al, "Optimization of Powder Core Inductors of Buck-Boost Converters for Hybrid Electric Vehicles", IEEE VPPC, pp. 730-735, 2009.

[2] F. Liffran, "A procedure to Optimize the Inductor Design in Boost PFC Applications", IEEE EPE-PEMC, pp. 409-416, 2008.

[3] J. Zientarski and al, "A Design Methodology for Optimizing the Volume in Signle-Layer Inductors Applied to PFC Boost Converters", IEEE COBEP, pp. 1177-1184, 2009.

[4] L. De-Sousa, B. Bouchez, "Rethink the Electrical Vehicle Architecture", Automotive Power Electronics International Conference and Exhibition, Proceedings Ref 2011-01, 2011.

[5] M. Rylko and al, "Revised Magnetics Performance Factor and Experimental Comparison of High-Flux Materials for High-Current DC-DC inductor", IEEE Transaction on Power Electronics, Vol. 26 N°8, pp. 2112-2126, 2011.

Richard Demersseman, Zaatar Makni
Valeo
2 rue André Boulle
94000 Créteil
France
richard.demersseman.ext@valeo.com
zaatar.makni@valeo.com

Keywords: power inductor, analytical calculation, finite element analysis, optimization, simulation

Wireless Charging: The Future of Electric Vehicles

A. Gilbert, J. Barrett, Qualcomm Inc. EID Europe

Abstract

The last few years have seen electric vehicle technology develop in leaps and bounds. EVs are now universally recognised as the future of the automotive industry, with cars such as the Nissan Leaf entering the market in 2011. However, there is no doubt that barriers to mass-market adoption still remain, primarily in terms of efficiency, cost and usability. It is these issues that are now being addressed by the development of wireless charging technology. Simplicity and minimum driver intervention are key features that win out time-and-time again and when these features are coupled with high power transfer efficiency, wireless charging is a winning combination. This presentation will outline how this technology works and the benefits it is set to bring to the electric vehicle industry.

1 Introduction

Wireless power is not new. Nikola Tesla demonstrated the illumination of phosphorescent lamps using wireless power at the 1893 World's Columbian Exposition in Chicago, and at a meeting of the National Electric Light Association in St. Louis later the same year. Tesla's work is behind much of the advancement in electromagnetic communications that we take for granted today, but it is only in the past 20 years or so that wireless power pioneers, such as Professor John Boys and Associate Professors Grant Covic and Udaya Madawala of The University of Auckland, have developed systems to transmit electric power efficiently across large air gaps without using wires.

The size of the air gap and the efficiency of a wireless power system, while important for the viability of any commercial deployment, are not the only aspects of a wireless power solution that need to be considered. Industry is littered with inventions that failed to address "human" aspects of adoption and this distinction is often the difference between success and failure of novel technology.

With burgeoning interest and increasing levels of investment by the automotive industry in Electric Vehicles, interest in wireless power is rapidly accelerating. Most major auto manufacturers either have electric vehicles on the market now or have firm plans to bring these zero emission vehicles to market in the next 12-18 months. However, there is no doubt that barriers to mass-market adoption still remain, primarily in the understanding of range, cost and usability.

2 Wireless Electric Vehicle Charging – How it works:

Instead of using a plug-in cable to charge the Electric Vehicle (EV), Wireless Electric Vehicle Charging (WEVC) technology uses the principle of magnetic induction to wirelessly charge the vehicle's battery. Power is transferred between a Base Charging Unit (BCU), installed either on or below the paved surface of the charging bay, and a Vehicle Charging Unit (VCU) fitted to the underside of the vehicle. A magnetic "Flux Pipe" couples power between the BCU and VCU charging pads and energy is wirelessly transferred between the two units to charge the EV battery. All the driver needs to do is park – the rest is automatic.

In principal, WEVC sounds quite simple. However, in practice, it is challenging to develop systems that offer flexible operation and easy parking for a broad range of vehicles. Many of the systems currently being developed for the automotive industry require precise alignment of the two charging pads, and in some cases the pads must be very close, even touching, to achieve the necessary power transfer rates and efficiency.

Interestingly, in a recent TREND industry survey[1], focused on EV charging, the opinion of many respondents was that plug-in EV chargers were 100% efficient. This is of course not the case. As with any electrical system, losses occur in transmission, power conversion and within connectors and components such as resistors, diodes and switches. Efficiency levels of plug-in EV chargers depend on the quality of components and connectors used, the length of cables along with input voltage, frequency of conversion and power factor.The efficiency of plug-in chargers is typically in the 85-95% range, but can be as low as 80% in some cases. Wireless charging therefore needs to deliver power transfer efficiencies above 90% to be comparable to plug-in. In the Qualcomm Halo WEVC system over 90% power transfer efficiency is achieved with efficiency increasing as the power rating increases from a 3.3kW system to 7kW and 20kW. Higher power transfer solutions are more efficient because the ratio of standing losses to power delivered improves.

Wireless power solutions also benefit from inherent isolation across the air gap – there are no physical connections obviating the need for the isolating circuits required in wired systems and the associated efficiency losses. The increase in wireless power transfer efficiency in the past few years has come about from improved materials, superior power electronics and from a better understanding of magnetic resonance and myriad system parameters that must be balanced to create two magnetically resonant circuits.

Qualcomm has been investing in wireless power for many years and recently acquired the wireless power technology of HaloIPT, a spin out company from The University of Auckland. The key differentiator that Qualcomm Halo is bringing with its wireless power technology addresses what we believe is a key requirement to enable mass-market adoption of EVs; the ability to maintain high efficiency even with misalignment of the charging pads.

3 Mass Market EV Adoption

As previously discussed, many of the highly efficient wireless power systems proposed today need precise alignment laterally (x,y plane), and close proximity in the vertical (z) plane. Any misalignment results in a dramatic drop off in power delivered and transfer efficiency. This means the EV driver needs to park very accurately in the charging station ensuring the on vehicle pad is precisely aligned with the ground based pad. We see two problems with this:

- First, this creates a need for complex and potentially expensive guidance and alignment systems built into the vehicle that assist the driver to accurately park and align the pads for charging.

- Secondly, we believe this new parking behavior would be difficult for drivers to adopt and would stifle mass-market adoption of EVs. Plugging in the EV to charge is a new behavior that automotive companies are suggesting could limit EV sales, and they are looking to WEVC to provide an answer to this problem. In one comment from an auto company executive it was expressed as, "When you purchase a € 100k electric car, you don't want to be constantly handling a dirty cable to recharge the battery, you want a wireless, cable free solution."

4 Tolerance

An aspect of the Qualcomm Halo WEVC technology results in the ability to be highly tolerant of misalignment of the charging pads in both the lateral and vertical planes while still maintaining required power levels and power transfer efficiency above 90%. This is achieved using a patented and unique magnetic coil architecture that creates a "flux pipe" between the two pads. This flux pipe provides improved return paths for the field and ensures maximum power transfer, even with a large air gap or with significant misalignment since the magnetic flux is effectively "pulled" towards the on vehicle charging pad ensuring minimal energy leakage.

5 Human Behavior

This Qualcomm Halo WEVC implementation means that drivers do not need to learn a new behavior. There is no need to plug-in a charging cable; no need to park precisely; in fact no need to park any differently. Providing the EV is parked normally within the charging bay, charging will be possible. We believe this ease of use of charging is a critical success factor for the mass adoption of EVs. WEVC has the potential to be hassle free and without any fuss. In addition, drivers are likely to charge little and often when compared with cabled solutions that require repeated effort to connect the charging cable, and then activate the charging. Drivers will even charge for 5 minutes at the local store, for 30-60 minutes at the supermarket or for longer periods at work or at home.

6 Street Furniture

One of the obvious benefits of wireless charging is that there is little or no street clutter since the BCU charging pad can be buried under the paved surface of the charging bay. A single power supply unit can power multiple charging bays and can be sited discreetly away from the roadside, sidewalk or kerb. With no moving parts and minimal maintenance WEVC is the ideal solution for public charging bays and there are no posts that can be vandalized and no cables that could pose a safety hazard or be stolen.

7 Trials Feedback

The CABLED trial in the UK that is using both wired and wireless charging, suggest that 77% of EV journeys lasted less than 20 minutes, vehicles are parked for 92% of the time, that drivers charge when the battery level indicated between 81-87% charged, and that charging was predominantly at home either early morning or late at night during off peak times.[1]

It is important though to remember that many EV trials are skewed towards a demographic that is an ideal EV driver; they have off street parking for home charging and probably park their car in a garage overnight. This leads to charging happening predominately at home, reducing public charging instances.

As the EV mass-market materializes we believe that many drivers will not have access to off road parking, as in city centers where accommodation is primarily apartments. Public charging, charging at work and charging at private charging points such as supermarkets, retail parks, cinemas and restaurants for instance will be needed.

If, as we believe, EV drivers will charge more often, when the opportunity arises, local communities and authorities will resist the deployment of a multitude of cable charging posts. Wireless charging will therefore become the preferred method for EV charging.

8 Wireless Trials

Wireless Electric Vehicle Charging trials have actually been around for some time, albeit on a small scale. In Turin and Genoa a total of 31 electric buses have been in operation since 2002 using technology that originated from The University of Auckland. The ground based charging pads are located where the buses stop to pick up passengers so they charge little and often. This has enabled the use of smaller batteries – a reduction from 135kWh to 45 kWh. The buses are promoted as electric, are smoother than the comparable diesel versions and people appear to like the experience.[2]

Rolls Royce Motor Cars are also using Auckland WEVC technology in their Prototype Phantom 102EX Experimental Electric Vehicle. This uses a 7kW charging system that has been tested extensively in the field and delivers a little over 92% power transfer efficiency.

In 2011, Qualcomm Halo announced a pre-commercial deployment of wirelessly charged passenger cars and light vehicles for London in the UK. The trial, will utilize public knowledge gained from the UK CABLED trial and will commence during 2012 with the first cars expected to be on the road by the summer.

The CABLED trial has been supported by the UK Government Technology Strategy Board (TSB), where two Citroen C1 WEVC vehicles have been successfully working for over 12 months.

Probably the most exciting use of WEVC has been announced at the MIA Low Carbon Racing conference in January 2012. Drayson Racing Technologies and Lola Group unveiled the much-anticipated electric-powered Le Mans prototype-racing car, the Lola-Drayson B12/69EV. Powered by 4 electric motors delivering over 850 horsepower it is expected to reset industry expectations for electric vehicles. The B12/69EV, which incorporates a slew of new technologies, including advanced Lithium Nanophosphate battery technology from A123 Systems, will be charged wirelessly using a 20kW WEVC system from Qualcomm Halo. Lord Drayson, one of the founding partners and a past American Le Mans winning team driver, has a vision of wireless charging built into the racing track so electric vehicle racing is not interrupted by pit lane charging.

9 Business Models

The success of WEVC, and potentially electric vehicles in general, will depend on new automotive business models that ensure the development and growth of a vibrant and innovative EV ecosystem of companies that deliver value to EV drivers. Delivering new technology into an existing and long-established industry though can be challenging.

One of the best examples of technology adoption and innovative growth can be seen in the mobile communications industry where a broad licensing based business model has powered the growth of 3G and now 4G technology.

Qualcomm spends over $2.9 billion in Research & Development (R&D)[3] annually, investing up-front in research programs and licenses out innovation to multiple companies. This creates a competitive and active supply chain that encourages additional innovation around a base standard that has been agreed collectively by the industry.

The complexity of WEVC technology and the relative infancy of the EV industry will lead to fragmentation of the supply and value chain unless robust business models and a single WEVC standard are agreed. For mass market EV adoption to materialize it is essential to take a long-term view of the EV industry and market. Early investment is essential to ensure EV adoption by consumers and businesses, while revenue will be realized some years in the future as more and more drivers purchase new electric vehicles.

10 Future Potential – Dynamic Charging

We believe stationary charging will be the focus of the initial implementation of WEVC for some time and will eventually replace the plug-in cable seen on most EVs today. It will take time to roll out the charging infrastructure, at people's homes, at factories, offices and eventually in public places.

Dynamic EV Charging (DEVC), where wireless charging pads are buried in roads and highways is a future technology that could totally eliminate range anxiety – though a commercial deployment is many years away.

11 Conclusions

Wireless power, while not new, is entering an exciting phase of development, especially in the automotive industry. Consumer ease of use and the need to avoid the creation of new driver behaviors positions WEVC as the ideal solution for EV battery charging, and also opens up the potential future opportunity of Dynamic EV Charging.

This is an exciting time for those companies that have EV programs in place and are helping to drive the industry and user adoption of EVs forward. The whole industry though needs to coalesce around a single standard for WEVC to ensure a vibrant ecosystem and competitive market place that has the potential to innovate and bring new ideas to the automotive industry.

Qualcomm Halo is one company that sees a very positive future for Electric Vehicles and will continue to invest in new technologies and solutions to make Wireless Electric Vehicle Charging a mass-market commercial reality.

References

[1] TREND EV Survey 2012.
[2] First years findings Press Release 18 July 2011.
[3] http://www.bbc.co.uk/news/technology-14183409
[4] Qualcomm financial results 2011.

Andrew Gilbert, Joe Barrett
Qualcomm Inc. EID Europe
Building 4, Chiswick Park
566 Chiswick High Road
W4 5YE, London
Great Britain
agilbert@qualcomm.com
jbarrett@qualcomm.com

Keywords: dynamic EV charging, wireless EV charging, inductive EV charging, energy transfer efficiency

Standard Proposal for Resonant Inductive Charging of Electric Vehicles

S. Mathar, RWTH Aachen
J. Bärenfänger, EMC Test
K. Baier, A. Heinrich, Daimler
W. Bilgic, IMST
V. Blandow, TÜV Süd
U. Blosfeld, Tyco
B. Elias, R. Peer, Audi
T. Eymann, Bosch
M. Hardt, ESG
J. Heuer, R. Knorr, Siemens
R. Heinstein, VDI/VDE Innovation + Technik
S. Heusinger, G. Imgrund, VDE DKE
S. Kiefer, Kiefermedia
S. Kümmell, IAV
D. Kürschner, U. Reker, Vahle
J. Mahlein, Brose-SEW
M. Mahrt, Eon
R. Marklein, Fraunhofer IWES
J. Meins, TU Braunschweig
R. Plikat, Volkswagen
C. Rathge, ifak
W. Schnurbusch, Conductix-Wampfler
O. Simon, SEW Eurodrive
P. Stolte, Bombardier
E. Stolz, Park & Charge

Abstract

Contactless energy transfer between power supply and electric vehicle as described in this proposal is based on the principle of resonant inductive coupling. A pair of charging pads, the stationary charging pad and the mobile charging pad, is used for this purpose. A charging pad may incorporate one or more coils and may be underlaid with material that is capable of conducting the magnetic field. As the stationary charging pad and the inverter are combined to form a unit within the stationary part (though they may be arranged at separate locations), no publicly accessible interfaces are required. The stationary part may be embedded in the ground and mounted flush with the road surface, or it may rest on the ground, e.g., for mounting in a garage. The inverter serves to convert the supply voltage and to set the frequency, current and voltage for the magnetic coil(s) of the stationary charging pad. In addition to the mobile charging pad, the mobile part of the charging equipment incorporates the on-board electronics.

1 Protection Targets

With this proposal, the following two protection goals are required to be achieved:

1) protection of persons, livestock and property against electromagnetic fields

2) protection against direct effects of electromagnetic fields, particularly with regard to heating and consequent risk of:

a) burning upon direct contact b) ignition and fire.

The following four protection areas are considered (see Figure 1). For each of the four areas described below the aforementioned two protection goals apply.

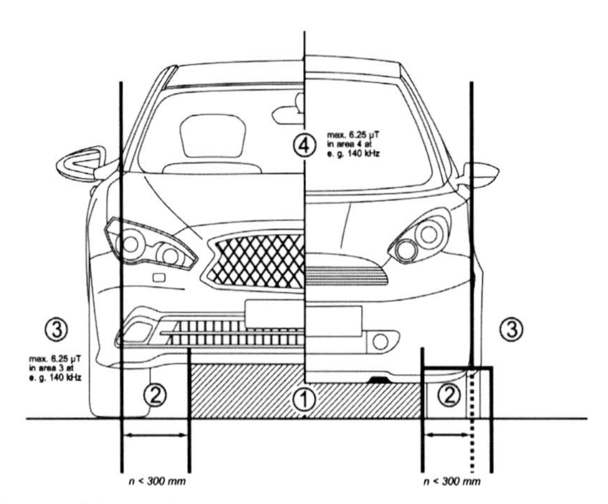

Fig.1. Areas of Protection

Area 1: Area of operation. Non-public area between the stationary and the mobile charging pad. For this area, protection target 2 is pursued with priority: to avoid inadmissible temperature rise of conductive parts and to prevent ignitions.

Area 2: Transition section. The non-public section between the area of operation (1) and the public area (3) is delimited towards the outside by the tangent line perpendicular to the vehicle's side sill. For this area, protection goal 2 is met based on area 1. Protection Goal 2 is met based on the assumption, that this is no public area.

Area 3: Public area. Area to the side, front and rear of the vehicle. For this area, protection goal 1 is persued with priority, in order to prevent users and uninvolved third parties from exposure to inadmissibly high magnetic fields.

Area 4: Vehicle interior. Protection goal 1 applies primarily to this area analog area 3: Users and uninvolved third parties shall not be exposed to inadmissibly high magnetic fields.

1.1 Avoiding Temperature Rise and Thus Ignition

In the protection area, the magnetic flux density shall be kept sufficiently low to avoid undue temperature rise of parts located within the area of operation. The maximum absolute temperature for parts that need not be touched during operation is specified on the basis of IEC 60364-4-42:2010-05 or DIN VDE 0100-420:

- metal parts with metallic surface: ϑ_{max} 80° C
- parts with non-metallic surface: ϑ_{max} 90° C
- these values apply at an ambient temperature of ϑ_{amb} 40° C

If the temperature of a test body exceeds the maximum permissible value, it shall be verified by other means, for example sensors, that any foreign material between the active surfaces does not exceed the maximum permissible values.

1.2 Field Limit for Public Area and Vehicle Interior

For protection areas 3 and 4, the maximum permissible magnetic flux density is limited with reference to the ICNIRP guidelines of 1998 [1]: e.g., 6,25 μT at 140 kHz. If the reference value according to ICNIRP is exceeded, verification shall be provided, for example, by simulation, that the basic limiting values (dosimetric values of tissue) are not exceeded. If compliance with the ICNIRP recommendation cannot be achieved, proper measures shall be taken. Proper measures for complying with the limiting values may be, inter alia, all-round monitoring, evaluation of current flow deviations and constructional provisions.

2 Efficiency Targets

For efficiency determination, the transmission distance from a.c. power supply – comparable with the energy supply point in conductive charging – to d.c. outlet on the vehicle side is considered. It is of no importance whether this output is coupled into a charging device or connected directly to a battery.

Efficiency calculation is based on the assumption that:

- the contactless transfer system is in the nominal position,
- the protection goals are met in protection areas 1 to 4,
- the maximum input power is drawn from the power supply.

Such calculated value is supposed to ideally be as follows: – efficiency $\eta B \geq 90\%$

The total power drawn from the power supply (e.g., in Germany) shall not exceed a current of 16 A at an a.c. supply voltage of, e.g., 230 V. In Germany, for example, this would be a maximum supply power of 3,68 kW. Taking into account the transfer efficiency, a calculated delivered power of more than 3,3 kW would thus be available on the vehicle. The maximum output power is to be stated by the manufacturer.

3 Geometrical Definitions – References and Dimensions

A nominal position is defined for the purposes of unification of general measuring methods, comparability and subsequent compatibility testing. It is characterized by the relative position of the charging pads with respect to the zero point on all three axes and with regard to their parallelism.

The height is defined as follows:

- ground clearance at nominal position: $d_{\text{mech, nom}}$ = 135 mm (Z axis)
- offset in the direction of travel: $d_{\text{offs-X-nom}}$ = 0 mm (X axis)
- offset transverse to the direction of travel: $d_{\text{offs-Y-nom}}$ = 0 mm (Y axis)

It is not important for the measurement whether the stationary coupling plate is in-ground or on-ground mounted.

3.1 Geometrical Size of the Stationary Charging Pad

Guideline dimensions for the functional surface of the stationary charging pad are as follows:

- in the direction of travel: $l_{x,\ primary} \leq 1\ 000$ mm (X axis)
- transverse to the direction of travel: $l_{y,\ primary} \leq 1\ 000$ mm (Y axis)
- in the height direction (reference height: road surface): $d_h = 0$ mm - 50 mm (Z axis)

The actual permitted size is determined by the magnetic flux density, the effects of which being verified by measurement of test bodies used on the test bench (see Clause 10) of the total system. For the size of the functional surface of the stationary charging pad, it must only be assured that the protection values are met in interoperable service. The square and the circular functional surface best provide positional tolerance with regard to rotations.

3.2 Position of the Stationary Charging Pad Relative to the Parking Position

The zero reference point is located centrally in relation to the parking position with respect to the X and Y axis, its position coinciding with the centre point of the stationary coupling plate. Absolute dimensions are not required for primary–secondary due to centre–centre arrangement. After consideration of many different parameters and utilization scenarios, this is the preferred arrangement.

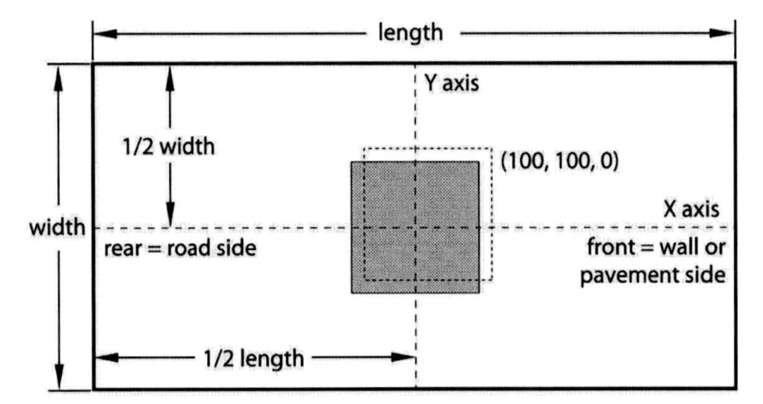

Fig.2. Position of the stationary charging pad in a standard parking space, maximum offset of mobile pad

The mobile coupling plate is preferably located at the centre of the vehicle in relation to the Y axis. For the X direction, its position shall be chosen such to allow a vehicle to be positioned in a parking space with reasonable safety in both X and direction.

3.3 Geometrical Size of the Mobile Charging Pad

The size of the mobile charging pad corresponds to the size of the stationary coupling plate. It is, however, not required to be exactly the same dimension. Guideline dimensions for the functional surface of the mobile coupling plate are as follows:

- in the direction of travel: lx,secondary ≤ 1 000 mm (X axis)
- transverse to the direction of travel: ly,secondary ≤ 1 000 mm (Y axis)

The size of the functional surface of the mobile coupling plate is such that the protection goals are complied with in protection areas 1 to 4.

3.4 Distance Between the Pads

For the design and electrical dimensioning of the contactless energy transfer system, the distance between the pads is of decisive importance.

For the stationary coupling plate, two mounting types are possible:

- in-ground mounting
- on-ground mounting

Values are given with reference to the mechanical surface of the charging pads.

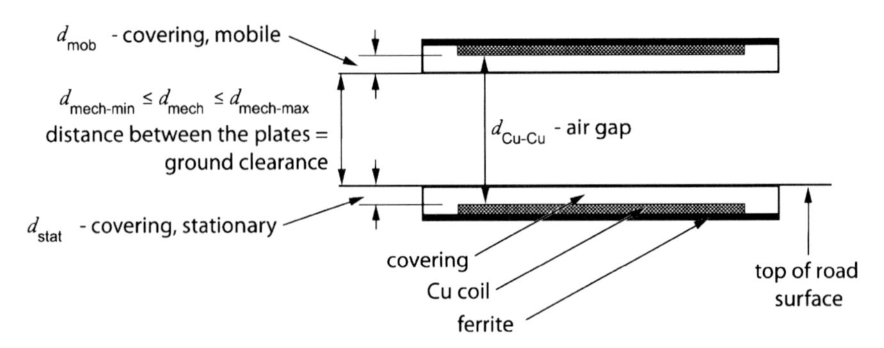

Fig. 3. In-Ground Mounting

The stationary charging pad is completely embedded in the ground and mounted flush with the road surface (e.g., ferrite, copper). The surface of the stationary coupling plate is located at zero on the Z axis. Thus the distance between the two coupling plates is equal to the ground clearance of the vehicle underneath the pickup (mobile charging pad).

The mounting height of the vehicle mobile charging pad is defined by specifying the following minimum and maximum distance values for the working area with reference to the system's zero point: ground clearance d_{mech} = 100 mm to 170 mm (Z axis).

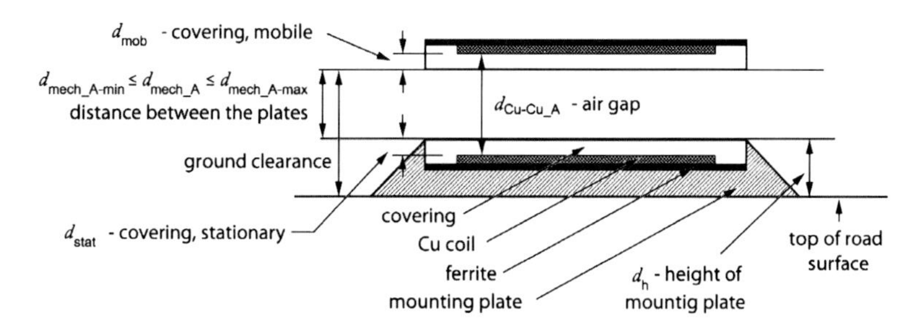

Fig. 4. On-Ground Mounting

The stationary charging pad is mounted in whole or in part on the ground up to a maximum height d_h.Thus the distance between the two charging pads is smaller than the vehicle ground clearance underneath the pickup (mobile charging pad): Max. mounting height above road surface d_h is defined by installation guide (Z axis, = zero point reference). A maximum mounting height does not only describe the maximum package integration options, but also specifies the critical minimum distance between the plates with respect to functional reliability. Defining a maximum mounting height is also reasonable in view of, for example, national specifications such as road construction regulations.

Consideration of the operating distance between the charging pads requires in each individual case that the actual onground mounting height d_h is to be deducted from the ground clearance (distance between plates: $d_{mech\_A} = d_{mech} - d_h$). The operating distance for on-ground-installations is defined at max. mounting height (e. g. 50 mm): d_{mech} A 50 mm to 120 mm (Z axis, i. e. cars with a ground clearance 100 up to 170 mm are covered). Accordingly other set ups can cover higher ground clearances, always meeting protection and efficiency targets.

4 System Frequency

Diverse distances between the stationary and the mobile coupling plate result in different inductance values. In order to meet the resonance condition, the feeding side must be able to track the frequency.

Frequency with which the energy tranfer is realized in the air gap:

- nominal system frequency at nominal position: f_{sys} = 140 kHz
- frequency range for resonance tracking (air gap and offset compensation): Δf = -20 kHz to +50 kHz

At frequencies above 150 kHz, more severe radio interference suppression and ICNIRP limits will have to be met.

5 Coil Construction

As a reference for the coil construction, a coupling plate is characterized by the following:

- number of phases: single phase (normal case, for exceptional cases see NOTE 3)
- characteristics of the magnetic field such as: single coil
- geometry of the assumed single coil: approximately rectangular

As the internal construction is not subject to this proposal, the physical coil number cannot be prescribed. On the other hand, it is required to determine that the magnetic field has to be configured assuming a single coil. If there wouldn´t be this minimum specification, it would be conceivable to use many single coils on both sides, this, however, almost entirely excluding any compatibility with one single coil.

The rectangular (ferrite) construction best ensures sufficient distance to the door sill. In the case of the same ferrite surface area, the round shaped design is broader. A round coil can, of course, also be used with a rectangular surface; this will, however, not be the optimum solution. On the other hand, an absolutely rectangular coil cannot (reasonably) be implemented. The phrase "approximately rectangular" is already a compromise. In deviation from the specified number of phases, multi-phase operation is permitted, if at the same time evidence is provided that the efficiency and protection requirements ar met as well as criteria for interoperability between single phase and multi-phase systems.

6 Circuit Topology

For simplification of the system and to provide assurance of compatibility, it is assumed that there is a series resonant circuit on both the primary side and the secondary side.

On the primary side the focus is on the following:

- Provision of a controlled load-dependent magnetic field, where applicable.
- Active tracking of the working frequency with position-dependent variations in the resonance condition.

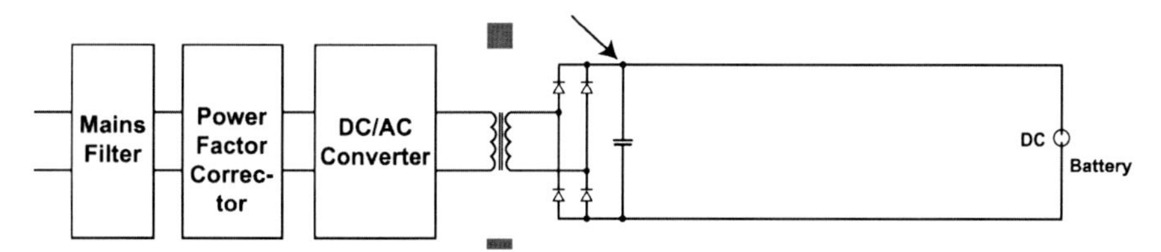

Fig. 5. Cicuit topology for maximum efficiency without DC/DC-module on the mobile side

Basic requirements are, of course, determined by the Battery Control Unit (BCU). In the resonant circuit, however, the primary side as the field generating side is the feeder (actio), whereas the secondary side is the collector (reactio).

Through its configuration and magnetic coupling with the primary side, the mobile charging pad affects the resonance condition of the total system. Energy decoupling is the task of the secondary side. As the design of the vehicle side interface is not subject to this proposal, other solutions are permissible, unless there are additional effects on the primary side.

7 Control and Communication

In the basic design, the mobile part is connected to the stationary part only via magnetic coupling. For basic operation, a control loop from the mobile part to the stationary part is not mandatory. A control loop is optional.

Unless a mobile part or a vehicle is positioned above the stationary charging pad, no magnetic field shall be generated that would offend the protection goals defined for protection area 3.

With the vehicle is approaching, the stationary part is activated by a signal from the vehicle and set into standby position. At the same time, magnetic fields are generated serving for precise positioning above the charge pad.

After the vehicle has been correctly positioned above the stationary part within the defined tolerance, power transfer is activated by an enable signal. Compliance with the required values is verified by test bench testing.

Primary power shall only be supplied, if all safety conditions are met on the secondary side. These include insulation, temperature and other monitoring procedures which are suitable, to detect safety critical situations coupled with vehicle specific requirements, where applicable.

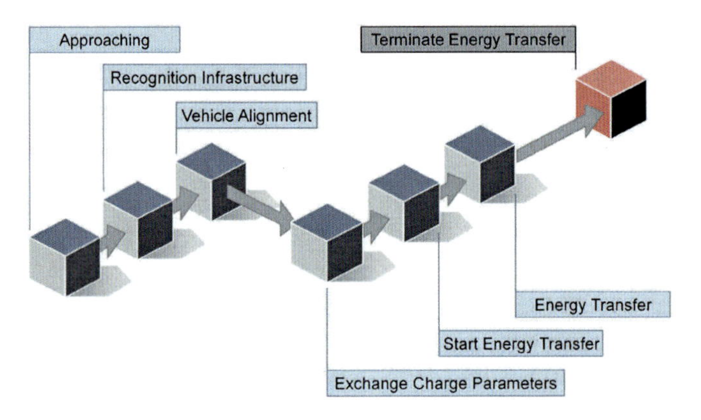

Fig. 6. Use Case description of inductive charging phases and their control requirements

7.1 Operating States and Handling

Despite of the phases describing the charging the process in general, different scenarios of system usage must be considered by the communication concept. As well the communication concept must show flexibility to cover and support different technologies, e.g. for vehicle alignment or energy transfer.

7.2 Communication Protocol

Taking under account that several EVs released in the next months are going to support different modes of charging it is desirable to provide an harmonised control communication where technical reasonable. In Figure 7 the principle approach of ISO/IEC15118 covering different charging modes is described. The majority of the specification comprised of the communication layers between physical layer and application layer, but also a core set of messages in the application layer are shared by all charging modes.

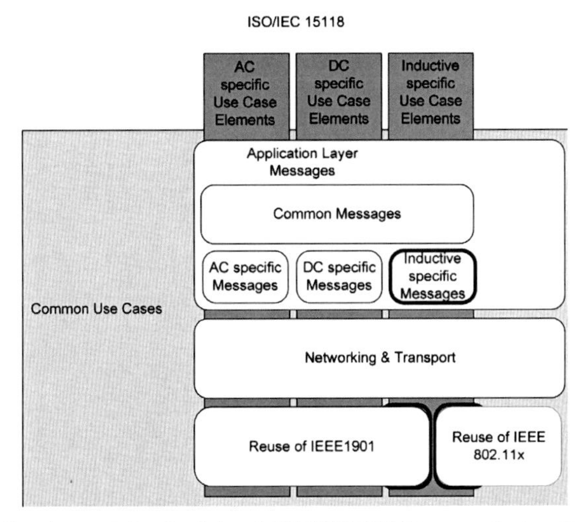

Fig. 7. Fundamental principle of ISO/IEC15118 to maximize the reuse of core parts of the specification to fulfill the requirements of the different charging modes

8 Test Bench

Verification of compliance with the values specified in this proposal with respect to:

- safety requirements,
- functional values and
- compatibility between the different systems

is provided using a standard test bench. The test bench incorporates receiving elements to accommodate:

- a stationary charging pad (charging plate) and
- a mobile charging pad (pickup),

their positions relative to the defined zero point being changeable within the range specified in this proposal. Connection of the appropriate inverters or on-board electronics is also needed for the test performance. Power measurement is done by load simulation.

The test body serves to measure compliance of foreign material located in the area of operation (area 1) with the relevant temperature limits. The temperature is measured under steady state conditions. In a first approach, materials available in Germany will be used.

References

[1] International Commission on Non-Ionising Radiation Protection (ICNIRP), Guidelines for limiting exposure to time-varying electric, magnetic, and electromagnetic fields (up to 300 GHz), Health Physics, vol. 74, n° 4, p. 494–522, 1998.

Sebastian Mathar
RWTH Aachen University
Institut für Kraftfahrzeuge
Steinbachstr. 7
52074 Aachen
Germany
mathar@ika.rwth-aachen.de

Samuel Kiefer
Kiefermedia GmbH
In der Spöck 1
77656 Offenburg
Germany
km@kiefermedia.de

Keywords: electric vehicle, automatic charging, smart grid, wireless charging, standardization, magnetic resonant inductive charging, wireless communication

Research Project E-Performance - In-Car-Network Optimization for Electric Vehicles

G. Gut, ForTISS GmbH
C. Allmann, Audi Electronics Venture GmbH

Abstract

In the automotive domain the permanent increase in functionality led to a vast number of electronic control units (ECUs) in today's cars, but packaging and network bandwidth demands became problematic in the last years. Thus it is vital to integrate more functions per ECU and shift in-car-networking complexity into software. To master this challenge, it is essential to find local and global optimization possibilities, which includes practical software component partitioning strategies while not overlooking the multiplicity of influence factors as well as smart software modules that help to reduce the energy demand wherever possible.

1 Introduction

The development of Electric Vehicles, in some facts, is a contradictory challenge that ought to combine customer expectations, quality and price. Knowing that the prospects on the cruising range of comparable combustion-engine vehicles cannot be realized contemporarily on battery-electric vehicles, a large variety of other automotive characteristics must be redefined. The redesign of these characteristics, going along with totally new approaches in vehicle characteristics, is indispensible for increasing the customer's benefit. The resulting layout criteria are not yet sustainably verified because of their mostly unknown, multi-variant interaction.

Therefore, the joint research project „e performance" is meant to demonstrate technical and economical feasibility on the basis of a re-usable module kit (see Fig. 1). This publicized, integrated development approach on the basis of „E-Modules" will be exemplarily demonstrated and evaluated using a collaborative vehicle concept called „e-tron research car 2012" [1,2,3].

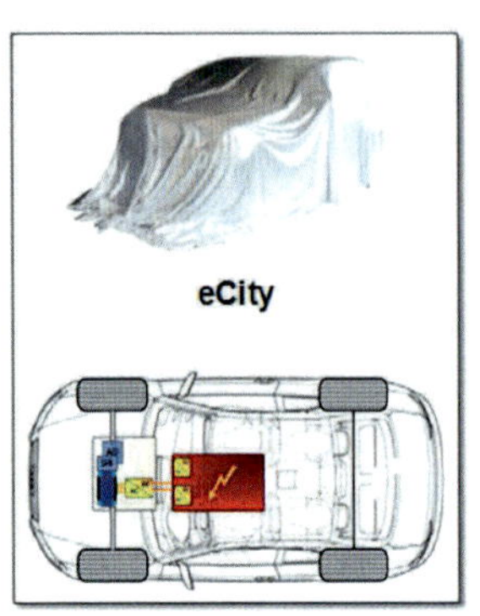

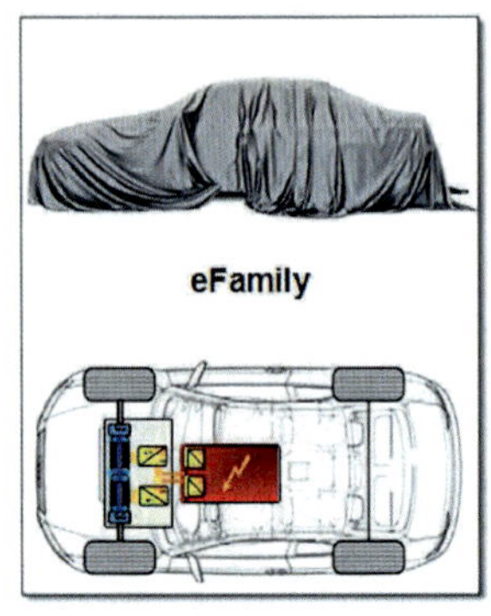

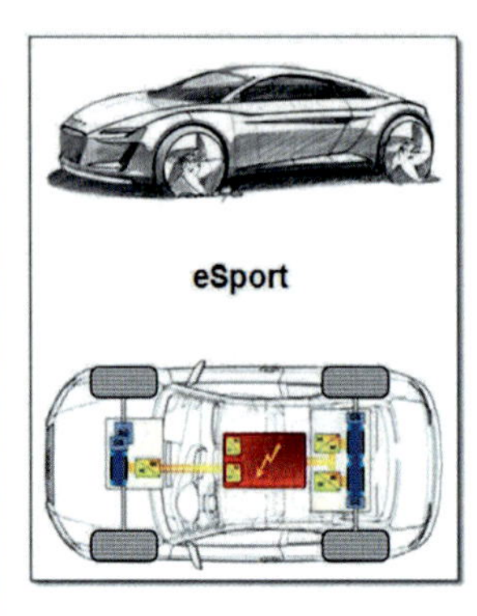

Fig. 1. Modular system kit for EV

2 Networking Issues

One of the mentioned "E-Modules" on the car is the network architecture including the software components. Nowadays modern vehicles with combustion engines comprise 80 electronic control units (ECUs) or more. This comes as a result of the rapidly increasing number of vehicle functions and thus in-car communication demand.

The continuous increment of driver assistance functions in the last years is likely to be the main contributor to the rising demand in network bandwidth. This includes adaptive cruise-control (ACC), traffic sign recognition, night vision, automatic parking and many more. What all these functions have in common is their background in image processing. As current network architecture layouts can be seen as a distributed system in this area this leads to massive communication of image data between sensors and control units. A second contributor is the infotainment system, where in the latest generation of premium cars, it is even possible to stream music and movies from the central system in the dashboard to the rear seats. Also in the last years, vehicle dynamics functions have consequently grown and need high speed real-time communication between the components. Torque vectoring can be seen as an example here.

To handle all these and also future functions, and at the same time avoid a further increase of the number of ECUs and bandwidth demand, new concepts have to be developed and prototypical tested.

3 Specific Characteristics of Electric Vehicles

The development of electric vehicles, seen as an alternative to cars with combustion engines, becomes more and more important due to rising gas prices and governmental funding. To cope with the difficulties involved, the development and introduction of a lot of new technologies is essential. Route- and cruising range prediction, adaption of charging times according to energy pricing, the energy store in the cloud as well as time and state dependent activation and deactivation of software functions like charge management are a few examples for this challenge.

Furthermore, cooling and heating of the traction battery is essential, as it is only safely operational in a limited temperature range. This is important for driving as well as for charging. Therefore and given the fact, that lost heat of the combustion engine isn't available any more make the thermal management and conditioning of the vehicle evolve from a one of many to a central and at the same time very complex system.

While taking out the combustion engine and gearbox, a lot of new and mostly high voltage (HV) components like traction inverters, charger, DC-DC converters, battery management, air conditioning compressor, sound actuators, HV safety and energy manager are added.

Yet distinguishing for electric vehicles is the new vehicle state "Charging", which is more complex than it looks like at first sight. This is due to the mentioned battery-bound issues, but also functional safety plays a major role here. An example is that the traction inverters have to be locked while charging, so that the vehicle cannot move during the charging process.

In summary there is a lot of new functionality that is dependent on the vehicle state and highly distributed.

4 Networking Goals

The permanent increase in functionality led to a vast number of electronic control units (ECUs) in today's cars, but packaging and network bandwidth demand became problematic in the last years. Thus it is necessary to not have a further increase and at best reduce the number of ECUs and at the same time the demand for communication between ECUs.

Besides that and due to the low energy density of batteries in comparison with gas the main objective in the development of electric vehicles is to save energy wherever possible [6]. Hence this is also a focus in the area of in-car-networking, especially as the functions needed for charging are spread over a lot of electronic control units (ECUs) and vehicle buses.

Thus it is vital to shift in-car-networking complexity into software and integrate more functions per ECU. To master this challenge, more processing power is needed, which often comes with increased power dissipation. To address this problem semiconductor companies came up with multicore ECUs, which provide more computation power with less energy consumption. Nevertheless, other solutions are also thinkable, e.g. ECUs additionally equipped with specialized processors like DSPs or ASICs for certain tasks.

5 Highly Integrating Vehicle Functions

An example in the domain of in-car-networking is the approach of highly integrating vehicle functions onto a single ECU. Thus it is possible to reduce the total number of ECUs significantly. There are currently many discussions ongoing on how to group the Software Control (SWC). Possibilities include grouping of SWC in a way to reduce communication demand between ECUs or such to reduce the latency between SWC by eliminating the communication over busses or by safety-level to reduce certification demand of lower safety-level SWC [4,5]. Aside these goals on software and bus level, the superior goal on a vehicle-base is to save construction space and energy, which leads to indirect goals of less cabling, ECUs, weight and energy consumption of components and cabling. [8]

In electric vehicles there are increased communication needs due to more complex networked functions. One of many examples here is the (temperature) conditioning of the traction battery. To master this, HV components like DC-DC converters and battery management systems are needed as well as comfort components like climate control, and also central energy management [7,9]. Thus we propose partitioning functions with high communication demand among each other on the same ECU as this can help to reduce the needed bandwidth on vehicle busses dramatically and thus can also save costs in the development for new bus systems as well as in the car for more expensive cables and plugs.

5.1 Vehicle Dynamics

Longitudinal and lateral dynamics functions have high real-time communication demand between each other. As these functions have grown separately they each bring their own ECU and are connected via FlexRay or a dedicated CAN bus. We propose to integrate these functions into a single ECU to reduce the busload dramatically or, in case of a dedicated CAN bus, to reduce the number of vehicle busses.

What can be seen as an extra here is that the control response time can be drastically reduced and thus may allow better driving dynamics.

5.2 Charging, Pre-Conditioning and Idle Battery Monitoring

These electric vehicle specific features are newly introduced and not used while driving. They need additional power that is not or cannot be used for driving, thus they have to be taken into consideration when calculating available range or even environmental impact [7].

To ensure a high level of energy effectiveness here, ECU selective wakeup and sleep is necessary. When using Gateways, a topology where the needed ECUs can communicate without the Gateway (e.g. connected to the same bus) should be preferred. Additionally, the implementation of intelligent algorithms for energy saving during these processes is advisable.

5.3 Loading and Unloading of SWC

Taking the possibility of (half-)dynamically activating or deactivating SWC into account can help integrate even more vehicle functions onto a single ECU by taking advantage of the fact that a lot of these functions are in the electric vehicle exclusively used in only one of the vehicle state "Drive" or "Charge". There are vehicle functions and thus SWC that are only needed exclusively in one them. We propose a partitioning SWC of both vehicle states, that are only active in exclusively one state, onto one same core of an ECU. For instance the control of the sound actuators and the charge management / charging device can be seen as a use case.

6 Prototypical Realization

Throughout the research project "e performance" it was a major goal to realize some of these concepts in the demonstrator vehicle. The in-car network in this vehicle consists of three main busses, representing "functional groups" or domains (see figure 2). The HV CAN-Network is hosting – mostly newly introduced for electric vehicles – HV-related components like a battery management system or charging management. HV can thus be seen as a new and additional domain associated with electric vehicles. The Comfort CAN-Network is mostly unchanged in comparison with combustion engine driven cars and is hosting components like central locking or wiper control. The third bus, here called Vehicle Control and Active Sound CAN-Network, is hosting components from multiple domains like active sound, which is required by law for electric vehicles, or traction inverter control. The infotainment unit is connected via WLAN and not part of the figure shown below. Aside the ECUs and the associated functions on these busses most of the more complex control functions have been highly integrated on a central ECU, which also serves as a Gateway between the vehicle busses. Without this high integration the newly needed and increased functionality would otherwise need even more ECUs, which means more and – due to more hardware components in electric vehicles like DC-DC converters – not available package space as well as increased energy consumption and weight due to more ECUs and cabling. Highly integrating functions with a high communication demand limits a global increase over all in-car networks and also has the benefit of a much faster data exchange between the integrated functions. The integration of many of the complex functions onto a highly integrated ECU can be seen as a further step to a centralized control and decentralized intelligent actuator/sensor system.

Over a dozen complex functions were highly integrated on a single ECU using a powerful 900MHz PPC single-core processor. Using Matlab Simulink models and code generators, vehicle functions of 6 project partners were put together. This took a lot of efforts to synchronize and needed detailed planning not only of interfaces, but also processor and memory resources as well as component run times and execution intervals (control response times).

Besides central vehicle functions like energy management, charge management and thermal management, also the longitudinal and lateral dynamics functions were integrated to run on this same hardware. Combining the vehicle dynamics functions reduced the busload by approximately 90 kbit/s and thus made bus capacity available for new functionality. To confirm functional benefits of a now combined control response time of 1ms – that can be experienced by a potential customer – first tests have to be waited for.

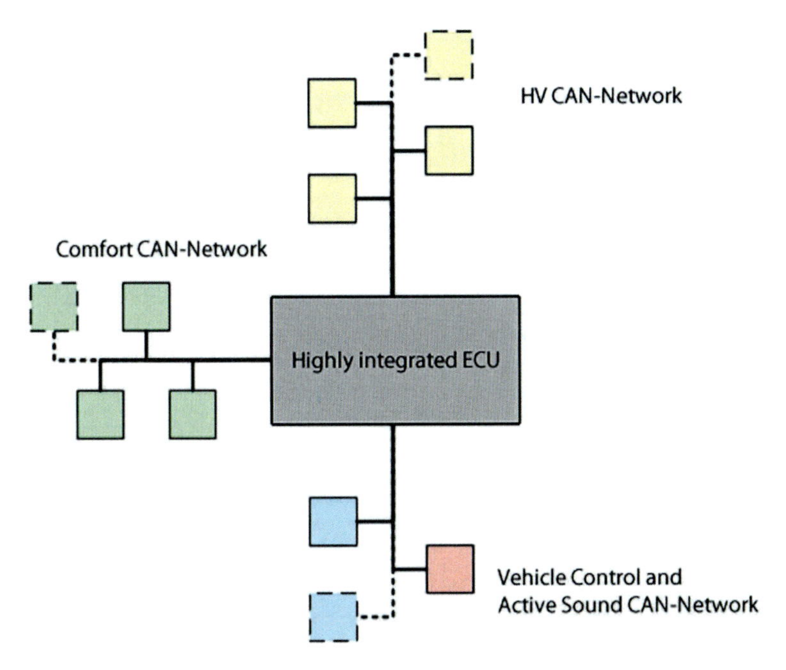

Fig. 2. Realized network architecture

The infotainment system is realized with an Apple iPad, solely building the central console where it is mounted to. It serves as Internet client with E-mail, browsing and social network integration, audio- and video-player as well as a display for the vehicle status and input device, e.g. for thermal and vehicle dynamics control. The iPad is connected to the vehicle via WLAN and further to the 3G network via a WLAN-UMTS-Gateway as well as to the in-car-network via a WLAN-CAN-Gateway (see figure 3). The iPad can also be taken out of the car and then used to gather the current vehicle status or configure charge times from remote locations via UMTS.

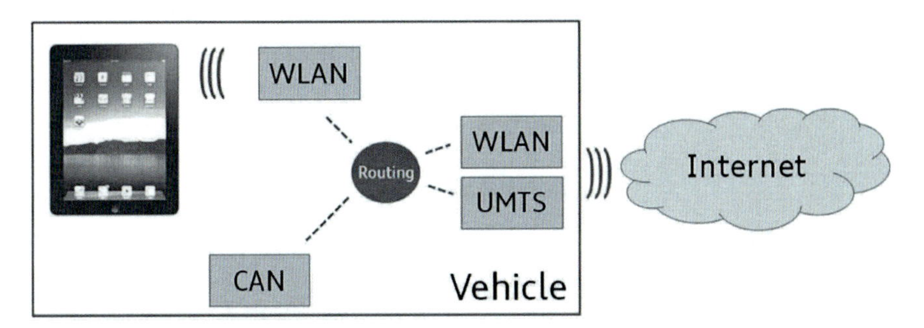

Fig. 3. iPad communication paths

The components needed for the charging process are connected to the same CAN bus, as far as no other functionality is affected, e.g. this is the case with the traction inverters, as in our case they only have one CAN interface and need high speed communication on another CAN bus, where most of the drive train ECUs are connected to. In these cases an electrical clamp – here called "charge clamp" – was used, so that not-needed ECUs aren't powered and energy consumption is kept at a possible minimum. When CAN transceivers that support selective wake-up are available for broader use they would be the logical replacement for these clamps by further reducing cabling and circuit complexity.

When the vehicle is turned off, there are two cases in which it has to perform a self-contained wakeup – pre-configured charging processes and battery condition monitoring. Even with the inactive vehicle, battery monitoring has to take place in intervals of approx. 10 minutes to ensure that the condition is not changing in an unexpected way. The battery management – lacking a low power mode – can perform these operations, but has to be awake. The charging management works like today's park heating, multiple timers can be set and charging is then automatically done when connected to a power source. Thus a low power ECU supporting this is needed here too. To solve these issues, an additional small-sized ECU – called "Ringer" – with a very low power consumption (less than 5mA) has been introduced. It provides a central timing source for the vehicle over a radio controlled clock and can also by regularly synchronized to a GPS clock. Furthermore it stores and manages the configured charging times that were set over the iPad. The Ringer is waking up the battery management system in a cyclic manner, notifies it to perform its operation and then controls the power down. This whole operations normally only takes seconds which makes it possible to save energy here by only starting up every 10 minutes. When the next configured charging time less the worst case time needed for charging is reached, the Ringer wakes up the charge management, which then calculates the actual needed time for charging according to battery status, temperature and further factors. If it lies in the future – which is mostly the case – it communicates the time to the Ringer and then powers down again. The Ringer will then wake up the charge management again in that point in time. Otherwise charging would start immediately. This process helps to gain an effective charging process.

7 Conclusions

We propose to highly integrate SWC on ECUs in a communication and processing-time oriented manner, always keeping the high number of individual factors of influence in mind. Besides a lot of to-be-solved issues in the fields of timing- and safety planning and analysis, integration and testing, this promises to reduce overall costs, weight, energy consumption and installation space. We look forward to a detailed analysis of the effects that our measures will have in practice on the viewpoint of the full vehicle.

References

[1] C. Allmann, Vorsprung durch Technik – Projekt e perforamance bei Audi, 1. VDI-Fachkonferenz Elektromobilität – Automobilindustrie trifft Energiewirtschaft, VDI Wissensforum, Düsseldorf, 2010.

[2] M. Schüssler, et.al., Forschungsprojekt e performance, Aachener Kolloquium, 2010.

[3] C. Allmann, M. Schüssler, Research Project e performance, Solar Summit, Freiburg, 2010.

[4] J. Schoening, Challenges of Changing a Common Vehicle Architecture – Evaluation Process for Functional Integration, 2006 SAE World Congress, Detroit, Michigan, 2006.

[5] K. Schmidt, et.al., Entwurfsaspekte für hochintegrierte Steuergeräte mit unterschiedlichen ASIL-Stufen, ATZ elektronik 01/2012, 2012.

[6] L. Brabetz, P. Hochrein, Robuste und energieeffiziente Bordnetzarchitekturen. at – Automatisierungstechnik, Vol. 60, No. 2, 2012.

[7] F. von Borck, et.al., Die Traktionsbatterie - Schlüsseltechnologie für den Durchbruch elektrischer Fahrzeugantriebe, ATZ elektronik 01/2008, 2008.

[8] M. Broy, et.al., Architekturen softwarebasierter Funktionen im Fahrzeug: von den Anforderungen zur Umsetzung, Informatik Spektrum, Vol. 34, 01/2011, 2011.

[9] A. Kallfelz, Battery Monitoring Considerations for Hybrid Vehicles and Other Battery Systems With Dynamic Duty Loads, Battery Power Products & Technology, Vol. 10, Issue 3, 2006.

Georg Gut
ForTISS GmbH
Guerickestr. 25
80805 Munich
Germany
gut@fortiss.org

Christian Allmann
Audi Electronics Venture GmbH
Sachsstr. 18
85080 Gaimersheim
Germany
christian.allmann@audi.de

Keywords: partitioning strategies, network optimization, high integration, software component allocation, energy consumption, charging process

Automotive Ethernet, a Holistic Approach for a Next Generation In-Vehicle Networking Standard

P. Hank, T. Suermann, S. Müller, NXP Semiconductors Germany GmbH

Abstract

For the next-generation in-vehicle networking infrastructure beyond CAN and FlexRay, the automotive industry has identified Ethernet as a very promising candidate. Ethernet being an IEEE standard and commonly used in consumer and industry domains provides a high re-use factor for components, software and tools. In addition, Ethernet has the bandwidth capability that is required for new driver assistance and infotainment systems, for example. However, to become a success story, solutions for the automotive industry have to be further optimised in terms of scalability, low cost, low power and robustness. The first optimisation steps on the physical layer level have already been taken but more innovation needs to be focused on automotive use cases. This paper discusses new network topologies and components and describes an evolutionary path of Ethernet into automotive applications.

1 Introduction

Communication and bandwidth requirements increase as more and more new, complex applications appear in the car, for example for enhanced safety and entertainment solutions. End users expect the same level of entertainment functions in the car as known from their home environment. Furthermore, existing vehicle control networks like the LIN, CAN and FlexRay standards are not designed to cover these increasing demands in terms of bandwidth and scalability that we see with various kinds of Driver Assistance Systems (DAS). Future networking technology should re-use as much as possible from consumer and other non-automotive domains, while taking into account the automotive-specific requirements. This includes hardware components as well as software stacks. Communication solutions for higher bandwidth, such as MOST, are available but expensive for a broader usage in automotive networking systems. Today, in-vehicle networking architecture appears as a heterogeneous system as a result of its historically grown nature (see the left schematic in Fig. 1.).

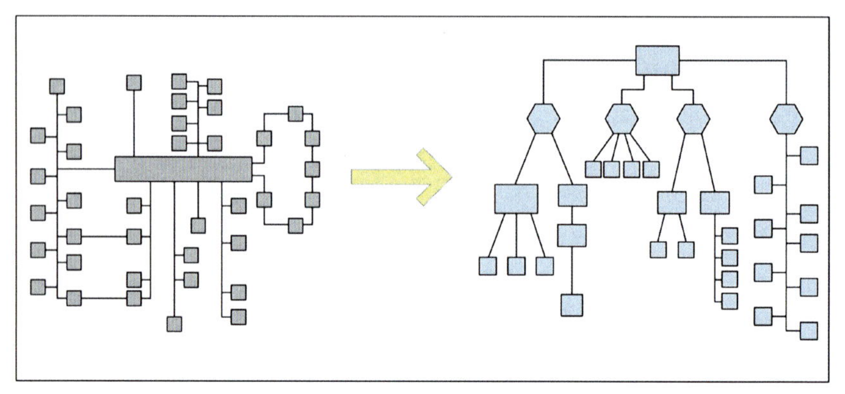

Fig. 1. Domain architecture, today and in the future

A new in-vehicle networking system built from scratch and without legacy would most likely have the architecture as shown in the right schematic of Fig. 1. Here, ECUs are structured in a hierarchical architecture where application domains are connected through a data highway. Ethernet provides all the prerequisites for such a holistic approach. It could be used as a backbone bus to connect the various application domains as well as for sub-networks that just require higher bandwidth. Today, switched Ethernet networks rely on a point-to-point communication where the available bandwidth is more efficiently used compared to broadcast systems like CAN or FlexRay. The switching concept can be advantageously applied to bridge the domain boundaries without time-consuming packing and re-sorting of the transmitted messages or packages as needed in a complex gateway.

The usage of Ethernet in the car means a paradigm shift in the design of next-generation in-vehicle networking systems: connecting different domain networks, transporting different kinds of data (control data, streaming, etc.) and fulfilling the stringent robustness demands in terms of extended temperature range and EMC performance.

2 Evolution towards Automotive Ethernet

Ethernet is an open LAN standard and defines the two lower layers of the OSI reference model. Over the past decades, the IEEE 802 standardisation committee has specified several physical layers from 10Mbps up to 10Gbps. IEEE 802.3u (100Base-TX) is widely used in consumer and industrial domains and has recently been selected for car diagnostics over IP as described in ISO 13400.

Higher layers have to be taken into account besides data link and physical layer, as shown in Fig. 2. The Audio/Video Bridging (AVB) Task Group of AVnu alliance [1] has defined IEEE standards covering higher-level services for IEEE 802-compliant networks. The protocols IEEE 802.1 AS, QAT, QAV and BA cover address timing and synchronisation, streaming reservation, forwarding and queuing in Audio/Video Bridging systems. In addition, transport protocol layer standards like IEEE 1722 facilitate stream-time-sensitive audio and/or video across Ethernet AVB networks and interoperability between end stations.

Driver Assistance Systems are increasingly using video cameras for applications like surround view. Activities for the standardisation of the communication protocol and physical layer have recently been started in ISO 17215 called "Video Communication Interface for Cameras". In order to further develop the Automotive Ethernet technology, an industry-wide Special Interest Group named **O**ne **P**air **E**ther**N**et Alliance has been formed for the physical layer. The so-called "OPEN Alliance" will work on standardisation for components and compliance tests based on the Broadcom BroadR-Reach technology [2]. A further intention of the interest group is to gather requirements for future technologies such as "Reduced Pair Gigabit". Finally AUTOSAR addresses Automotive Ethernet in their software layer stack.

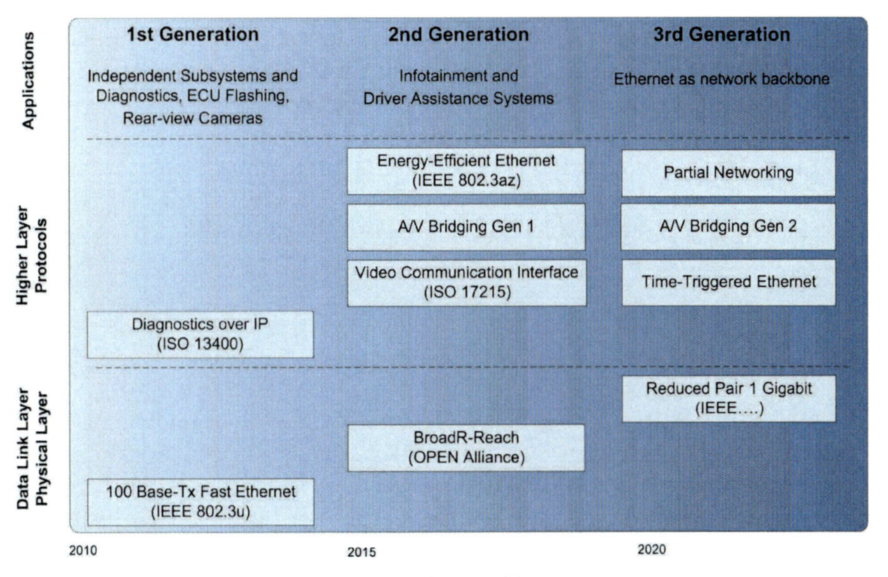

Fig. 2. Ethernet roadmap, generations and layers

Ethernet has not been developed for TDMA networking and one topic that still needs to be addressed for automotive systems is a suitable solution to achieve the required real-time performance and Quality of Service. AVB already includes measures to ensure the timely delivery of media streams.

For the Automotive AVB Gen2 Work Group, latency time improvements are one of the major targets. First applications for Time-Triggered Ethernet took place in Avionic systems [3] with highest safety level requirements. Defined in SAE AS6802 and different to Audio/Video Bridging, TTEthernet is based on a distributed clock-synchronisation algorithm that finally results in an exact schedule with deterministic behaviour.

Although the integration of both AVB and TTEthernet is possible [4], further investigations are needed to allow its use in automotive applications where multimedia streams, real-time control data as well as diagnostic information and software updates will be transmitted on the same network.

2.1 Generation 1: Diagnostics Over IP

The first Ethernet applications for the automotive industry are On-Board Diagnostics (OBD) and the update of ECU flash memories. For the reading-out of diagnostics data and the updating of software within a given time frame, the performance of existing in-vehicle bus systems such as CAN and FlexRay was not sufficient. Ethernet 100Base-TX with CAT 5 has been chosen for the interface between the vehicle and the diagnostics test equipment. The higher bandwidth of Ethernet saves costs in service and production. ISO 13400 and ISO 14229 utilise existing industry standards and define a long-term stable state-of-the-art diagnostics standard. Products and components are already available from versatile suppliers as automotive requirements such as robustness and temperature tolerances have been relaxed for this specific use case.

2.2 Generation 2: Driver Assistance Systems and Infotainment

The second generation of Automotive Ethernet will address infotainment and camera systems for surround view applications. Today's rear-view camera solutions often use LVDS for the transfer of video data, which works well for single cameras. In future systems, this will consist of more cameras and fuse with sensor data from short/long-range radar, for example (Fig. 3). An LVDS-based system becomes inefficient in terms of wiring harness and expensive cables and connectors. Switched Ethernet allows video cameras to be connected to a central control unit for synchronisation and further processing.

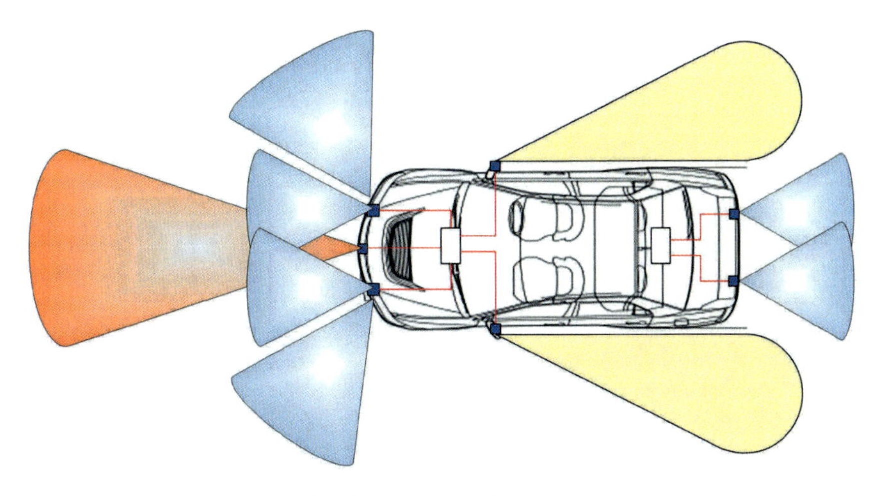

Fig. 3. Camera and radar for Driver Assistance Systems

Ethernet cameras can further benefit from "Energy-Efficient Ethernet" (IEEE 802.3az), which introduces a Low- Power Idle (LPI) mode and wake-up functionality to save energy when the cameras are not being used. Additionally, solutions for Power over Ethernet (PoE) are preferred to further reduce wiring harness. The demand for higher bandwidth and lower latency is obvious. Multiple high-resolution cameras for uses such as object detection require uncompressed data transfers, for example to avoid compression artefacts for obstacle detection, and are a strong driver of higher bandwidth.

Recent infotainment solutions are mainly based on proprietary and non-scalable technologies. Automotive Ethernet addresses this emerging application field in a cost-effective manner by making use of the AVB standard. Synchronised transmission of video and audio data with guaranteed latency can already be achieved with existing AVB Gen 1 Ethernet components. However, both applications fields will benefit from current AVnu standardisation activities and the most recent Switch and PHY developments based on the BroadR-Reach physical layer.

2.3 Generation 3: Ethernet as Network Backbone

While for generation 1 and 2 Ethernet remains confined to subnets of certain applications like infotainment and driver assistance, with generation 3 Ethernet will become the backbone of the in-vehicle network. A typical backbone is illustrated in Fig. 4. The design of such a network will be a paradigm shift in the way communication between ECUs has to be organised by the network management.

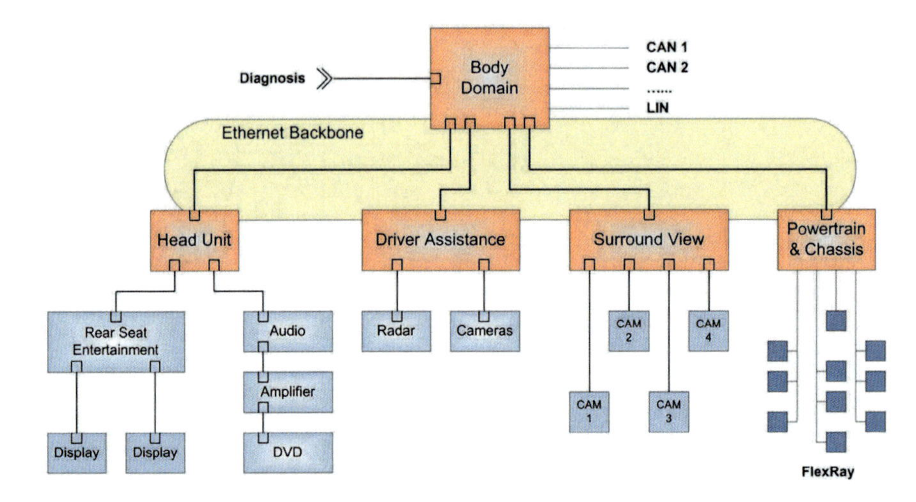

Fig. 4. Ethernet backbone in domain architecture

The communication network will be organised in a highly hierarchical way with the main domain controllers connected via an Ethernet backbone. Sub-networks below the domain controllers may be Ethernet-based, too, with switches bridging between the network levels. This structure provides a scalable solution as each port of a switch can in general be implemented as 10Mbps, 100Mbps or 1Gbps without any changes to higher protocol layers. The paradigm shift also becomes visible in how a message is transferred via possible domain boundaries to its destination. While in today's networks complex and network-dependent gateways realise this function, known and mature IP-based routing within switches and routers are proposed for backbone networks. The advantage is that IP routing is fully independent of the underlying network implementation, thus allowing a homogeneous addressing concept for the whole in-vehicle network. Moreover, the IP concept enables the straightforward connection of the car infrastructure to the Internet [5], a trend which is mainly driven by end users expecting the same access to services provided by the Internet as known from home.

A further characteristic of the new architecture is that in principle there will be one backbone network technology only, namely Ethernet, which has to accommodate different data communication classes like diagnostics, video/audio streaming and highly dependable control data. While AVB Ethernet and TTEthernet can already provide different levels of Quality of Service (QoS) combined with real-time performance, further R&D activities are needed to validate the secure coexistence of these different data communication classes on the same network.

3 BroadR-Reach – a 100Mbps Automotive Ethernet Solution

Though 100Mbps Ethernet and IP technology has been available since the early 1990s, it took Ethernet almost 20 years to attract the interest of the automotive industry as a next-generation in-vehicle networking standard. This was partly because of the lack of a physical layer suitable for use in vehicles. Fig. 5 illustrates the technical principles in more detail and explains why the consumer-oriented physical layers of Fast and Gigabit Ethernet did not make it into the car. Fast Ethernet is based on a MLT-3 signalling scheme (distinguishing three levels +/0/-) and features a uni-directional communication on two twisted pairs of cable. Gigabit achieves 10x higher data rate by introducing a bi-directional communication on four twisted pairs of cable and a PAM-5-based signalling scheme. To compensate for the 6 dB signal-to-noise loss of the PAM-5 versus the MLT-3 signalling, a costly error forward correction code (Trellis) has to be employed.

The high symbol rate of 125 MBaud for both Fast and Gigabit Ethernet makes a significant contribution to electromagnetic emissions in the critical FM radio band and thus rules out the use of inexpensive unshielded twisted pair cable in an automotive environment. The BroadR-Reach technology has managed to almost halve the symbol rate to 66.6 MBaud and paved the way for the use of unshielded twisted-pair cable. In principle, BroadR-Reach can be regarded as a "light version" of Gigabit, adopting the bi-directional communication scheme while requiring only one pair of cables. Thanks to the PAM-3 signalling, a bit error rate of less than 10^{-10} can be achieved even without error forward correction.

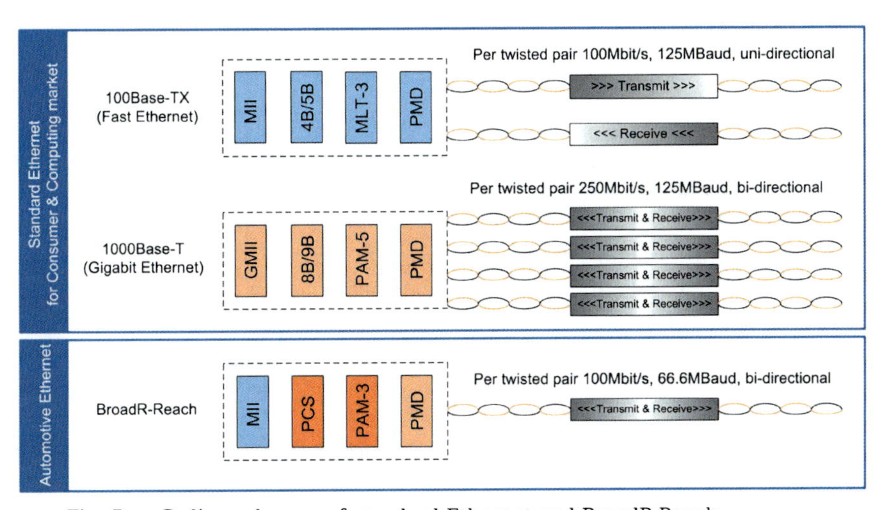

Fig. 5. Coding schemes of standard Ethernet and BroadR-Reach

Automotive applications impose considerably higher requirements on electronic systems and their components compared to the consumer world, mainly in terms of EMC [ISO11452] and environmental conditions.

First investigations indicated that BroadR-Reach is suitable for use in the automotive environment. However, to achieve the robustness needed for a next-generation in-vehicle networking standard, new optimised components need to be developed.

The system diagram in Fig. 6 reveals the main components of an automotive BroadR-Reach link. Compared to standard Fast Ethernet, the bill of material could be significantly reduced. Applying a capacitive coupling instead of a conventional transformer, the physical layer hardware effort – consisting of PHY, common mode choke (CMC), coupling capacitors, connector and UTP cable – is very similar to a FlexRay or CAN link.

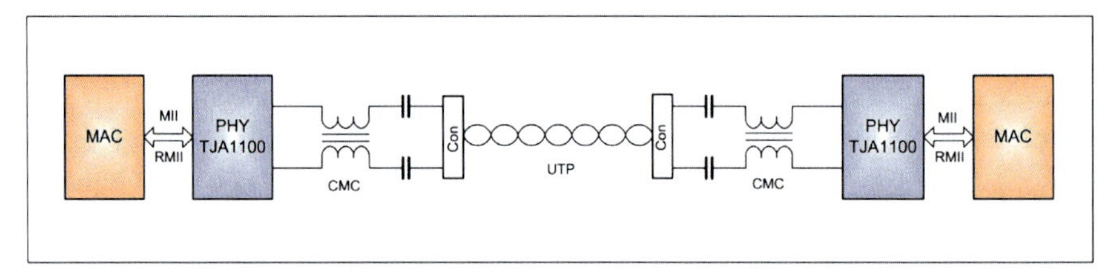

Fig. 6. BroadR-Reach system diagram

The PHY as the interface between the analogue transmission medium and the digital MAC controller largely determines the robustness and emission performance of a link. While a consumer PHY is optimised to support a cable length of more than 100 m, automotive PHYs typically have to deal with a link length of less than 10 m.

The challenge here is to find a pulse shaping and receiver equalizer implementation optimised for such a cable length that can meet the stringent automotive emission and immunity requirements. Fig. 7. shows the signal spectrum of the NXP transmitter concept with optimised pulse shaping, indicating that the automotive emission requirements can be met without the use of expensive low-pass filter components. For comparison, the corresponding signal spectrum of a Fast Ethernet transmitter reveals higher signal energy in the critical FM radio band.

The BroadR-Reach technology allowing the usage of unshielded twisted-pair cable makes Ethernet cost-competitive for automotive applications. Since FlexRay cable is able to cope with harsh automotive environmental conditions

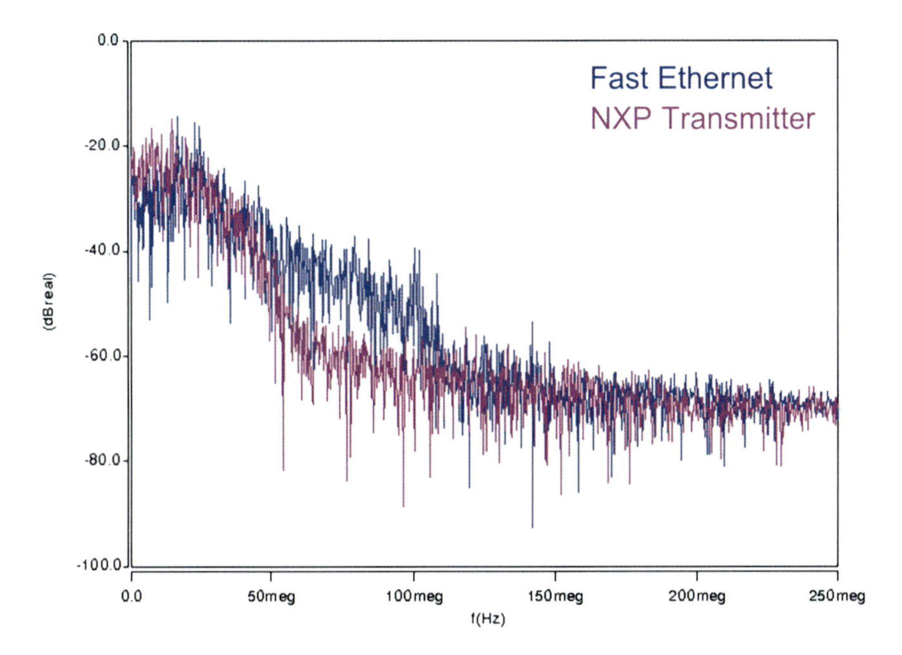

Fig. 7. Signal Spectrum

it is the first choice for Automotive Ethernet, too. In contrast to consumer applications, a careful study of the impact of environmental conditions (mainly temperature) on the Ethernet signal integrity is needed. Fig. 8 compares the simulated eye pattern for a 20 m FlexRay cable at room temperature with that at an elevated cable temperature of 105°C.

The smaller eye opening for high temperatures indicates rising Inter-Symbol Interference (ISI) as a result of cable bandwidth limitations. As long as the cable length is kept below 20 m, the reduced signal-to-noise ratio can be typically compensated by adaptive equalisation in the receiver without compromising the bit error rate.

With BroadR-Reach, a physical layer suitable for use in the automotive field is available which is cost-effective and can meet the main automotive requirements in terms of EMC, robustness and environmental conditions. In order to deploy the full potential of the BroadR-Reach technology, the development of dedicated automotive PHYs and common mode chokes is encouraged.

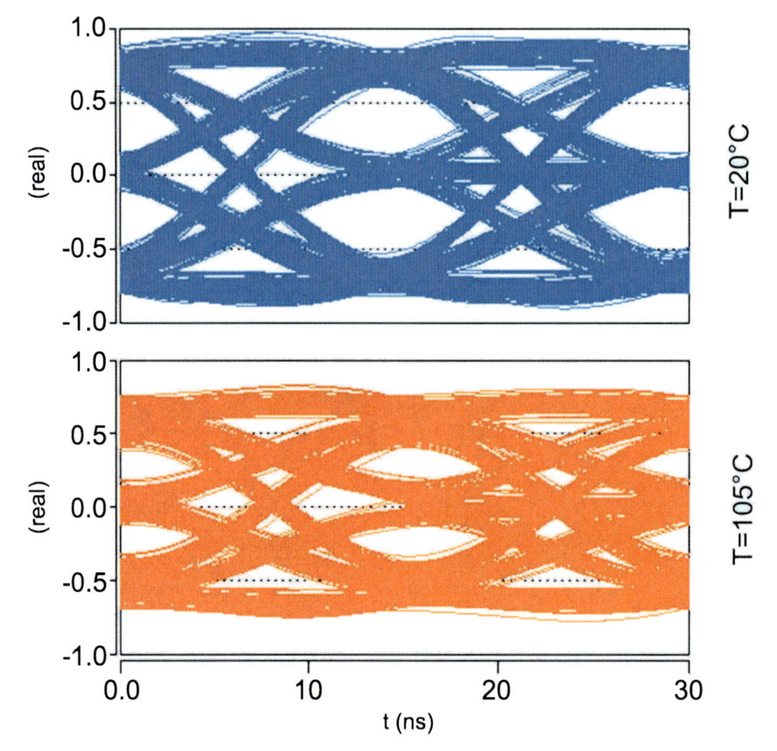

Fig. 8. Signal eye pattern

4 Conclusions

- Increasing communication bandwidth needed for driver assistance and infotainment systems
- Network topologies change from decentralised domain-specific architectures to hierarchical architectures that need a backbone
- Ethernet provides scalability and flexibility for next-generation in-vehicle networking architectures
- Over the next decade, CAN and FlexRay will remain for body domain and safety-critical communications
- Further R&D activities are needed to validate the secure coexistence of different data communication classes on the same Ethernet network

- New automotive-optimised components are required, mainly affecting Ethernet switches and PHYs – first promising steps have been taken with the BroadR-Reach technology

- OPEN Alliance and AUTOSAR are driving further standardisation on the hardware and software levels

References

[1] http://www.avnu.org
[2] http://www.opensig.org
[3] T. Streichert, Daimler AG, IP-basiert Netze für eingebettete Systems, SEIS TP2 Übersichtsvortrag, Munich, Germany, page 3, 20.09.2011.
[4] M. Plankensteiner, TTTech, Vienna, Austria, Ethernet Learns to Drive Hanser Automotive 12/2011, page 14, 2011.
[5] C. Eckert, IP-basierte Kommunikationen im/zum Fahrzeug, Fraunhofer München und Darmstadt, SafeTRANS Workshop Präsentation, Frankfurt, 05.05.2010, page 6, 2010.

Peter Hank, Thomas Suermann, Steffen Müller
NXP Semiconductors Germany GmbH
Stresemannallee 101
22529 Hamburg
Germany
peter.hank@nxp.com
thomas.suermann@nxp.com
st.mueller@nxp.com

Keywords: network systems and components, domain architectures, IEEE 802.3, Ethernet, gigabit Ethernet, SWITCH, PHY, BroadR-Reach

Ethernet-Based and Function-Independent Vehicle Control-Platform: Motivation, Idea and Technical Concept Fulfilling Quantitative Safety-Requirements from ISO 26262

M. Armbruster, L. Fiege, G. Freitag, T. Schmid, G. Spiegelberg, A. Zirkler,
Siemens AG Corporate Technology

Abstract

This paper presents the outline of a new system architecture for future electric vehicles. It is designed to simplify the development of advanced assistant functionality (e.g. ADAS) and is based on highly integrated smart actuators. A platform approach is chosen to meet functional as well as non-functional requirements outlined in this paper. A logically centralized platform computer is used as cross-domain runtime environment. All sensors and actuators are accessible from this platform computer. A middleware encapsulates the communication to physical hardware and provides mechanisms for functional safety and security. These mechanisms are fully transparent to vehicle control functions and mask platform failures up to ASIL-D functions. Moreover, platform mechanisms even allow for fail-operational behaviour of these functions and support them in a mixed criticality environment. A key characteristic is "plug-and-play" capability (PnP) for software and hardware, which is supported by OS and middleware even for safety-critical functions. This paper does focus on selected communication mechanisms based on standard Ethernet hardware. Safety assessments are just rudimentary and for the sake of completeness.

1 Introduction

Future mobility is mainly driven by the three megatrends climate change, urbanization and demographic change. Out of those, three development goals can be derived. The first goal Zero Emission means energy resources have to be used efficiently and effectively to satisfy the required and increasing demand on mobility. The second one, Intelligent Mobility, enables the reduction of traffic density based on intelligent traffic control using a high degree of connectivity and inter-modality. While the third, Zero accidents, represents the significant increase in traffic safety, especially with the aim to ensure individual mobility into advanced age. E-Mobility thereby plays a central role addressing and implementing the previously mentioned goals. Therefore the lighthouse project eCar at Siemens AG Corporate Technology does focus on information and communication technology (ICT) concepts to integrate smart sensors and acutators (aggregates) as well as on highly integrated aggregates.

The key paradigms of the ICT platform (Fig. 1.) are:

- A logically centralized platform computer runs coordinating applications that calculate target values in overlaying control loops, provide sensor fusion, situation detection, and strategy planning of driver assistant functionality.

- A Ethernet-based network connects sensors, actuators (aggregates - general term describing a sensor/ actuator node connected to the Ethernet-network that offers one or more sensor- and/or actuator functions) and the logical central platform computer.

- Sensors and actuators operate with local "smart" intelligence, processing raw input and controlling physical processes locally, offering a software-based interface to the remaining vehicle network.

Smart actuators run control and analysis loops necessary for their local operation, like a wheel-hub engine. They do not process control data on vehicle layer, i.e., from other actuators. Controllers working on vehicle layer run on the central platform computer and comprise all control algorithms that cover more than one aggregate. The central platform-computer consists out of several Duplex-Control-Computers (DCC) to ensure a very high degree of integrity. The Ethernet backbone provides redundant, reliable communication between the nodes in a ring topology, which is explained in chapter 3 in more detail. This Ethernet topology offers redundant paths over a single physical ring and is optimized for a platform approach. An inner highly consistent and redundant core ring interconnects all DCCs. Aggregates are attached to this core via breakout networks. For the latter any topology may be used but rings are preferred.

The platform computer is constructed as general purpose computer to execute any vehicle-function and connects to aggregates via the Ethernet-based communication backbone only. Performance and availability requirements determine number and size of the DCCs. They host applications of all the original automotive domains, like passenger management, assistance and drive-train control. They provide uniform access to sensor data, pre-processed and sent

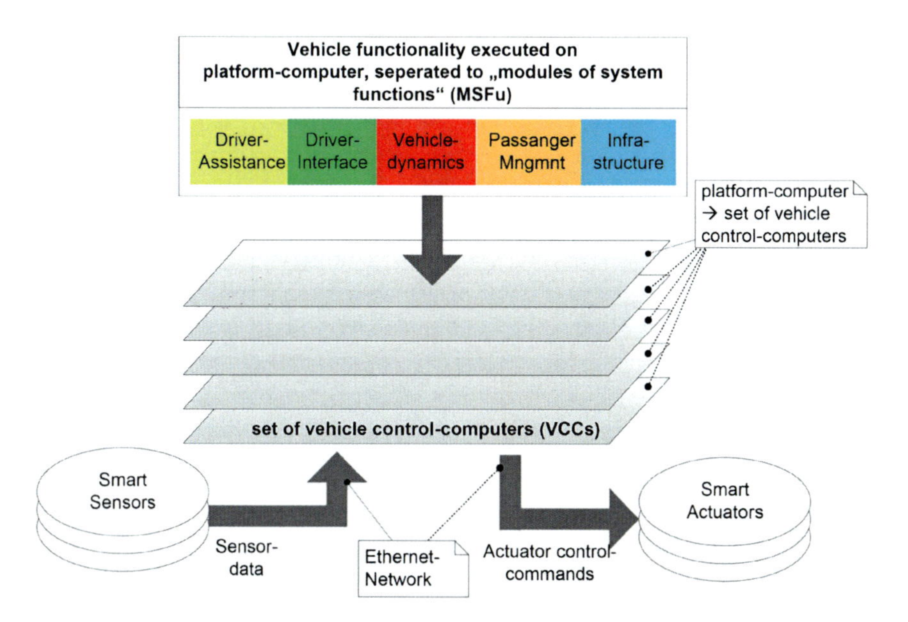

Fig. 1. A (logically) centralized computer as core of the platform

as object lists. The platform computer allows for mixed criticality operation in order to host, and separate, applications from different domains and safety classes on the same hardware.

The ICT platform cannot guarantee functional safety of the vehicle by itself, but it ensures integrity and availability of the computing and communication hardware. This does not eliminate the need for process-conform development, but relieves the application developers of dealing with hardware-specific detection and failure-handling mechanisms. This modularizes functionality necessary for fail-operational behaviour, PnP capabilities, and dynamic resource sharing to cut hardware costs. Duo-duplex redundancy is used to detect and unanimously isolate faulty Duplex Control Computers (DCCs) and to provide a second DCC (slave) to guarantee availability. If an application is configured to run redundantly, it is executed on two lanes of a DCC to ensure integrity. A second, slave DCC can be configured as hot-standby to ensure fail-operational behaviour.

The definition of "platform" thereby comprises the following items of a full vehicle control system:

- A central (fault-tolerant) platform computer, which consists of multiple duplex control computers (DCC), with access to all sensors and actuators via an Ethernet network.

- An operating-system and a middleware providing means to
  - schedule software-based system functions and enable plug and play for software and hardware (sensors and actuators) for safety- and non-safety-relevant items,
  - extend known basic-software mechanisms by safety and security mechanisms,
  - extend known basic-software mechanisms by platform-specific mode (health) management functionality, covering platform reconfiguration due to faults, master-slave switchovers to ensure fail-operational behavior.
- A communication and power distribution network.

Smart sensors and actuators are not part of the platform. The same holds for vehicle functions running on the central platform computer: the platform connects both.

The following, second chapter derives core requirements from ISO 26262 that are the motivation for the designed exchange of data between DCCs of the central platform computer. Chapter 3 presents the network design of a fault-tolerant Ethernet network capable of handling ASIL-D fail-operational functions. The paper closes with an outlook and links the presented concept to the BMWi-funded project RACE – Robust and Reliant Automotive Computing Environment for Future eCars.

2 Key Requirements and Architectural Implications

The platform concept shall be able to execute any software-based vehicle function that is classified from QM, ASIL-A up to ASIL-D, and even ASIL-D plus fail-operational behaviour with respect to ISO 26262. This chapter derives a set of core requirements on which the design of the communication network and the chosen redundancy handling is based. Such "core" requirements influence the architectural design significantly (hardware, communication links, and failure-monitoring functionality), but are, of course, not complete and rather an out of context consideration in terms of ISO 26262 (called "ISO" from now on).

2.1 Considered Requirements of ISO 26262

The platform is treated as a "Safety Element out of Context"; for details concerning a more formal way of handling "platform safety" please refer to [1] or [2]. Systematic and probabilistic faults leading to hazards can be avoided or at least reduced to a tolerable degree based on the measures defined by ISO. Systematic faults have to be "handled" mainly by an adaption of the development process. The architecture is influenced by technological measures, which are considered in this paper.

Regarding the core steering function, the hazard and risk analysis identifies the first core requirements. In this paper, we assume that any invalid or missing control output passed on to the actuators will lead to an hazard being classified to ASIL-D without any safe state and a fault-tolerance time of 50ms (based on C3-controllability, S3-Severity, E4-exposure for the scenarios highway or secondary road or country road).

According to part 5 of ISO, the assumed worst case hazard leads to the following quantitative requirements:

- the single-point fault metric target value for the platform-computer is $\geq 99\%$,
- the latent fault metric target value is $\geq 90\%$,
- the random hardware failure target value is $< 0{,}4 * 10^{-8}$/h.

The random hardware failure target value is reduced by 60% compared to the data given in the ISO as we assume within this paper that only 40% of the hazard's safety budget (10^{-8}/h in case of an ASIL-D hazard) can be assigned to the platform computer.

Those numbers already motivate a high degree of fault-detection coverage within the platform computer. Furthermore the fail-operational requirement will strengthen this motivation and extend the high demand on integrity by a demand for availability based on fault-tolerance as shown in the following two subchapters.

2.2 Impact of Fail-Operational Requirement

In addition to above requirements, fail-operational behaviour poses additional quantitative requirements concerning fault-detection coverage of the DCCs. The question addressed here is: will a SPFM (single point fault metric) of 99% be sufficient to implement a properly working fail-operational behaviour given by the maximum random hardware failure target value. Duo-duplex architectures are able to implement fail-operational behaviour with random hardware failure target values of about 4*10-9/h if fault-detection coverage of about 99,998% (The calculation is based on typical λ_{DCC} going from $5 * 10^{-5}$ to 10^{-4}) is achieved [1].

Such coverage means (nearly) all failures within a duplex computer have to be detected. Especially regarding the PHYs of the physical communication interfaces this is difficult to achieve and in turn motivates a multi-path data exchange between duplex computers. The two parts of a duplex computer (DCC lane A and B) send their data to aggregates as well as to a second DCC 2 on different network paths, which ensures the very high degree of fault-detection.

Redundant DCCs are needed not only to increase fault-detection capabilities, but also to ensure availability for fail-operational behaviour. Multiplicity and operation mode of the DCCs is not known in aggregates: only DCCs store configuration data and mode data of the vehicle, whereas sensors and actuators need a local view onto their operation within the vehicle only. Dividing responsibilities in this way facilitates adaptability and reusability of the platform. Thus, sensors and actuators only consider control commands received from central DCCs consistently, i.e., same/similar data received on redundant paths. Conversely, sensors and actuators are not able to control central DCCs, i.e., passivation, restart, and switchover.

Each DCC determines the vehicle status in every software cycle (10ms). Status includes function-specific information, like aggregate health status, and platform-specific information, like health of DCCs and allocation of functions to DCCs. Based on this status exchange, each DCC decides on its own which vehicle function to execute and whether to execute vehicle functions as master or slave. The platform may consist of more than two DCCs to increase functional and/or operational scalability.

The key cornerstone for correct platform behaviour is that all status information is exchanged unambiguously, even in case of "any" ("any" means that only a very small probability of failure per hour leading to an inconsistent exchange of data between DCCs can be tolerated). This probability is significantly lower than 4E-9; further details are shown in [2]) fault that might disturb the exchange of data between DCCs. Unambiguous status information available in every DCCs is called platform data consistency. Referring to above risk assessment, faults within any DCC or faults disturbing the data exchange between at least two DCCs must be detected, isolated and recovered within 50ms.

The stated requirements and implications of fail-operational behaviour lead to the assumption that single path data exchange between DCCs is not sufficient to fulfil the quantitative safety requirements.

2.3 Performance Requirements

The performance of a communication network depends on the number of nodes traversed by a message. To estimate the performance of our concept, we assume – based on experience made in former projects such as SPARC [3] - the platform-computer will consist out of less than 10 DCCs and less than 40 aggregates will be used within a vehicle. For the communication relations between DCCs, we consider 40 bits describing the vehicle status, 10 bit state data per vehicle function, and 20 functions per DCC. For the communication to the 40 aggregates we assume 2500 data values of 16 bits each. The sum of this data is to be transferred within one application cycle of 10ms.

Given a communication structure with an inner ring for DCC-DCC-communication and an outer ring or branch lines for communication to aggregates (see chapter 3 for details), we calculate a network load of about 12 Mbit/s in the inner ring and 11 Mbit/s in the outer ring. The latter is split up if multiple outer rings are used. Those values are not very demanding for a 100 Mbit/s Ethernet network, which allows us to allocate bandwidth to other, non-critical traffic and thus use one homogenous, mixed criticality network for all purposes. For future applications like video streaming the concept can be upgraded to Gigabit Ethernet.

Allowing non-critical traffic on the same network impacts the timing of the critical traffic. Without preventing collisions between critical and non-critical traffic, e.g., by using time-based preemption like in PROFInet IRT, TT-Ethernet or as planned in AVB Gen2 [4], a critical packet can be delayed by other packets of lower priority in every hop. For the longest possible path through the network, 40 aggregates, 10 DCCs and a connecting switch, this would accumulate to more than 6ms in the worst case (with packets of maximum size). With a reasonable limit of packet size of 200 bytes the maximum delay comes down to less than 2ms, which is acceptable with respect to the cycle time of 10ms. Future improvements, e.g., based on AVB Gen2, will lead to a transport delay below 1ms.

To summarize, bandwidth and latency requirements are important and also manageable, but the key driver for the Ethernet backbone shown in this paper is the fail-operational availability. The following chapter will show a concept to reach these requirements with standard Ethernet components.

3 Ethernet-Based Communication System

In the last chapter it has been argued, that from a performance point of view, it will be possible to use Ethernet to connect all nodes of the vehicle-network, see Fig. 2. The Ethernet-based network infrastructure described in this chapter can be built from standard Ethernet hardware and is fully compatible to standard Ethernet devices. Standard devices can send non-critical traffic over the same mixed-criticality network infrastructure as used for the critical control traffic, even though there might be some limitations regarding available bandwidth and allowed packet size. Within this paper no time-aware network scheduler will be considered, but the concept is open to do so in future activities. In the following, the data exchange between DCCs and between aggregates and DCCs will be described in a detailed way.

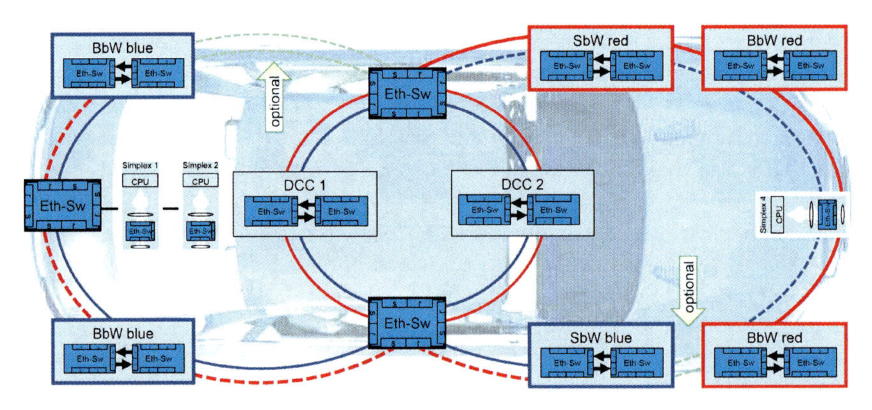

Fig. 2. Inner ring with branches; optional redundancy over green links

3.1 Data Exchange to Ensure Integrity

The redundant data exchange as sketched in the Fig. 3 is used to detect failures caused by a sending lane immediately. Therefore each lane of DCC 1 sends out its data (control data, application status, and platform status) cyclically to DCCs 2 and 3. As long as DCC 1 is fault free, the output data of DCC lane A and lane B is identical. In case of disparity, the receiver can immediately detect the failure and discard the received data. Discarded data is not passed on to upper software layers. Instead, for a certain time, the last valid data that has been received is used. After a well-defined confirmation time, a fault is indicated and a system reconfiguration is initiated.

Fig. 3 depicts the data packets sent out by lane A using yellow lines and orange lines indicating the output data of lane B. Both packets are transmitted via the "red" network to the receivers DCC 2 and DCC 3. Packets of lane B are coded before being sent out in order to detect falsifications in links or switches. Both packets from lane A and B are sent via the same Ethernet network and such via identical switches. Without coding, a network switch forwarding the data from lane A and B may corrupt both packets identically. The failure would not be detectable within any receiver and the platform must be considered to behave out of control. By using a different coding for lane A and lane B packets, the probability of a fault corrupting the uncoded and the coded packet in a way, that in both cases the packet looks correctly, i.e. the CRC is correct, and the falsified coded and uncoded data look still the same, will be negligible.

Please note: Fig. 3 and all following figures show a snapshot of the communication from DCC1 to DCC2 and DCC3. All the DCCs are the same and do the shown communication themselves once in a cycle. Furthermore it is not a restriction of the concept, but a simple example, that three DCCs are shown.

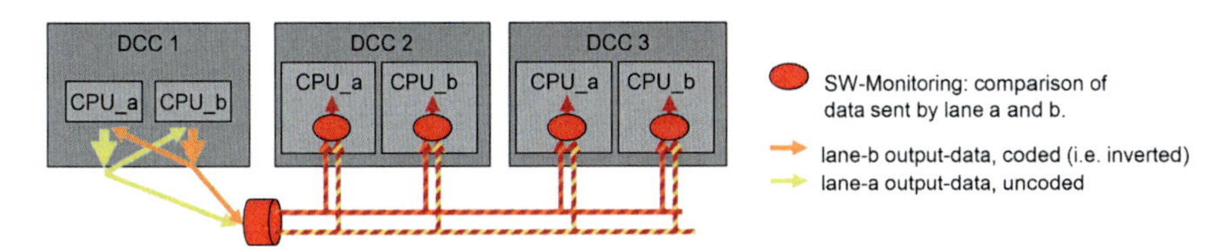

Fig. 3. High-integrity X-DCC data-exchange concept

Each lane of a DCC sends its output data to its neighboring lane within the same DCC. Each lane monitors its opposite lane and may passivate itself, and thus the DCC, in case of inequality. This monitoring is called wrap-around monitoring. DCCs are recognized as passive if at most one lane keeps sending data; proper operation of the remaining platform is not endangered because only consistent DCC output is processed and all nodes unanimously identify the failed DCC.

3.2 Data Exchange to Ensure Availability

Fig. 4 shows the "extended" logical communication relations between three DCCs with exchange of data on two paths. As described above, lane A and lane B of DCC 1 send out non-coded (yellow) and coded (orange) data to all other DCCs. Thus, data consistency amongst non-faulty DCCs can be ensured even with one faulty data path. Therefore, data of lane A and B sent out via the "blue" data-path is monitored. The same is true for data sent out via the "red" path. Software mechanisms indicated by the blue ellipse titled "voting" in Fig. 4 ensure consistency as long as the system will have less than any two faults.

3.3 Physical Network Topology

To save cabling effort, the physical topology differs from the logical topology. Instead of two dedicated media for the "blue" and "red" packets, we use two directions of an Ethernet ring (ring redundancy instead of parallel redundancy). Fig. 5 shows the physical topology, the striped arrows show the packet flow which is identical to the logical topology in Fig. 4.

3.4 Network Protection Using the Concept of Network Fuses

The price for the lower cabling effort of the ring structure is the following: As shown in Fig. 5, "red packets" and "blue packets" use the same Ethernet switches. Thus, a faulty switch can influence both directions, i.e., the physical independence of lanes is broken by the Ethernet switches. This is not a problem for a dead switch, it breaks both ring directions at one point, resulting in a single line of Ethernet segments. After such a fault, the system will work without a loss of functionality, but with a loss of redundancy. The same is valid for a switch corrupting some or all packets: A faulty packet will be detected by the mechanisms illustrated above, thus a faulty packet is as good as no packet. But there is something worse: A so-called babbling idiot.

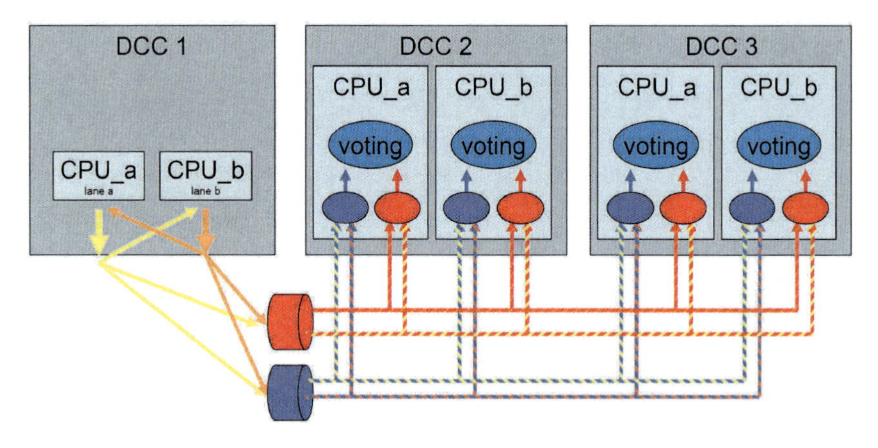

Fig. 4. The logical topology

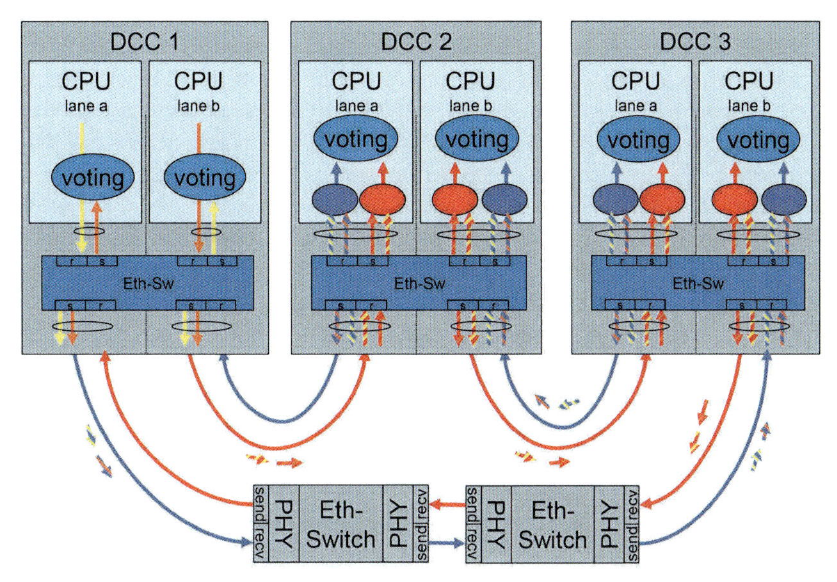

Fig. 5. The physical topology

A babbling idiot is a node that consumes big amounts of bandwidth in both directions, i.e., sends out a lot of unplanned packets not expected by any receiver. This is not a problem as long as the babbling idiot does not send data packets with high priority, so that the useless packets steal the bandwidth needed for the critical traffic. If it does, it could break the critical traffic in both directions, i.e., all communication in the network. If the babbling idiot is a CPU, the issue is easy to solve: we know, that every CPU is sending a relatively small amount of data during normal operation, our estimations came out to something around 1 Mbit/s. Limiting the ingress rate at the ports coupling the CPUs to the ring with an adequate measurement interval thus can reduce that failure to a waste of 1% of the overall bandwidth.

The situation gets a bit more difficult, if a switch is the babbling idiot: Every switch in the ring is not only sending and receiving data from one end station – as known from star topologies – but is also forwarding all the ring traffic. Rate limiting does not help us here, because if the babbling idiot is sending packets of the same traffic class as the critical traffic, we would arbitrarily drop critical traffic without knowing if it was inserted by the idiot or by someone else. Even stream sensitive rate limiting does not help, as we do not know if the babbling idiot is sending packets with a faulty source addresses. The solution we propose for this issue is working similar to the electrical fuses we know from the breaker box at home: If any device consumes more energy than the circuit can provide, the fuse fires and the circuit is dead. Similar to the maximum current, we can estimate a maximum bandwidth, which will occur in the network. This is possible, as we know the traffic pattern of the concerned devices. We set up a rate monitoring at any ingress-port connected to the ring, and block that port, if the bandwidth limit is exceeded. The trick is that we do not limit the traffic to the rate limit, but completely block the port. Thus, if any node exceeds the rate, it will get cut off the rest of the ring, and we get a situation similar to a dead switch. In difference to the electric fuse at home, we can automatically re-engage the fuse, after the bandwidth returns back to normal values. In Fig. 6, the turquoise switch in DCC 2 is the babbling idiot, the turquoise lines are transporting lots of useless packets. If the other switches would forward that traffic all lines would be turquoise, i.e., the whole network would be down. But the "fuses" in DCCs 1 and 3 have recognized the exceeded traffic limit and switch to blocked mode, so that DCC2 is excluded from the ring and DCCs 1 and 3 can keep the critical functions alive.

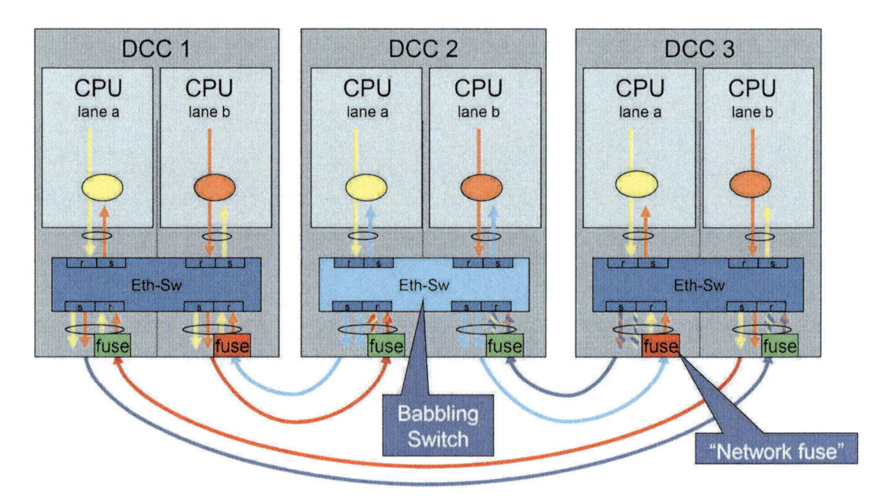

Fig. 6. Network fuses

3.5 Integration of Aggregates into the Platform-Computer Network

Next, the focus broadens from the backbone network ("inner ring") to the sensor | actuator network ("outer ring"). In principle, the aggregates could be directly integrated into the inner ring, but for an increased reliability of the platform computer it is preferable to use one or more outer rings or branches to connect aggregates.

Due to the limited space, only two DCCs are shown here, furthermore the CPUs are not shown, even if the DCCs are just the same as the ones above. Fig. 7

shows the communication from DCC1 to a Steer-by-Wire actuator (SbW, blue). In this example, SbW blue has a highly available logical connection to the platform computer. The logical connection applies the same communication patterns as described above for the inter-DCC communication. It receives coded and uncoded data over two independent paths, depicted green and turquoise in Fig. 7. This way, it can decide if the received data are correct by comparing the coded and uncoded data, and it can hark back to the redundant data, if they are not.

In the complete setup two redundant steering actuators are installed, see SbW red in Fig. 7. Highly available connections are not mandatory for a reliable operation if one steering actuator can steer the car alone. That is, the ring structure provides highly available connections and gives us additional availability of data, but based on redundancy designed into system architectures, a single connection between aggregate and platform computer will suffice.

The same holds for sensors: Use highly available connections sending the sensor's data pair wise over two paths, or connect two sensors to branches, each sending a data pair over one disjoint path, or just connect four simple and unreliable sensors using at least two different paths and evaluate data consistency within data fusion middleware. Consequently, there are different approaches to gather sensor data reliably with simple sensors when using the platform computer's intelligence.

Depending on cabling effort and bandwidth requirements one could consider installing multiple outer rings or branch lines in a car. Fig. 2 shows an example for the cabling of four brake-by-wire actuators (BbW) and two steer-by-wire actuators (SbW). In this example, the blue branch connects brake actuators for the rear axle, the red branch connects brake actuators for the front axle and one steering actuator, while a second blue branch connects the second steering actuator. The dotted lines indicate that those links would be ready to close the ring with very low additional effort, as switched Ethernet is always full duplex, which is anyway needed for the feedback from the actuators. With only two additional links, depicted as green dotted lines in Fig. 2., the three branches are connected to two outer rings, which provide redundant paths to any aggregate for rather low costs. In this topology, a link failure, e.g., caused by a defect cable or plug, does not degrade system functionality, whereas the same failure can lead to the loss of half of the actuators without the green ring links.

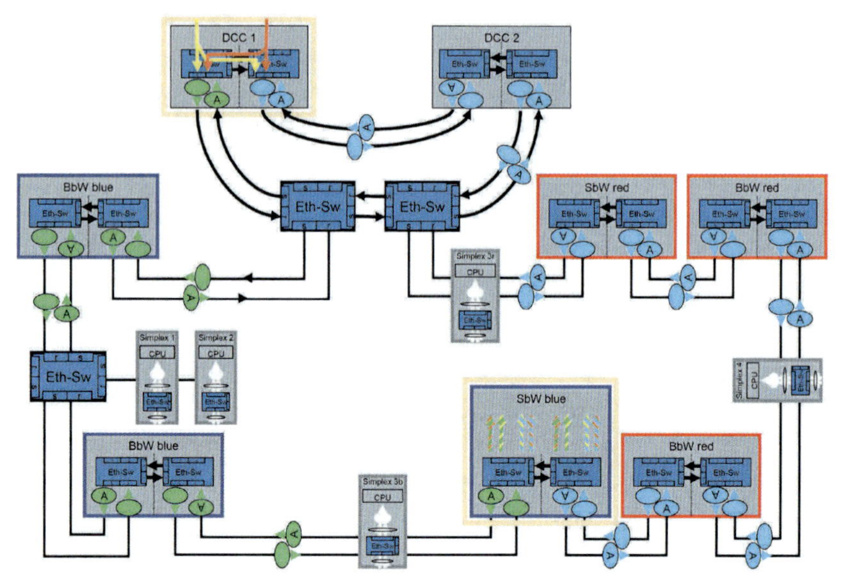

Fig. 7. The Outer Ring

3.6 Network Protection Using VLANs

With the design presented up to this point, we do not even need a babbling idiot to waste all bandwidth for useless packets. We designed a multiple ring topology – a meshed Ethernet network – in which broadcast messages would multiply and circulate, if we do not care about it. In office Ethernet networks, the rapid spanning tree protocol (RSTP, IEEE 802.1D) helps us out of this issue, but that would open the ring and give us reconfiguration times of hundreds of milliseconds. In industrial requirements, several ring protection protocols are used, but all protocols that break the ring and reconfigure after a detected failure need a few hundred milliseconds for reconfiguration. An interesting alternative would be HSR (High-availability Seamless Redundancy, IEC 62439-3), where all frames are really sent in both ring directions and the duplicate is filtered at the receiver. This is what we need, but HSR requires special hardware.

To stick to standard Ethernet, the middleware sorts out the duplicates based on source address, destination address and cycle counter, and the network uses VLANs (Virtual Local Area Network, IEEE 802.1Q) to protect the ring from circulating packets. As Fig. 8 shows we need two VLANs per DCC to enable the DCC to send to both lanes of any other DCC from both of its own lanes. Additional VLANs are necessary to reach outer rings. Luckily, we do not need a number of VLANs per aggregate, but two per outer ring.

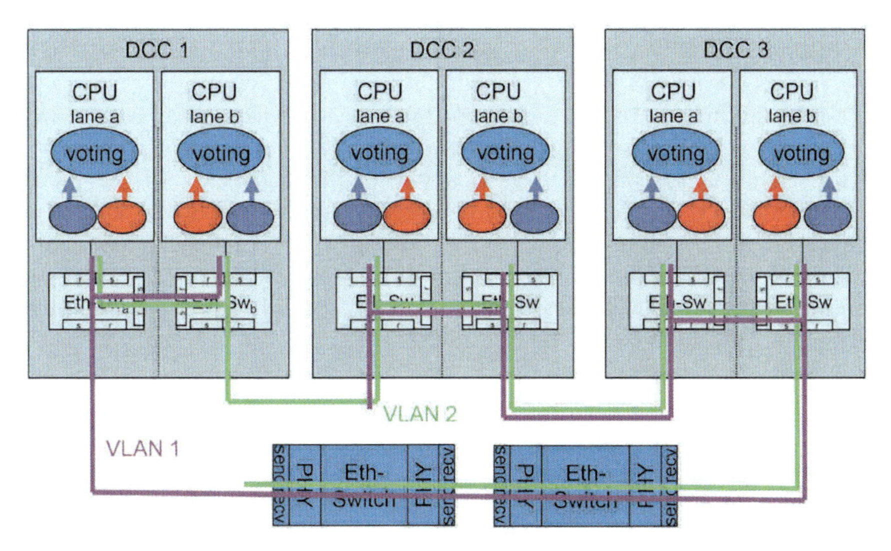

Fig. 8. Two VLANs in the DCC inner ring

For the first one, any device in the whole car network is member of this VLAN, but one of the coupling switches. For the second one, we do the same for the other coupling switch. Now, any DCC can communicate to any aggregate over two paths, refer to the green and turquoise paths of the example in Fig. 8.: The green packets use the first VLAN, where any device but the right coupling switch is member, the turquoise ones are using the second VLAN, where only the left coupling switch is excluded.

For non-critical traffic, we set up an extra VLAN, which gives a loop-free connection of all nodes. Redundancy is not needed for that VLAN, as it is meant for diagnosis, multimedia and integration of legacy aggregates. That traffic will see a limited bandwidth and possibly packet size, but apart from that a "normal" standard Ethernet with IPv6 communication is possible.

For critical traffic, the concept foresees to use Layer 2 Ethernet packets, as the IP header does not give us any advantage, but additional overhead. With or without IP headers, the applications as well as the aggregates need to know the address, to where they should send their information. For the applications, the middleware shall help to find the proper destination addresses. For that purpose, we define a set of function-specific multicast addresses. Thus, an application just has to send or request data to or from, for example, steering actuators or sensors. The middleware knows the standardized multicast addresses for steering actuators and sensors and just has to send the data to the multicast address for steering actors respectively register for the multicast address for steering sensors.

4 Conclusions

In a conventional vehicle all safety-critical subsystems are controlled by the mechanics and only assisted by the electronics to prolong its functionality. However the last responsibility classically bears the mechanics. The approach presented here replaces mechanical backups by highly reliable, fail-operational electronics. The n-duplex approach to redundancy leverages scalability with respect to functional performance and operational availability. The platform concept is based on a clear separation of responsibilities, where intelligent aggregates are connected to power and Ethernet network only without using designated signal lines.

On the way to more convenient, assisted and autonomous driving there are three aspects to be developed in future vehicles. Obvious is the need of ADAS functions that will assume duties and responsibilities of the driver. In the end, these functions will and have to take corrective actions to the drivers command. Although some people decline it, this is necessary to achieve "Zero Accidents". Second is the need of smart sensors and actuators that complement such functions. Highly integrated wheel-hub e-motors will be the low-cost, high function starting point for such a development. Mechanical and thus human control backups are then no longer possible, because the driver is hardly able to take over control in assisted driving situations. The third and main aspect covered in this article is the high-available communication backbone necessary to facilitate above innovations, supporting a fail operational system behavior. The approach described within this paper will be transferred to the BMWi-funded project RACE which aims at the development of an ICT as shown before. Consortium partners are Siemens, AVL, TRW, fortiss, TUM (RCS and SSE), RWTH Aachen (EONERC and ISEA) and Fraunhofer AISEC.

References

[1] M. Armbruster, Eine fahrzeugübergreifende X-by-Wire Plattform zur Ausführung umfassender Fahr- und Assistenzfunktionen, Verlag Dr. Hut, München, 2009.

[2] R. Reichel, M. Armbruster, X-by-Wire Plattform – Konzept und Auslegung, Automatisierungstechnik, at9/2011, 583-596, 2011.

[3] SPARC – Secure propulsion using advanced redundant control, http://www.transport-research.info/web/projects/project_details.cfm?ID=36021

[4] Audio/Video Bridging is an IEEE 802.1 Task Group, which specified time-synchronized low latency streaming services through 802 networks. In AVB Generation 2, industrial and automotive requirements will be met. see http://www.ieee802.org/1/pages/avbridges.html

Michael Armbruster, Ludger Fiege, Gunter Freitag, Thomas Schmid, Gernot Spiegelberg, Andreas Zirkler
Siemens AG Corporate Technology
Otto-Hahn-Ring 6
81739 Munich
Germany
michael.armbruster@siemens.com
ludger.fiege@siemens.com
gunter.freitag@siemens.com
thomas.ts.schmid.ext@siemens.com
gernot.spiegelberg@siemens.com
andreas.zirkler@siemens.com

Keywords: Ethernet, duo-duplex, vehicle control platform, out of context, integrity, reliability, fail operational, plug-and-play

Design of a Robust Plausibility Check for an Adaptive Vehicle Observer in an Electric Vehicle

M. Korte, G. Kaiser, V. Scheuch, F. Holzmann, Intedis GmbH & Co. KG
H. Roth, University Siegen

Abstract

With the increasing number and complexity of Advanced Driver Assistance Systems (ADAS) and rising control facility by individual controllable drives in electric vehicles (EV) the reliability of sensor signals becomes more and more important in nowadays vehicles. In order to enhance the safety, the estimation of vehicle states and parameters gets more relevant. In most state of the art functions a vehicle observer secures the correctness of the delivered states. As the performance of observers depends on their input signals a novel plausibility check is implemented. In this paper the checked signals serve the designed adaptive vehicle observer, based on Extended Kalman filtering technique, as input signals. Thus the integrated vehicle functions can control the electric actuators with more precision in order to improve the driving performance and a minimization of energy consumption by an optimal use of the available road traction. The complete system, existing of plausibility check and observer is validated by simulation and will be implemented in an electric vehicle within the EU funded project eFuture.

1 Introduction

The current development of vehicle functions, especially ADAS, increases the influence of control algorithms on the actuators and thus on the vehicle dynamics. Especially in electric driven vehicles the control possibilities rise by individual assessable torque of the electric machines [1]. As this raises new requirements to the vehicle safety demands the currently published norm ISO 26262 [2] specifies guidelines for necessary software safety mechanisms at the software architecture level. To fulfil these defined requirements, a novel robust plausibility check is presented in this paper with regard to the work of Versmold and Saeger [3].

Through the close connection of the plausibility check to the vehicle observer the robustness of the estimation results is upgraded. In detail the number of vehicle states delivered by standard sensors, which are mounted on the vehicle, are extended by the vehicle observer based on Extended Kalman filter as described in [4]. Here the sideslip angle, vehicle mass, road friction coefficient and effective tire radius are estimated and built together with the measured and filtered vehicle states the basic inputs for the Vehicle Dynamics Control (VDC) systems. In the project eFuture [5], comprising the building of a save and efficient electric vehicle, the functions Electric Stability Control (ESC) and Torque Vectoring (TORVEC) [6] have great benefit through this robust and adaptive vehicle state and parameter estimation.

In this paper the development of a robust plausibility check for an adaptive vehicle observer in an electric vehicle (EV) will be presented and simulated. After a compact function description in the next section the functionality and architecture of the plausibility check will be shown in section 2. Simulation results in a 14 degree of freedom vehicle model based on recorded measurements will be introduced in section 3. Finally, conclusions are given in section 4.

1.1 Function Description

The function plausibility check as the first receiver verifies the incoming sensor signals and detects faulty or missing signals. The signal error detection mechanisms are composed of a single signal check, a redundant signal check and a model-based signal check. Moreover in this function some error handling mechanisms are included. Here signals that are detected as faulty are corrected in the redundant check, if possible, and missing signals are replaced temporarily. Additionally an offset compensation is implemented during stand still. The function is completed by the calculation of the confidence and an activation decision of the vehicle observer algorithm. The arrangement of the plausibility check is shown in Fig. 1. The precise functionality of each block will be explained in the second section.

2 Plausibility Check

In this chapter the four single subfunctions of the plausibility check (see Fig. 1) shall be explained in detail after the equipped vehicle sensors are introduced in the following subsection.

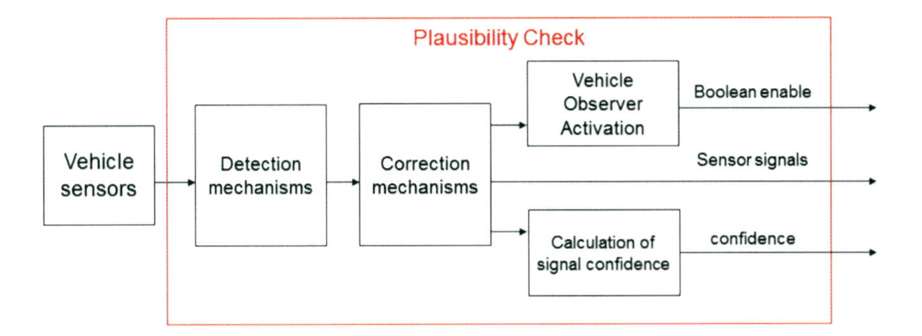

Fig. 1. Structure of the plausibility check

2.1 Vehicle Sensors

The vehicle prototype is equipped with the following sensors that deliver the input signals for the plausibility check:

- Yaw rate sensor that additionally measures the longitudinal and lateral acceleration
- Steering angle sensor mounted on the steering column that delivers the steering angle
- Active wheel speed sensor mounted on every wheel that measures the angular wheel speed and is a component of the Anti-Lock Braking System (ABS)

2.2 Detection Mechanisms

The detection mechanisms block consists of the single signal check, the redundant signal check and the model based signal check. These functions are executed in parallel and are independent from each other. The detailed functionality of each detection mechanism is explained in the following.

2.2.1 Single Signal Check

The single signal check contains the range check of all sensor signals and the check of gradients. The range check is done by comparing the current signal with the range given by the respective datasheet. The check of gradient for every signal is done in the second subfunction. The limits for the gradient value were evaluated by analysis of measurement data from the prototype in driving situations with high dynamics.

The value for the confidence calculation is greater zero whenever the range and gradient check detect no failure. If the signal value or rather the gradient does not exceed a specific lower threshold, the confidence value is not affected. As soon as the property value exceeds this lower threshold the confidence value decreases linear until it reaches its minimum with crossing the upper threshold. The process of the confidence value calculation can be seen in Fig. 2.

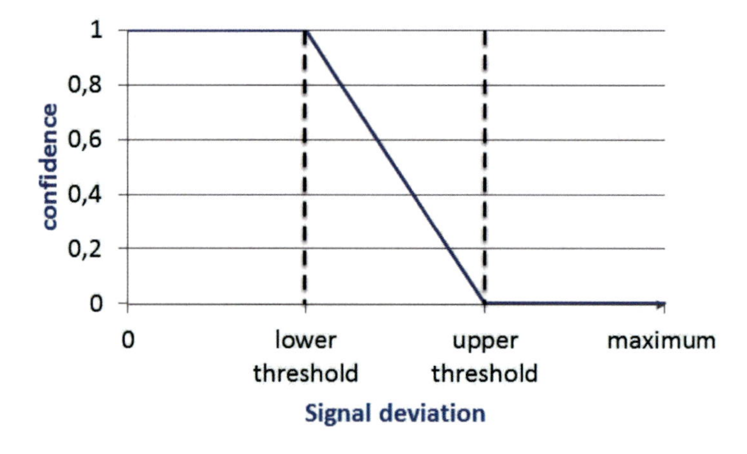

Fig. 2. Single signal check: confidence calculation

2.2.2 Redundant Signal Check

Compared to the single signal check the redundant signal check offers a much higher detection potential. It uses redundant sensor signals for analysis and calculates the confidence value from the difference of them. The obvious disadvantage of this method is the need of redundant sensors which stands in conflict to the cost optimisation in the automotive production.

As the prototype won´t be equipped with redundant sensors the different angular wheel speeds are seen as redundant signals. The analysis of the signals is arranged in two different steps: The detection of the deviation between each wheel and the determination of the deviating signal.

Additionally a check of axle speed is implemented. Here the front and rear wheel detection results are summarized. If an axle deviation is detected the mean of the rear wheels speeds is assumed as reference signal. Especially in acceleration situations when the wheel slip of the driven axle is great, this method shows good performance.

The confidence of the redundant signal check depends on the number of signals detected as deviated. The confidence is one if no wheel speed is detected as deviating. With the increasing number of deviating signals the confidence decreases linear until it is zero when all wheel speeds are detected as deviated.

2.2.3 Model-Based Signal Check

In contrast to the redundant signal check the model-based check examines the connections between different signals for the confidence assessment. Here analytical connections based on mathematical description and vehicle behaviour modelling constitute the basis for the evaluation of the signal plausibility.

This model-based signal check was developed on the fact, that in driving manoeuvres with yaw rate unequal zero, e.g. curve driving, the outer wheels spin with higher velocity than the inner ones. Through empirical determination in the nonlinear 14 degree of freedom (DOF) simulation model based on [7] the following formula which describes the dependence between wheel speed and yaw rate could be formed:

$$|\dot{\psi}| * 0.3 - \Delta\omega_{max} \geq 0 \tag{1}$$

Here the maximum deviation between the different wheel speeds $\Delta\omega_{max}$ should not be greater than 0.3 times the absolute yaw rate. As this equation was evaluated in a vehicle model it's range of validity is limited due to parameter variations during dynamic driving situations. So this formula is valid as long as the side slip angle of the vehicle is below 10 rad.

2.3 Correction Mechanisms

The correction mechanisms block is consists of the replacement of signals and the offset compensation. Just as in the detection mechanisms the blocks are executed in parallel and are independent from each other. The functionality of this block bases on the results of the detection mechanisms. The detailed functionality is explained in the following.

2.3.1 Signal Replacement

The signal replacement is designed to catch two possible failures: Delayed or missing and faulty sensor signals. In the first case signals that could not be identified by the range check will be replaced by their last measured value. Since this is not suitable for permanent sensor malfunction, the maximum time for signal replacement is set to 100 ms. The replacement of faulty sensor signals uses the results of the redundant signal check described in 2.2.2.

2.3.2 Offset Compensation

As any of the equipped sensors has offset problems over lifetime or due to incorrect mounting, the detection and correction of that deviation is very important for a robust and reliable state estimation. The offset is estimated from the average sensor signal during valid standstill.

Here standstill is seen as valid when the angular velocity of all four wheels is below a defined threshold near to zero. To protect the concept against fatal sensor malfunctions, the estimated offset value is limited to the respective data of each sensor. The concept was validated with recorded sensor signals from test drives with the prototype (see section 3).

2.4 Calculation of Confidence

In general the confidence represents the reliability on the incoming sensor signals. As explained before the maximum value of the confidence is one. Due to the fact, that the sensor signals serve as inputs for the vehicle observer the calculation of the confidence has to consider that. Since all signal inputs have impact on the estimation of the vehicle state there is only one confidence calculated for all signals.

2.5 Vehicle Observer Activation

In order to save calculation capacity, the vehicle observer algorithm should not be executed when there is no need. As the VDC won´t act on the vehicle dynamics near standstill the vehicle observer is activated when the vehicle velocity rises above a defined threshold. Because this decision should not be affected by wheel slip the two angular velocities of the non-driven rear axle are taken into consideration. Finally the algorithm is executed whenever both rear wheels have an angular velocity greater than 1 rad/s.

3 Results

The verification of the plausibility check was done in two different ways: In a nonlinear vehicle simulation model with 14 DOF, where the tire characteristics are calculated with a modified Dugoff model [8], and with recorded data from drive tests with the prototype. As the presentation of all results would overrun the length of this paper the most representative were selected.

3.1 Simulation Results

In Fig. 3. to 6. the results of a straight line acceleration manoeuvre on slippery road in the simulation model are displayed where the redundant signal check detects deviation and the signal replacement corrects the deviated signals. In this scenario the driver accelerates with maximum gas pedal while there is no acting on the steering. In Fig. 3. the original angular wheel speeds and in Fig. 4. the corrected original angular wheel speeds by the plausibility check are displayed. In Fig. 5. the observed lateral velocity of the vehicle for both inputs and reference and in Fig. 6. the confidence value is shown.

It can easily be seen that the plausibility check detects and corrects the deviation between the front and rear axle wheel speeds during periods of great wheel slip on the driven axle. The improved accuracy of the longitudinal velocity estimation by the vehicle observer with plausibility check is apparent even without detailed analysis. The confidence value decreases for the period of signal replacement from 1 to 0.8.

3.2 Offline Verification

With the data taped at the driving tests the performance of the offset compensation could be verified since the mounted sensors had a high offset level. Fig. 7 illustrates the results for the lateral acceleration during a straight line braking scenario. In this scenario the driver decelerates from a certain velocity by

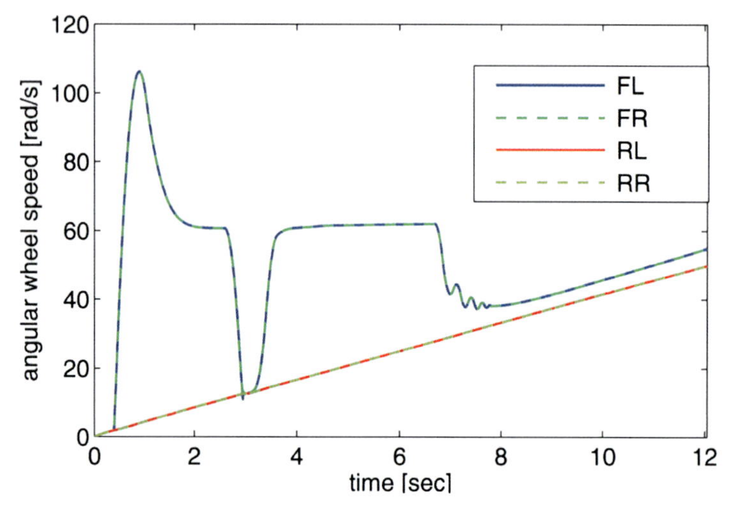

Fig. 3. Original angular wheel speeds

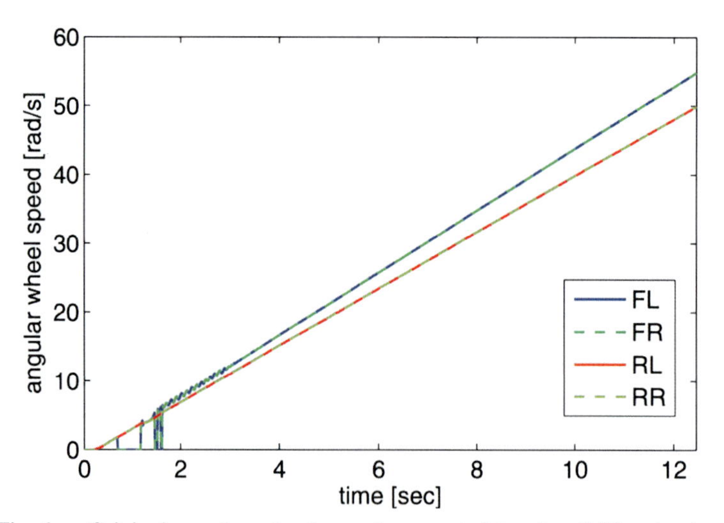

Fig. 4. Original angular wheel speeds corrected by plausibility check

full brake without steering. It is assumed that the lateral acceleration in such a driving manoeuvre should be zero mean. It can be seen that the offset compensation lowers the measured sensor signal so that the corrected signal becomes nearly zero-mean during standstill periods. The offset compensation method was used for the longitudinal and lateral acceleration as well as for the yaw rate and showed good performance for all recorded scenarios.

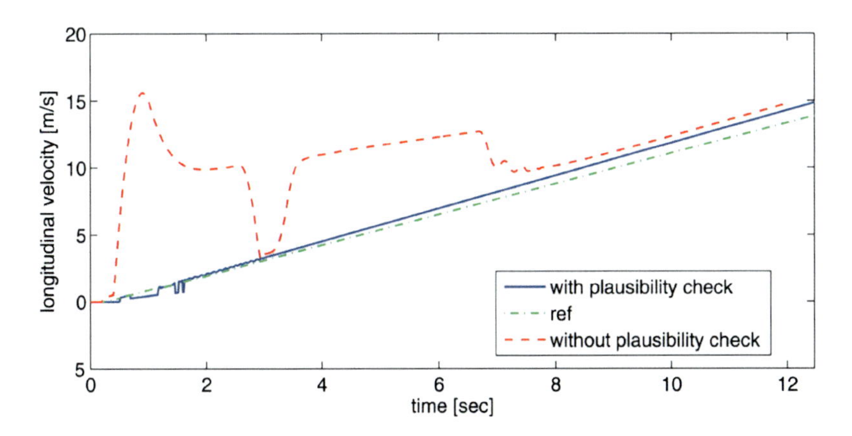

Fig. 5. Observed latent velocity

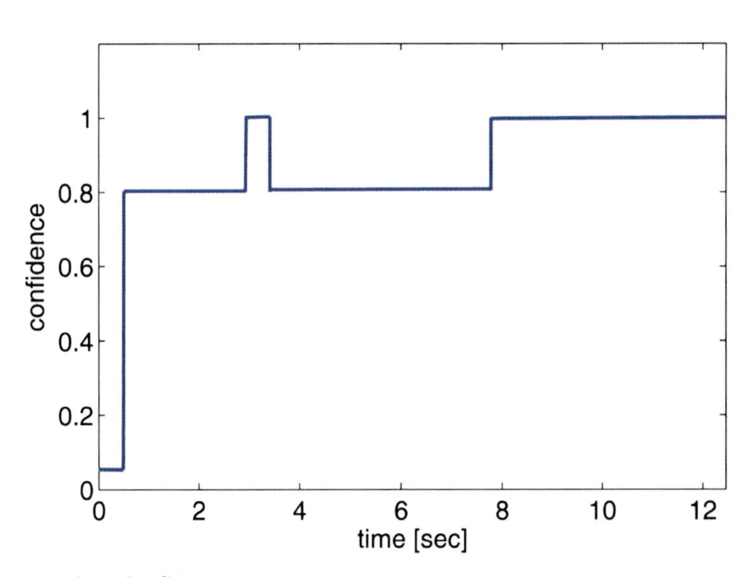

Fig. 6. Confidence value

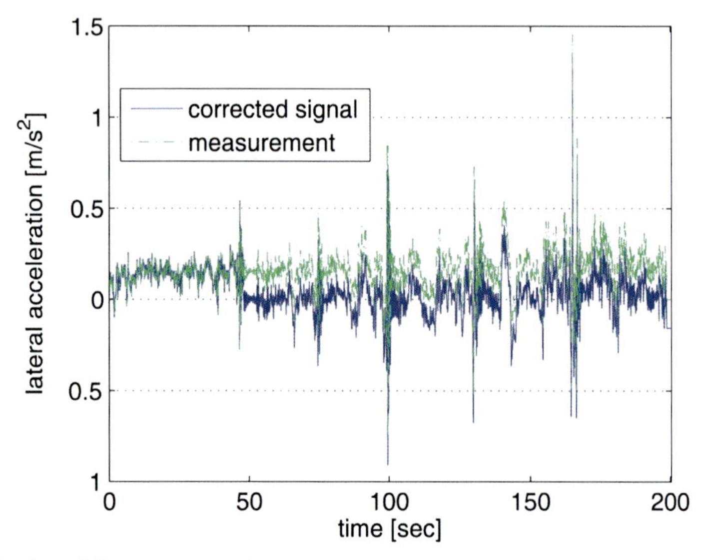

Fig. 7. Offset compensation

4 Conclusions

The presented plausibility check improves the overall performance of the vehicle observer towards its robustness and confidence by dealing with faulty or missing sensor signals. So the safety demands that arose by new VDC systems, especially in electric driven vehicles, could be fulfilled with regard to ISO 26262. Here detection and correction mechanisms could avoid a violation of technical safety caused by wrong sensor signals, without any special sensor systems. Next steps for the development of the introduced plausibility check are the offset compensation for the steering wheel, the design of a new concept for the replacement of missing sensor signals based on the Markov chain theory, the algorithm optimisation concerning its real-time ability and the final tuning when the prototype hits the road.

References

[1] VDE, Elektrofahrzeuge – Bedeutung, Stand der Technik, Handlungsbedarf, VDE, 2010.

[2] ISO 26262:2011, Road vehicles – Functional safety – Part 6 Product development at the software level, ISO, 2011.

[3] H. Versmold, M. Saeger, Plausibility Checking of Sensor Signals for Vehicle Dynamics Control Systems, Forschungsgesellschaft Kraftfahrwesen Aachen, 2006.

[4] M. Korte, F. Holzmann, V. Scheuch, H Roth, Development of an adaptive vehicle observer for an electric vehicle, European Electric Vehicle Congress, 2011.

[5] V. Scheuch, E/E architecture for battery electric vehicles, ATZelektronik worldwide edition, 2011-06, 2011.

[6] G. Kaiser, F. Holzmann, B. Chretien, M. Korte, H. Werner, Torque Vectoring with a feedback and feed forward controller – applied to a through the road vehicle , IEEE Intelligent Vehicle Symposium, 2011.

[7] R. Rajamani, Vehicle Dynamics and Control, Springer, New York, 2006.

[8] Z. Shiller, Optimization Tools for Automated Vehicle Systems, 1995.

Matthias Korte, Frédéric Holzmann, Gerd Kaiser, Volker Scheuch
Intedis GmbH & Co. KG
Max-Mengeringhausen-Str. 5
97084 Würzburg
Germany
matthias.korte@intedis.com
frederic.holzmann@intedis.com
gerd.kaiser@intedis.com
volker.scheuch@intedis.com

Hubert Roth
Universität Siegen
Department Elektrotechnik und Informatik
Hölderlinstr. 3
57076 Siegen
Germany
hubert.roth@uni-siegen.de

Keywords: sensor plausibility, vehicle state observation, parameter estimation, Extended Kalman filter, vehicle dynamics, electric vehicle

Simplified Architecture by the Use of Decision Units

V. Scheuch, G. Kaiser, F. Holzmann, Intedis GmbH & Co. KG
S. Glaser, French Institute of Science and Technology

Abstract

E/E architectures become more and more complex and thus hamper the introduction of new functions which in turn are essential for the development of electric vehicles. Consequently, new concepts are needed to enable an easy introduction of new technologies along with an interconnected but still manageable architecture. The proposed functional architecture approach aims at a safe operation of electric vehicles with a well defined decision path from the driver and the ADAS functions in the command layer to the actuators of the execution layer. The backbone of this architecture consists of two dedicated decision units. The command decision unit mitigates between requests coming from the driver and the ADAS functions whereas the execution decision unit selects the best actor signals for the current vehicle status. One benefit of this hierarchic approach is the ability to control the validity and the transition between driving modes by a set of rules which can be designed and adjusted to meet the vehicle requirements individually. The concept is not limiting the number of driving modes but is scalable for future applications, different market needs, and vehicle configurations.

1 Idea

The development of lean E/E architectures suffers from the evolutionary growth of the number of functions and thus, control units. The growing complexity leads to rising costs and an increased susceptibility of failures. In the early days of automotive industry, functions had a one to one assignment to the corresponding components and networking was not established. By introducing new functions like ABS (Antilock Braking System) which requires information by several sensors the information exchange via data busses became an integral part of every vehicle's architecture. As the connection of control units via communication busses turned out to be very easy, robust, and safe, the number of control units and components attached to the busses grew with the years, resulting in a rather complex structure of control units and busses. This grown E/E architecture still operates the complete vehicle fleet in the world but many approaches have been published to overcome the dilemma of deteriorating the situation by postponing a change which could cure it [1,2,3].

The trend of electro-mobility is even enforcing this evolution by the introduction of new functions and components. Energy saving functions are broadly introduced and influence the driving behaviour, battery and charging management control the new high voltage net and inverters operate the electric motors. The limited energy resources on the other hand require new strategies to operate the vehicle at a low consumption level despite new functions. Consequently, new architecture concepts need to enable an easy introduction of new functions and simultaneously account for an efficient operation.

2 New Architecture Approach

The proposed functional architecture aims at a safe and efficient operation of electric vehicles but is not limited to this kind of vehicles. The need for efficient energy usage was escalated by electric vehicles due to a limited battery performance but of course can and should also be applied to hybrid and internal combustion driven vehicles.

The functional architecture is shown in Fig. 1. The approach uses a well defined decision path from the driver and the command layer to the actuators in the execution layer. The backbone of this architecture consists of two dedicated decision units, located in the command and execution layer, respectively. The command decision unit mitigates between requests coming from the driver, the ADAS (Advanced Driver Assistance System) functions, and the energy management whereas the execution decision unit has to choose the best actor signals for the current vehicle status (e.g. steering vs. braking while sliding in a curve). Both units follow the goals of safety, efficiency and comfort. The decision units act as state machines also enabling smooth and intelligent transitions between requested driving modes, e.g. when multiple ADAS functions are activated.

Currently each assistance function operates without taking into account other assistance functions. The activation / deactivation is performed by the driver and a specific validity check within the function. The interconnection with other assistance functions often is not foreseen and is also not calibrated as a system in the vehicle. In the case two functions are active or operate in the transition regime from one function to the other the functional behaviour is not well understood and even not predictable in all cases.

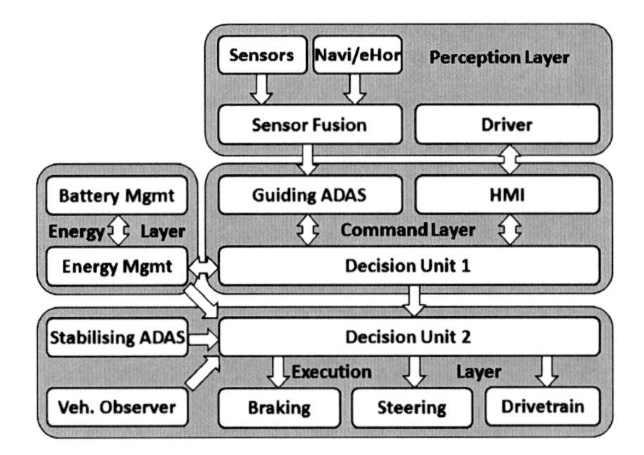

Fig. 1. Proposal of a functional architecture with decision units as the backbone.

In an example the driver uses ACC (Adaptive Cruise Control) and LKAS (Lane Keeping Assistant) simultaneously to achieve maximum driving comfort on a highway. The preceding vehicle might decelerate stronger than the ACC is able to follow so the driver might want to change the lane to avoid hard braking. Assuming the indicator has not been used in this critical situation, LKAS would recognize an unintended steering and reacts with a counter steering. As a consequence, the lane change could not be executed as expected leading to an even more critical situation where the vehicle stays on its lane and valuable time for braking is lost.
The benefit of the decision unit concept is to migrate the activation/deactivation intelligence and the validity check from the ADAS functions to a supervising function which can be extended to more assistance functions all being addressed in parallel (figure 2). The validity check and the transition strategy can then be designed and calibrated individually according to the vehicle's purpose or even to the actual driving situation. While choosing safety as the highest priority, the balance between comfort / performance and efficiency can vary.

This concept holds for both the command layer and the execution layer where guiding and stabilising ADAS functions are located, respectively. The ADAS functions of the different domains have different aims. The guiding ADAS (ACC, LKAS, EBA (Emergency Brake Assist), ...) assist in keeping a safe trajectory and react on changes in the road infrastructure whereas the stabilising ADAS (ESC (Electronic Stability Control), TCS (Traction Control System), ABS) assist in keeping the vehicle controllable in critical situations by individually controlling the wheels. The different functionalities require different decision units with properties adjusted to the specific tasks in the command and execution layer. Nevertheless, the overall structure and idea described above is valid for both functions.

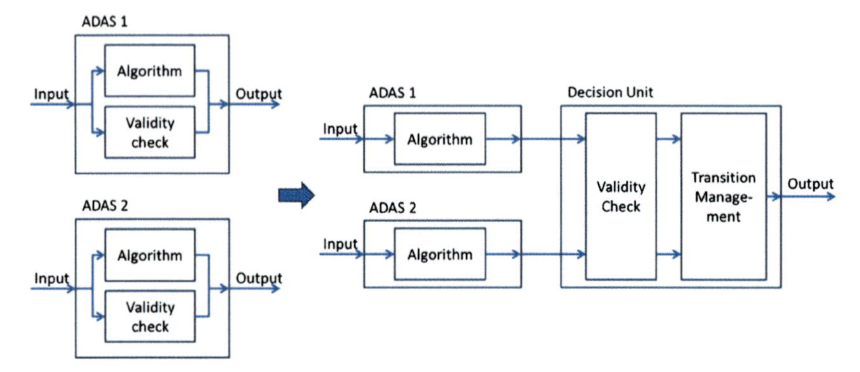

Fig. 2. Migration of the ADAS validity check sub function to the decision unit. Left: conventional parallel ADAS functions without connection. Right: decision unit as supervising function performs connected validity check and the transition from one to the other ADAS function.

3 Decision Units

The supervising function has been split into two decision units to account for the different functional domains and to reduce the complexity of each function. They comprise their own specific sub functions and parameters and need to be manageable during the development process. Both functions will be described and discussed below.

The command layer decision unit (DU1) is meant to mitigate between the motion requests of the driver and the different ADAS functions and to pass the most suitable motion vector to the decision unit in the execution layer. The term most suitable can be defined by fixed parameters during the development or can also be set by the driver via energy modes.

The inner structure of the decision unit 1 is shown in Fig. 3. The two paths responsible for guiding the vehicle, the driver and the ADAS functions, are assessed separately in the sub functions manoeuvre recognition and stability assessment before they will be merged in the transition management module which decides on the compatibility of both motion vector requests and which one shall be passed on to the decision unit in the execution layer. The output is a two dimensional vector which consists of a longitudinal and lateral component: the acceleration defines the longitudinal whereas the direction angle defines the lateral movement. The translation into actuator commands is done by the decision unit 2.

To assess the motion requests with respect to the current motion vector, the vehicle observer delivers the current vehicle state (longitudinal and lateral acceleration/velocity, yaw rate, side slip angle) to the single modules for reference.

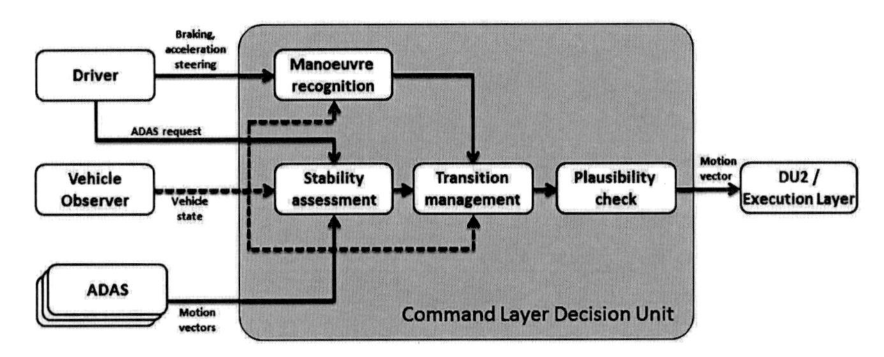

Fig. 3. Structure of the decision unit in the command layer (DU1).

The manoeuvre recognition sub function aims at foreseeing the driver's intention in the very next time interval. On the base of all operations of the driver (pedals, steering wheel, indicators,...) the module extrapolates the probability for acceleration, deceleration, a lane change or any combination of these. This time dependent probability matrix is the input for the transition management sub function and describes the driver's behaviour.

In the stability assessment sub function the output of the single ADAS functions are compared to the current vehicle state. Each ADAS function can operate within a limited regime of parameters, its stability range. As an example the stability range for the ACC function is shown in figure 4. The current vehicle state delivered by the vehicle observer marks one position in this diagram. As long as the current vehicle state is within the safe range (long distance/low velocity) the function operation will be passed by the DU1 to the execution layer. When the critical range has been entered, the braking capability of the ACC is close to the limit of stopping the vehicle before a collision. In this situation the DU1 will start the driver warning cascade by visual and acoustical messages. By further reducing the distance or increasing the velocity the border to the unsafe range will be crossed and an alarm message will be sent to the driver to operate the brakes. This border is chosen with respect to a proper driver reaction and a safe emergency manoeuvre. Parallel to the alarm message the function will be deactivated by the DU1 as the stability criteria for ACC are no longer valid. In case an emergency brake assist is implemented, this function can step into its operating regime after ACC deactivation and assist a safe stopping of the vehicle.

The stability ranges of the ADAS functions are fixed tables defined during the vehicle calibration by the ADAS function owner. The DU1 uses these stability diagrams/tables to assess the operational validity for the current vehicle state. With this concept the parallel operation of ADAS functions is properly manageable and extendable. For the current project several ADAS functions have been implemented to investigate the transition behaviour and the parameterisation of the stability diagrams: four different ACC types, EBA, AEB (Autonomous Emergency Brake), LDW (Lane Departure Warning), LKAS, LCS (Lane Change System).

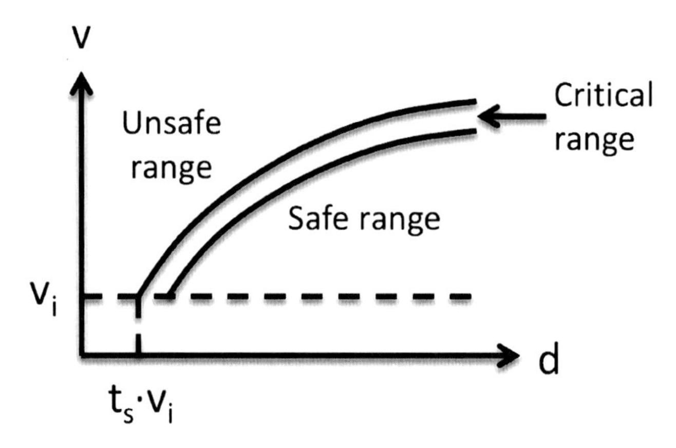

Fig. 4. Stability range diagram for the ACC function. v is the ego vehicle velocity, v_i is the velocity of the leading vehicle, d is the distance between ego and leading vehicle; t_s is the minimum allowed time to collision.

The final decision for activation and deactivation of an ADAS function will be performed in the transition management sub function. The information from the stability assessment module will be transformed into smooth transitions to avoid any discontinuities in the driving behaviour. If necessary, driver messages will be triggered by this module and passed on to the HMI (Human Machine Interface) functions. The final output is a superposition of the motion vectors from all active ADAS functions and the driver. With this concept it is possible to achieve safe transitions between ADAS functions and the driver while the ADAS algorithms purely concentrate on their intrinsic task and do not have to take into account the interference with other functions.

For the sake of functional safety, the final motion vector will be assessed with respect to the current vehicle state and the current velocity / acceleration limits set by the current energy mode. This is achieved by the sub function plausibility check.

After the motion vector on the command layer has been selected and verified it will be passed on to the DU2 in the execution layer. The main task of this function is to select the best actuator commands for the steering, braking and drivetrain, i.e. the electric motors. As the DU1 in the command layer, the DU2 shall assess the validity ranges of the ADAS functions (here the stabilising ADAS) and decide how the requested motion shall be realised. In the present project most of the state of the art functions have been implemented: ABS, ESC, TCS, TVS (Torque Vectoring System, TorVec).

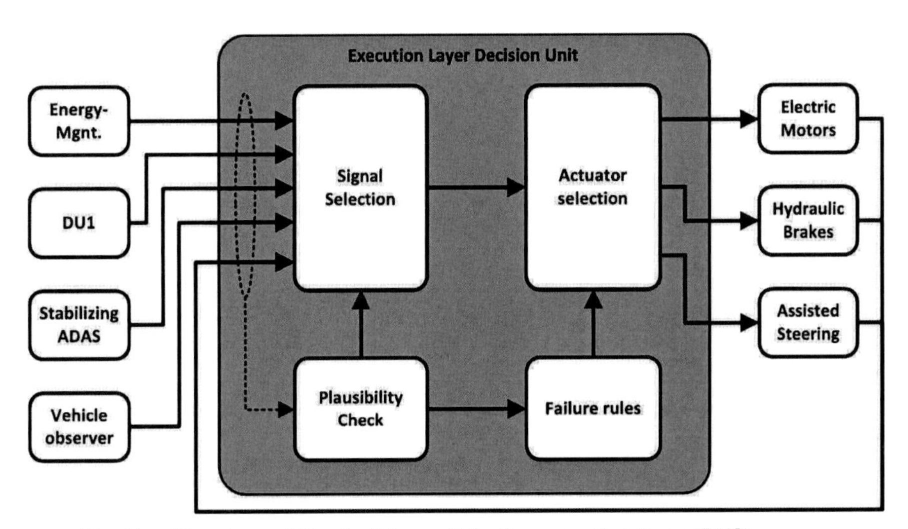

Fig. 5. Structure of the decision unit in the execution layer (DU2).

The task of the execution decision unit can be divided into three main tasks (see figure 5). First the validity of the input signals will be validated by a plausibility check. In the second step, an appropriate drive signal from the DU1 and the stabilizing ADAS functions will be selected. Finally, in the third step, the appropriate actuator commands to fulfil the drive requests will be chosen.

The first step in the DU2 is to validate the correctness of all input signals by checking the value range and the rate of change. The inputs are received from the DU1, the stabilizing ADAS functions, the energy management, the vehicle observer or the vehicle actuators. After checking the validity of every input signal the failure probability of the input result will be analysed, determining the status of all signals. As soon as at least one signal is incorrect a specific rule based process will be followed. Depending on the type of failure detected the driver will be warned accordingly and the vehicle operation will be turned into a state for safe driving. For example if the DU2 detects that the angular velocity signal of one wheel is corrupted, several decisions will follow. The ABS and TCS functionality of the wheel in question will be switched off because these functions rely on the wheel speed information. The possible torque of all electric motors will be limited to reduce the probability of wheel spinning or blocking. A minimum of electric torque is still available to give the driver the possibility of driving the vehicle to a safe place.

If no failure is detected the DU2 proceeds with a standard process, which is to determine the operational range of the vehicle (accelerating, braking, understeering, oversteering or standard operation) and the wheel (spinning, blocking or normal rotation). After calculating the actual operational range, the appropriate signals will be considered. For example if the vehicle is braking and the wheels block the ABS commands will be selected.

Figure 6 shows an overview of the general stabilizing ADAS functions. If the DU2 detects that the vehicle has a high positive longitudinal acceleration the activation the TCS is possible. If high lateral accelerations are measured the activation of ESC is possible, etc. After selecting the appropriate signal the corresponding actuator commands will be determined and applied to the actuators. According to the balance parameter of comfort vs. efficiency, the DU2 arbitrates the braking command between recuperation and hydraulic braking. For maximal efficiency the full recuperation capability will be used while the remaining braking force will be applied via the hydraulic brakes. For full performance, only a minimum recuperation is allowed.

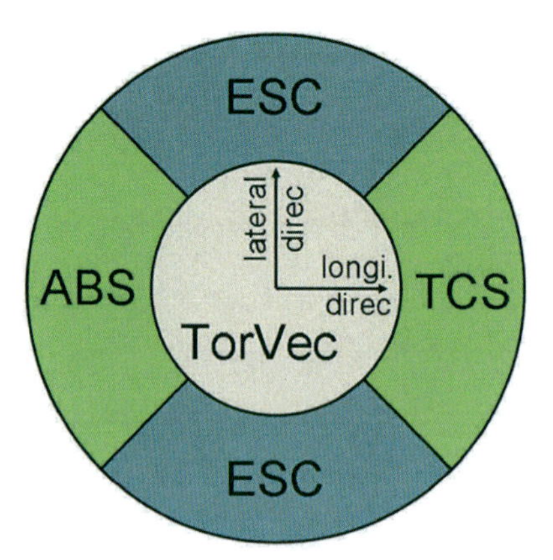

Fig. 6. Validity diagram for stabilising ADAS functions. In the centre region (normal range of longitudinal and lateral forces), the torque vectoring function establishes a safe torque distribution on the driving wheels. For high longitudinal accelerations, the TVS is active while for strong braking ABS becomes active. High lateral forces will engage ESC.

So the DU2 follows the same concept as the DU1 by selecting the best signal/ vector from a scalable set of ADAS functions by applying simple rules in a state diagram and the decision which signal shall be forwarded can be parameterised to meet different driving behaviours.

4 Conclusions

The concept of decision units as central intelligence functions for electric vehicles offers the potential to easily add assistance functions to the vehicle architecture (scalability) and to balance between diverging behaviours like efficiency vs. comfort by simple rules and parameters (configurability). The concept has been set up for electric vehicles as the need for energy efficiency and the inclusion of new functions is vital for the success of this technology. The architecture concept is being implemented into a full electric vehicle in the project eFuture which is funded by the European Commission in the 7th Framework Programme [4,5]. Beside the decision unit architecture several recent function will be implemented to show the capabilities of the concept. These include the above mentioned ADAS functions as well as the vehicle observer, torque vectoring, and a human centred energy management system with a driving style evaluation and comprehensive driver feedback functions. The concept will be applied to an electric vehicle with two front motors.

References

[1] M. Jaensch, P. Prehl, G. Schwefer, K. D. Müller-Glaser, M. Conrath, Modellbasierter E/E-Architekturentwurf 2.0, ATZ Elektronik 2011, Nr. 4, S. 28-33, 2011.

[2] M. Weinmann, K. Elshabrawy, B. Baeker, Development of a flexible E/E-System architecture for conventional and electrified powertrains, 14. Internationaler Kongress Elektronik im Kraftfahrzeug der VDI-Gesellschaft Fahrzeug- und Verkehrstechnik, Baden-Baden, 7. und 8. Oktober 2009.

[3] E. Kelling, T. Raste, Trends in der Systemvernetzung am Beispiel einer skalierbaren E/E-Architektur für die Domänen Vehicle Motion und Safety, 15. VDI-Konferenz „Elektronik im Kraftfahrzeug", Baden-Baden, 2011.

[4] www.efuture-eu.org

[5] V. Scheuch, E/E architecture for battery electric vehicles, ATZ Elektronik 2011, Nr. 6, S. 28-33, 2011.

Volker Scheuch, Frédéric Holzmann, Gerd Kaiser
Intedis GmbH & Co. KG
Max Mengeringhausen Str. 5
97084 Würzburg
Germany
volker.scheuch@intedis.com
frederic.holzmann@intedis.com
gerd.kaiser@intedis.com

Sébastien Glaser
French Institute of Science and Technology
for Transport, Development and Networks
14, route de la Minière, Bâtiment 824
78000 Versailles – Satory
France
sebastien.glaser@ifsttar.fr

Keywords: architecture, electric vehicle, efficiency, safety, driver assistance, decision unit

Plug-In Hybrid Electrical Commercial Vehicle: Energy Flow Control Strategies

S. Agostoni, F. Cheli, F. Mapelli, D. Tarsitano, Politecnico di Milano

Abstract

Nowadays, the greatest part of the efforts to reduce pollutant emissions is directed toward the hybridization of automotive drive trains. Plug-in Hybrid Electric Vehicle (PHEV) seems to be a good short term solution for replacing the conventional combustion engine propelled vehicles, in order to improve fuel economy and reduce pollution emissions. Such topic has a particular relevance while looking at vehicles that operate in urban environment, like light commercial vehicles used for goods delivering even in limited traffic areas. In order to obtain a wide range, full performance, high efficiency vehicle and, at the same time, reduce pollutant emissions, the most feasible solution, at present, is the PHEV, which combines batteries (that can be charged during the night or enough long stops directly from the electric power grid) that feed electrical drive together with a standard Internal Combustion Engine (ICE). In fact today Full Electric Vehicles can not assure the basic requirements of driving range, performance and load capability needed for a commercial vehicle operating in urban environments, mainly because of the low energy density of actually available batteries. Considering the average daily mission of a commercial vehicle delivering goods in urban environments, PHEV can cover even long distances from the hub to the city centre, exploiting the hybrid driving mode (which can increase the efficiency with respect to standard ICEVs) and then use its pure electric driving range (30-60 km) to deliver goods inside the city centre. Since the PHEV has two on-board engines (electric and endothermic) and two energy storage systems (the electrochemical batteries and the fuel tank), energy control strategies have to be developed and introduced in order to find out the most efficient one. The full energetic model of a Plug-In Hybrid Electric Commercial Vehicle, presented in previous papers [1] and already validated exploiting experimental tests performed on a prototype developed at the Mechanical Engineering Department of Politecnico di Milano, will be used in this paper. It will be used to develop energy flows control strategies able to allow the commercial vehicle to perform its daily mission in hybrid and pure electric driving modes.

1 Introduction

Vehicles moved by an Internal Combustion Engine (ICE) are nowadays spread worldwide, and are actually considered to be one of the most relevant causes of air pollution and greenhouse gas emission. One possible solution to such problem is the wide distribution of pure electric vehicles which are moved by an electric motor fed by a battery charged in different ways, using different sources of energy. One of the most significant limits to the distribution of such kind of vehicles is their limited driving range (generally near to 120 km), which is still far from that of a common ICE propelled vehicle (up to 1000 km).

Future development of electrochemical accumulators, such as battery and super capacitors, will probably allow the reaching of higher performances but, at the moment, the most feasible and promising solution in order to obtain a vehicle with both the good characteristics of an electrical and an ICE vehicle seems to be the Plug-in Hybrid Electrical Vehicle (PHEV)[2].

The present work particularly deals with Plug-in Hybrid Electric Vehicles (hereunder referred as PHEV): in this configuration the vehicle can run without any gas emission using only the electrical drive exploiting the energy stored inside the battery and coming directly from the electric power grid. This solution increases the overall efficiency of the vehicle and will reduce pollutant emission. The idea of PHEV is here applied to a commercial vehicle which is daily used to deliver goods in urban areas. The possibility to identify an average daily mission for such vehicle can be exploited in order to optimize the usage of energy stored inside the battery. In the next sections of this paper the numerical model used for the simulation will be presented, the average daily mission of the commercial vehicle will be described together with the strategies adopted to manage the power and the energy coming from the two propulsion systems of the vehicle (ICE and electric drive). Finally the benefits obtained with the usage of a PHEV with respect to a standard ICE vehicle will be shown.

2 The Model of the Plug-in Hybrid Vehicle

As already specified, a PHEV can be schematically represented as a parallel hybrid in which the energy for the battery recharge can come directly from the ICE motor or, more efficiently, from the AC net. The vehicle model is based on the structure represented in Fig. 1.

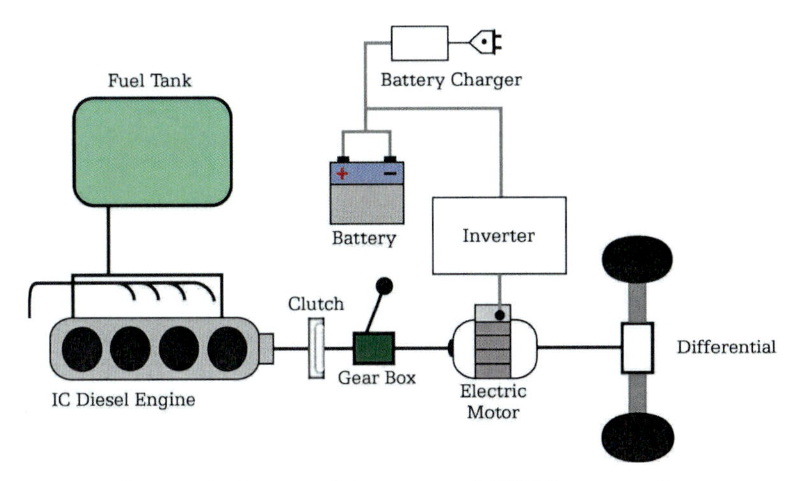

Fig. 1. Schematic diagram of the commercial PHEV presented.

In particular the model used to perform the simulations shown in the next sections, comes from the one presented in [1], which is based on a standard diesel commercial vehicle with a maximum full load weight of 3,5 tons. The choice of such kind of vehicle is due to the fact that this category of vehicle is one of the most diffused for goods delivery in urban areas of the city of Milan. In [1] the realization process of the prototype is described together with the model-validation procedure. That model has been slightly modified before to run the simulations presented in this paper.

In the modeled PHEV the driver can select which drive traction mode is used:

- ICE mode: the drive power is given from the conventional ICE motor which is supplied by diesel fuel. In this mode, the electrical motor can be activated only for regenerative braking and for actuate the battery charging strategy while the vehicle is running. Such running mode will be mainly used in extra-urban areas.

- Electric mode: the drive power is supplied to the electrical motor through the inverter by the battery pack. The ICE motor is turned off. In this configuration the vehicle becomes a Zero Emission Vehicle (ZEV).

- Hybrid mode: the vehicle is moved by both ICE and Electric Motor (hereunder EM). The control strategy of the electric system works in order to optimize fuel consumption recharging batteries, using regenerative braking and allowing the ICE to work at operating conditions that correspond to its higher efficiency. Such behavior can be observed looking at Fig. 2.

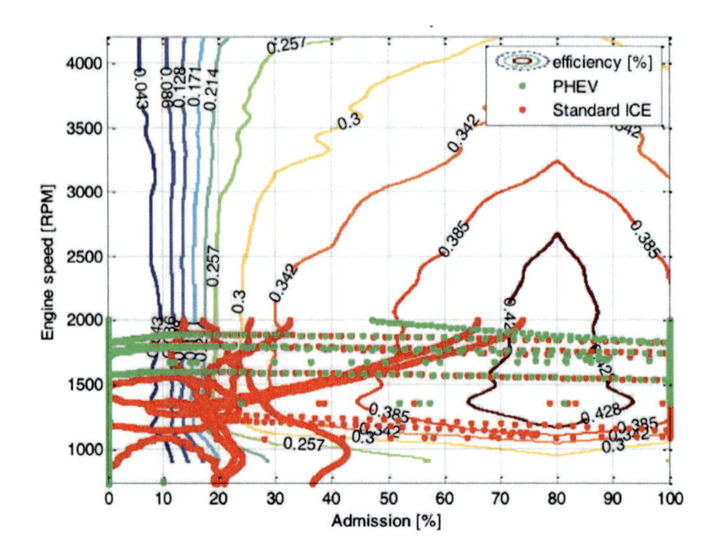

Fig. 2. Comparison of working points between the standard ICE vehicle and the commercial PHEV in "Hybrid mode".

In Fig. 2 it is shown how the working points at low load and low speed are avoided by the "Hybrid mode".

The simulation model has been developed using the object oriented approach ([3]-[10]): every single device has been modeled as an object connected to the others by means of input signals and output state variables.

The objects set represents the whole vehicle power train model. All the single devices represented in Fig. 1 have been considered (Lithium-ion battery, inverter and electric motor, gear box, ICE motor) together with driver and longitudinal dynamic of the vehicle.

Fig. 3. shows a schematic representation of the main components of the model. The detailed description is reported in the following sections.

The traction battery pack used for the prototypal vehicle is composed by 78 Li-Fe elements connected in series with 90 Ah of capacity (C_{batt}).The model of the battery is based on data provided by the manufacturer from which a map of battery voltage as function of the charge / discharge current and the state of charge can be obtained. Tab. 2. shows some technical data about the battery pack chosen for the vehicle model.

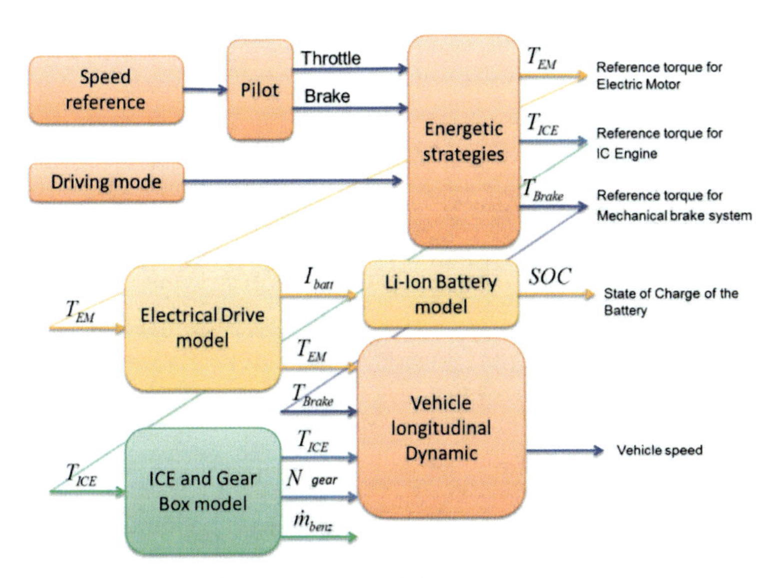

Fig. 3. Vehicle's object oriented model.

Tab. 1. shows some technical data about the electric motor chosen for the vehicle model.

Electric Motor Data	
Characteristic	**Data/Values**
Electric Motor Type	Asynchronous 3-Phase Motor
Motor Control	PWM-Field Oriented Control
Rated Power	22 kW
Maximum Power	55 kW
Rated Torque	260 Nm
Rated Speed	785 RPM

Tab. 1. Electric motor data

Battery Data	
Characteristic	**Data/Values**
Battery Type	Li - Fe
Rated Capacity	90 Ah
Number of Cells	78
Rated Voltage	250 V
Max./Min. Voltage	312 V / 218 V
Max. Charge/Discharge rate	5C / -5C (1C = 90A)

Tab. 2. Battery data

The ICE motor model is based on torque and fuel consumption maps [1].Tab. 3 shows some details about the engine of the prototype vehicle on which the model is based (see [1]).

Standard ICE Vehicle Data	
Characteristic	**Data/Values**
ICE Type	Diesel, Common Rail, VTG Turbocharged, Euro 4
ICE Maximum Power	65 kW @ 3500 RPM
ICE Maximum Torque	220 Nm @ 2000 RPM
Maximum Weight	3500 kg

Tab. 3. Standard ICE vehicle data

The torque produced by the EM and by the ICE meet before the differential. So the difference between the total force supplied by the two motors and the total resistance forces allow to obtain the longitudinal vehicle's acceleration and by its integration the vehicle's speed. The overall resistance force is due to vehicle inertia, tire rolling resistance, aerodynamic resistance and, if present, road slope.

3 Driving Modes and Energy Control Strategies

The definition of the average daily mission of a commercial vehicle used to deliver goods in urban areas has been made dividing such mission in three different parts. For each of these parts, a specific energetic control strategy has been defined with the final aim to optimize the usage of the energy stored inside the battery and to minimize pollutant emissions in urban areas. Below the description of the three parts of the mission with their specific energy control strategies is given.

3.1 From Hub to the City Center

The vehicle starts its daily mission with the fully charged battery (it is supposed it has been charged during the night exploiting the Plug-In function of the vehicle). Generally the hubs of the delivering companies are placed in sub urban areas. Starting from the hub, the vehicle get close to the city center exploiting the "Hybrid mode" drive. The control strategy developed for this specific part of the daily mission is represented in Fig. 4.

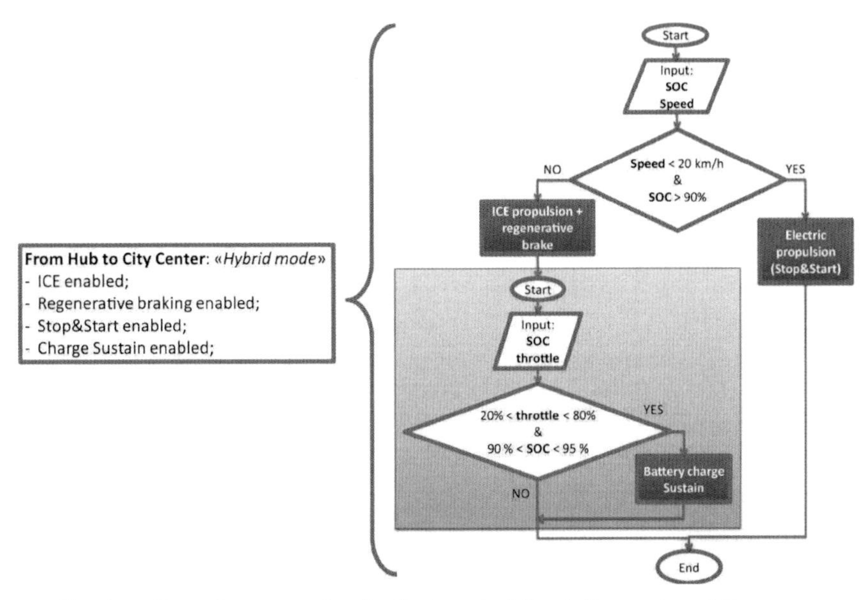

Fig. 4. Control strategy for the first part of the daily mission of the commercial vehicle.

In this mode the vehicle is moved by the standard ICE, which is helped by the electric motor during the vehicle startups and below a defined threshold of speed ("Stop&Start" function). Because the battery state of charge should be maintained high for the next part of the vehicle mission, both the "Regenerative Brake" function and the "Charge Sustain" function are activated: the first allows the recovery of the kinetic energy during each braking phase, while the second one allows the electric motor to absorb some power from the ICE while it is running with a relatively high efficiency depending on some parameters like the pilot throttle command. This allows the system to keep the battery SOC over a certain threshold that can guarantee the performing of the next part of the daily mission. Such threshold can be accurately defined, once the entire daily travel of the vehicle is known.

3.2 Inside the City Center

When the vehicle reaches the urban area the driving mode changes in "Electric mode". The logic scheme of this control strategy is shown in Fig. 5.

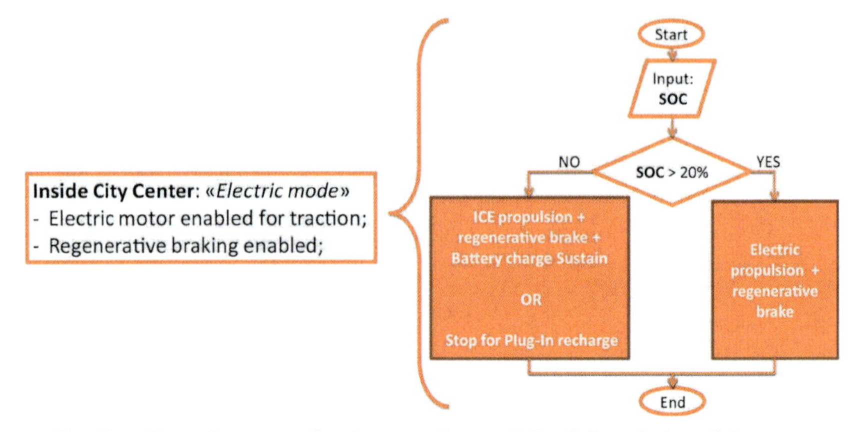

Fig. 5. Control strategy for the second part of the daily mission of the commercial vehicle.

In this case the ICE is shut down and the vehicle runs exploiting only the electrical powertrain with no pollutant emission. Only the "Regenerative Brake" function is now active in order to recovery as much energy as possible increasing the driving range. In fact in case the SOC of the battery would fall below a minimum threshold, the ICE should turn on in order to allow the vehicle to move and/or to recharge the batteries. In case this would not be possible (for instance if the vehicle is inside a Zero Emission Area), another solution could be to use the Plug-In function of the vehicle in order to charge the batteries.

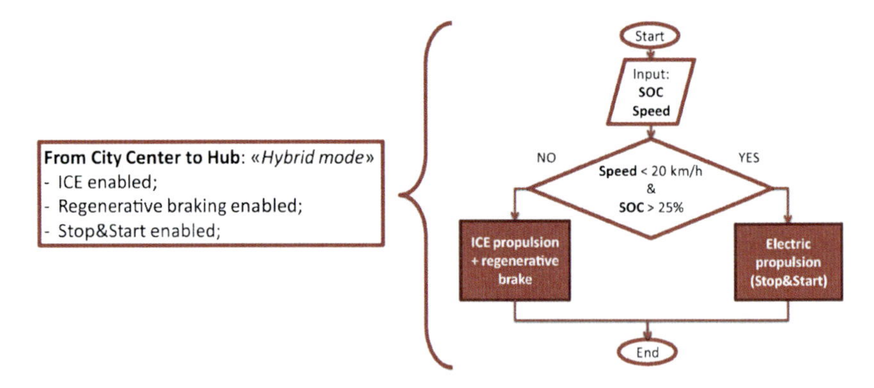

Fig. 6. Control strategy for the third part of the daily mission of the commercial vehicle.

3.3 From the City Center back to the Hub

Once all the goods have been delivered inside the city center, the vehicle changes again the driving mode to the "Hybrid mode". For this last part of the daily mission the control strategy for the hybrid powertrain is represented in Fig. 6.

The difference with respect to the first part of the daily mission is that the battery State Of Charge should no more be maintained as high as possible. For this reason only the "Regenerative Brake" function is activated. In fact the energy recovered during the braking phases can be exploited to activate the "Stop&Start" function assuring that the state of charge of the battery does not fall below its minimum value. The "charge sustain" function is now disabled in order to save fuel and to reduce the operational costs: in fact the cost of 1 kWh produced by an ICE is much greater than the cost of the same kWh coming from the AC net. A schematic representation of the average daily mission of the vehicle is reported in Fig. 7.

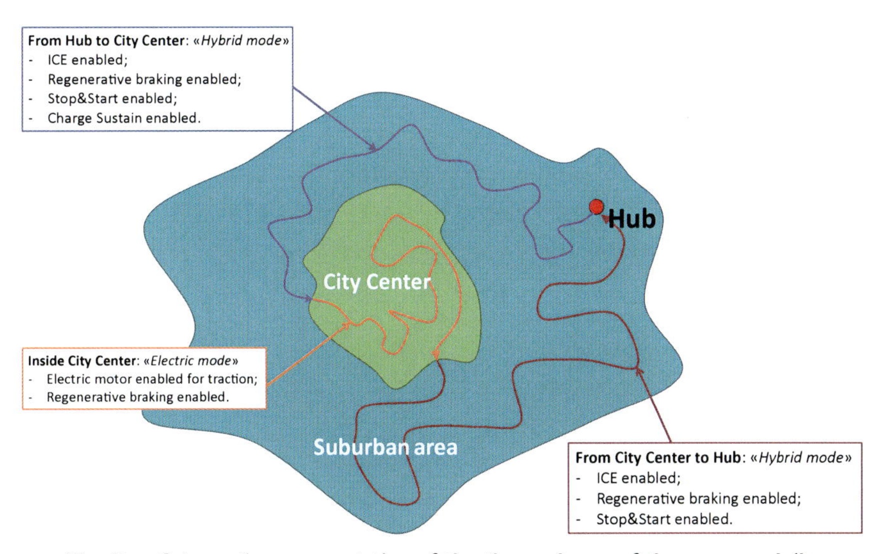

Fig. 7. Schematic representation of the three phases of the average daily mission of a vehicle used for goods delivery in urban areas.

Each one of the aforementioned energy control strategies need the definition of different parameters, like: the speed threshold for the intervention of the "Stop&Start" function, the power absorbed by the electric motor due to the "Charge Sustain" function, the maximum torque applied by the electric motor when the "Regenerative Brake" function is enabled, etc. Moreover the usage of the energy stored inside the battery can be modified knowing the exact route that the vehicle will run during its every working day: in fact the number of kilometers travelled in suburban areas and those travelled in the city center influence the definition of the energy control strategies.

4 Benefits of Energy Control Strategies

The model described in the previous sections has been finally exploited in order to evaluate the benefits that a PHEV with the aforementioned energy control strategies can provide with respect to a standard ICE vehicle while performing the same average daily mission.

The results obtained from the performed simulations will be shown in this paragraph. In order to obtain from each simulation results that could be as much as possible representative of the real working mission of a commercial vehicle running in an urban and suburban environment, all the simulations have been performed using a speed profile derived from an experimental urban drive cycle measured in the city of Milan. In particular such speed profile has to be able to represent both a suburban and an urban driving cycle. In fact the suburban part will be used to simulate the vehicle behaviour during the "From Hub to the City Center" and the "From the City Center to the Hub" part of the daily mission, while the urban profile will be used for the "Inside the City Center" part. Two sample parts of such speed profile are reported in Fig. 8.

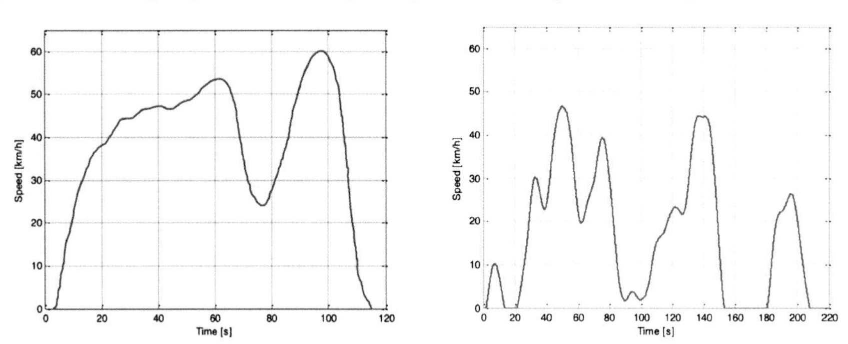

Fig. 8. Experimental speed profiles (sub-urban (left) and urban environment (right)).

From the comparison of the two speed profiles shown in Fig. 8, it is possible to note that their average speeds are different and that number and the length of the stops are higher for the urban profile. The total length of the entire daily mission of the vehicle has been set to 150 km, which is an average obtained from a statistical analysis provided by a delivery company of the city of Milan.

In the next table the results obtained from three different simulations will be shown. The first simulation (1-ICE) has been made with the vehicle running in "ICE mode"on the total distance of 150km. In this way the results obtained in terms of fuel consumption, CO2 emission and operational costs will be considered as reference values for the evaluation of the benefits provided by the hybrid propulsion system.

The second simulation (2-PHEV) has been performed exploiting the hybrid powertrain and the energy control strategies presented in the previous section. For this simulation, and in particular for the urban speed profile, some results are represented in Fig. 9.

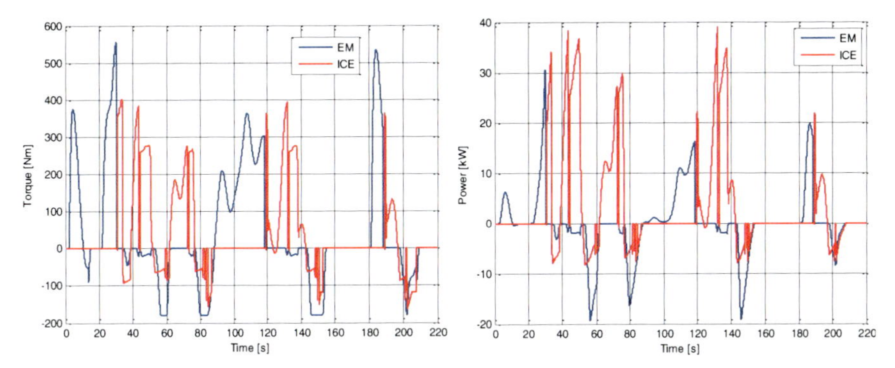

Fig. 9. Graphical results for simulation 2-PHEV: the torque applied by the two propulsion system to the rear differential (left), the power provided both by the ICE and the EM (right)

Finally a third simulation (3-Charge DEPL.) has been performed under the hypothesis of a vehicle running in "charge depletion mode": in this case the battery available on the vehicle has a smaller capacity (about 5 kWh) and it is continuously used by the hybrid system. In particular the entire daily mission is in this case performed by the vehicle in "Hybrid mode", with both the "regenerative brake" function, the "Stop&Start" function and the "Charge sustain" function enabled.

Simulation Results for Commercial Vehicle on its Average Daily Mission							
Strategy		From Hub to the City Center	Inside the City Center	From the City Center to the Hub	Daily Cons. / Emiss.	Daily Total Cost	Year Cost
Speed Profile		Sub-urban	Urban	Sub-urban			
1-ICE	Fuel [liters]	4.49	5.05	4.49	14.03	23.57€	4950€
	Energy [kWh]	--	--	--	--	--	
	CO_2 [kg]	10.75	12.10	10.75	33.60	--	
2-PHEV	Fuel [liters]	3.84	--	3.76	7.60	12.77€	3249€ (-1701€) (-1050€ tax saving)
	Energy [kWh]	1.55	13.55	1.75	16.85	2.70€	
	CO_2 [kg]	9.46	2.26	9.34	21.06	--	
3-Charge DEPL	Fuel [liters]	3.97	4.21	3.97	12.15	20.41€	4397€ (-553€)
	Energy [kWh]	0.85	1.60	0.85	3.30	0.53€	
	CO_2 [kg]	9.64	10.37	9.64	29.65	--	

Tab. 4. Simulation Results: Commercial Vehicle on Average Daily Missions

The emission of CO_2 has been calculated considering the stoichiometry of the combustion reaction for the ICE and the average European emission of CO_2 for each kWh of electrical energy produced (see [11]). The operational costs have been estimated considering the Italian average cost of one liter of diesel fuel proposed by ACI (www.aci.it) which is 1.68 EUR / liter, and the average cost of 1 kWh provided by one of the Italian power suppliers (www.enel.it) which is 0.16 / kWh.

Finally the operational costs extended over 1 year have been calculated considering the minimum number of working day specified by the Italian law, which is 210. The economical benefits referred to simulation 2-PHEV could also take into account the fact that it is possible to avoid the payment of taxes applied to pollutant vehicles inside the city center running in "Electric mode". For example, in case of the city of Milan, the aforementioned tax amounts to 5EUR / day. This means that with a commercial PHEV it is possible to save another 1050EUR / year.

5 Conclusions

An energetic model of a PHEV, based on the model developed and experimentally validated by means of a prototype HEV already presented in [1], has been exploited to implement energy control strategies.

The model has been used to investigate the benefits in terms of fuel and energy consumption that could be obtained driving the vehicle "Hybrid mode" with respect to the traditional vehicle moved by a standard diesel ICE.

In particular, the benefits of the presented PHEV has been observed on an average daily mission performed by a commercial vehicle which is used to deliver goods in urban areas. Sample simulations have been performed in order to show the reduction in terms of fuel consumption and CO_2 emission obtained by the usage of a PHEV. Because of the actual great difference between the price of 1 kWh of electrical energy and 1 liter of diesel (1 kWh costs about 10 times less than 1 liter of diesel) the obtained results show also significant benefits in terms of operational costs of the PHEV with respect to a standard ICE vehicle. Also the preview of the economic benefits on a whole year is highlighted.

The benefits provided by the proposed structure of PHEV have been also compared with a different hybrid configuration in which the commercial vehicle is equipped with a smaller battery and works in "charge depletion" mode.

The presented PHEV model of a commercial vehicle is available for further developments such as: the study of energy control strategies correlated to the real daily travel of the vehicle, optimization and the analysis of solutions with different power train components.

References

[1] F. L. Mapelli, D. Tarsitano, S. Agostoni, Plug-in Hybrid Electrical Commercial Vehicle: Modeling and Prototype Realization, IEEE International Electric Vehicle Conference (IEVC) 2012.

[2] C. Chan, The State of the Art of Electric, Hybrid, and Fuel Cell Vehicles, Proc. of the IEEE, vol.95, no. 4, pp. 704-718, 2007.

[3] S.M. Lukic, J. Cao, R.C. Bansal, F. Rodriguez, A. Emadi, Energy Storage Systems for Automotive Applications, IEEE Trans. on Industrial Electronics, vol. 55, no. 6, pp. 2258-2267, 2008.

[4] W. Gao, C. Mi, A. Emadi, Modeling and Simulation of Electric and Hybrid Vehicles, Proc. of the IEEE, vol.95, no. 4, pp. 729-745, 2007.

[5] T. E. Bihari, P. Gopinath, Object-Oriented Real-Time Systems: Concepts and Examples, Computer Vol. 25, Iss. 12, pp 25-32, 1992.
[6] A. Emadi, Y.J. Lee, K. Rajashekara, Power Electronics and Motor Drives in Electric, Hybrid Electric, and Plug-In Hybrid Electric Vehicles, IEEE Trans. On Industrial Electronics, vol. 55, no. 6, pp. 2237-2245, 2008.
[7] F. Mapelli, D. Tarsitano, and M. Mauri, Plug-in hybrid electric vehicle: Modeling, prototype realization, and inverter losses reduction analysis, IEEE Transactions on Industrial Electronics, vol. 57 n2, pp. 598 – 607,2010.
[8] M. Ehsani, Y. Gao, J. M. Miller, Hybrid Electric Vehicles: Architecture and Motor Drives, Proc. of the IEEE, vol.95, no. 4, pp. 719-728, 2007.
[9] F. Castelli Dezza, F.L. Mapelli, A. Monti, Sensorless induction motor drive for low cost applications, bled (slovenia), Proc. of isie – ieee 1999.
[10] R. Manigrasso and F. L. Mapelli, Design and modelling of asynchronous traction drives fed by limited power source, IEEE-Vehicle Power and Propulsion Conference, pp. 36-43, 2005.
[11] F. Cheli, F. Mapelli, R. Viganò, D. Tarsitano, Start&Stop energy management strategy for a Plug-In HEV: Numerical Analysis and Experimental Test, In: G. Meyer, J. Valldorf [Eds.], Advanced Microsystems for Automotive Applications, Smart Systems for Green Cars and Safe Mobility, Springer, Berlin, 2010.

Stefano Agostoni, Federico Cheli, Ferdinando Mapelli, Davide Tarsitano
Politecnico di Milano
Mechanical Department
Via G. La Masa, 1
20156 Milan
Italy
stefano.agostoni@mail.polimi.it
federico.cheli@polimi.it
ferdinando.mapell@polimi.it
davide.tarsitano@mecc.polimi.it

Keywords: plug-in hybrid HEV, commercial vehicle, energy control strategy, energetic mode

The MobicarInfo System: Tackling Key Issues in EV Range Anxiety

A. Monteiro, Inteligência em Inovação, Centro de Inovação
R. Maia, Critical Software
P. Serra, Instituto Pedro Nunes
P. Neves, Centro para a Excelência e Inovação na Indústria Automóvel

Abstract

The MobiCarin Vehicle Infotainment System addresses the key elements that give OEM systems an advantage over portable devices: seamless integration with the vehicle, hardware control interfaces, and vehicle configuration. The Visteon-based system, currently being developed by the MobicarInfo Consortium, is tailored to the needs of electric vehicle owners, offering access to a set of resources under a single umbrella. These resources include: battery status; up-to-date charging station map and station reservation; route planning; real-time information on public transport, and car-sharing. Vehicle configuration and control features include: remote access, in-vehicle and remote AC control, driving mode, power window control, speed and seatbelt notifications, information from temperature and parking sensors, and onboard diagnostics. The system is being developed by the members of the consortium, namely, Critical Software, INTELI, IPN and CEIIA. A functional prototype will be available in late 2012.

1 Overview of the Market for Infotainment Systems

According to the information company IHS [1], in 2011, automotive infotainment revenue amounted to $32.5 billion and is set to reach $41.2 billion by 2016. In regions like Western Europe and North America, infotainment systems are one of the few areas of growth in an industry that is, to a large extent, stagnant in terms of vehicle production and sales. On the other hand, the current price of these systems, make them a very attractive extra for OEMs to sell. How many other optional features are customers willing to buy at the current price of infotainment systems?

If past growth of the infotainment market is mainly a result of the need to extend the connectivity made available by mobile devices such as smart phones or tablets and changes brought on by factors such as social networks, future growth will benefit more from factors intrinsic to the auto industry itself. A good example of this trend is in-car infotainment systems for electric vehicles (EV). These systems are an essential element in overcoming drivers' range anxiety. With global EV sales expected to reach over 1 million in 2016, infotainment systems will clearly benefit from the growing number of EVs on the roads throughout the world.

Another example is vehicle customisation. If in the past vehicle customisation was achieved by combining different components and systems during final vehicle assembly or by installing after-market accessories, the growing electrification of the vehicle is shifting customisation downstream to the customer who can now effortlessly change the configurations. By simplifying this process, infotainment systems play an essential role in this transition.

2 Overview of Trends in Infotainment Systems

2.1 Current Infotainment Systems

Infotainment systems are starting to emerge in our daily life. The fact that a simple radio receiver isn't sufficient nowadays has caused a need to disseminate audio file players, to control your mobile phone while driving without the need of touching it.

Nowadays, an infotainment system is a mix of entertainment, with a navigation solution, Bluetooth features and in some cases, internet connectivity. Communication capabilities for the vehicle will soon become mandatory following the European Commission requirement for all new models of passenger cars and light vehicles from 2015 to be fitted with eCall devices (SIM-enabled) [2].With the passing of time, the internet connectivity will become more mainstream and with it, several new features will emerge. Especially regarding social network interaction, traffic information, route suggestion and other useful driving features.

2.2 Future Infotainment Systems

The future holds several challenges and open questions in the infotainment area. How will mobile devices coexist with infotainment systems? Will the mobile devices be an extension of the infotainment systems or the other way around? Will there be one or more solutions?

Where will the source of data be? The infotainment side? On the mobile devices side? On the cloud? Is it sufficient to store that music files, photos and other files on the storage device of a smartphone? Will it be easier to simply put all the user information on cloud storage and access it from the several user devices?

How about processing power? Cheaper handheld navigation solutions are broadly adopted nowadays, how can this be used in an infotainment system? How about cloud processing for route planning? Google Maps has one of the fastest route planning methods in the world, is it viable to use cloud-based processing in an infotainment system?

How will the user interface with the system? What will happen to the buttons that we so much enjoy? Will we be using touchscreens or voice commands? Will we have mind reading technology that will enable actions with a simple thought? How complex will the GUI in future systems? How will the information be displayed to the user? How much information is too much information? Where will the information be displayed? In a typical screen placed somewhere in the dashboard? Will it be displayed in the car window, right in front of the driver?

All these questions are still unanswered, the future, with the use of studies and tests, will show what is the road ahead for future infotainment systems.

3 The MobicarInfotainment System

3.1 MobicarInfotainment System Architecture

The MobicarInfotainment is composed of several abstraction layers. These abstraction layers serve the purpose of creating independence between layers. This way, it is possible to switch one of the layers for another and the system, if all the interfaces are maintained, will remain functional.

The first abstraction layer is the operating system, which is responsible for the system startup, initialization, hardware control (drivers) and network connectivity.

On top of the operating system, a hardware abstraction layer will provide, to the upper layers, a generic method of accessing the several hardware components, for instance, USB drives, GPS antenna, network connectivity and even the CAN network that will communicate with the car.

The next layer, the data processing layer, is responsible for accessing the hardware abstraction layer and serving processed data to the top layers.

The last two layers are responsible for visual presentation and user interface. The visual control layer will initiate one or more HMI layers so that the users can access and interact with the information available on the system.

This layered approach can enable operating system independence and hardware independence. Specifically in this case, Meego Linux and the Visteon Rapid platform (hardware and base software) were selected for this project.

3.2 System Features – General Features

The system will include the usual multimedia features like, FM radio receiver, audio files player, video player, photo viewer and aux in control. The user will be able to access any multimedia file available in USB storage. The system will always be connected to the internet, providing several online features such as facebook and twiter, which will be integrated inside the platform, and access to two mobility platforms. A remote control application will be able to control charging, V2G operations and air conditioning from a smartphone. Using Bluetooth connectivity, the user will control and interface with any compatible mobile phone enabling the execution of phone calls.

More focused on EV features, the user will be able to create and select charging schedules. This way, the car owner may enter his home, connect the charger to the car, and schedule a charge on a more economical schedule.

3.3 Integration with Vehicle

The Mobicarinfo system is integrated into the vehicle, allowing it to gather information from the vehicle's electronic modules and issue commands to its actuators. The integration between these two systems allows the Mobicarinfo to present to the user real-time measured parameters, perform calculations and present data such as average values and statistics and also issue control commands automatically or by user's interaction.

The Mobicarinfo is connected to the vehicle by power, sound system, antennas for internet connectivity, FM radio tuning and GPS for positioning and it also provides an USB port to allow the user to connect an external storage device. The connectivity between the Mobicarinfo and the onboard electronic modules is achieved via the CAN (Controller Area Network) bus. The bus networking topology allows the vehicle's measurement and actuator modules that interact with the Mobicarinfo to be distributed throughout the vehicle. Fig. 1 represents some of the modules on board the vehicle that exchange information with MobicarInfo.

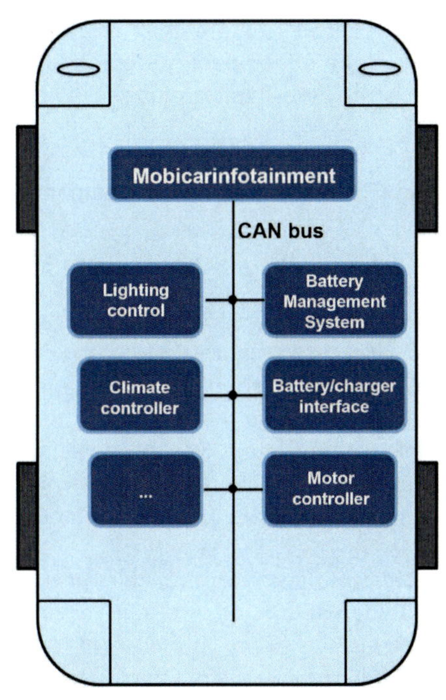

Fig. 1. Vehicle's communication architecture[CEIIA]

These represented modules work in a two way stream with the Mobicarinfo. The lighting control system, for instance, communicates its status and also allows the user to turn the lights on or off. The battery charger interface exchanges information about charging status and allows the user to control it by setting a start time after leaving the vehicle unattended or simply turning it on when connected to a charging station.

Integration of Mobicarinfo with the vehicle allows remote control and survey of the system as well. The user may choose to turn on the climate control system when the vehicle is plugged to a charging station, thus using mains power without wearing out the vehicle's batteries.

Integration with the motor controller makes it possible to have automatic management over the driving mode. The Mobicarinfo accesses data about destination, traffic conditions, charging stations location and battery status. With all the information combined, it may be configured to automatically restrain the vehicle from pulling too much power, wearing out the batteries too quickly and keeping the driver from reaching its destination or the nearest charging station. A warning is also issued to the driver explaining why the vehicle isn't pulling less power.

The vehicle can also communicate maintenance related data such as malfunctions or remaining battery life which makes Mobicarinfo a diagnostics tool as well, allowing a technician to access information for vehicle maintenance.

3.4 Integration with Mobi.E Electric Mobility Management System

Range anxiety is one of the main barriers identified by many potential EV buyers. In order to mitigate this barrier, various initiatives have been put in place by governments and regional authorities around the world. Most initiatives include the creation of a public EV charging infrastructure with strategically placed stations. Nonetheless, the current 160 km range of most EVs is still a real and psychological barrier to buying and using an EV in longer trips. In-car infotainment systems play an important role in EV range management as they combine vehicle information (e.g. current kWh/km) with infrastructure information (nearest charging station, availability and remote reservation).

In the case of the MobicarInfo system, from the very beginning, EV-specific features were seen as fundamental. The fact that Portugal has one of the largest networks of public charging stations in the world (1300 normal and 50 fast charging stations for the pilot project alone), made Mobi.E a natural partner for developing and testing the system's EV-specific features. Moreover, the Mobi.E system exhibits a series of characteristics that are expected to dominate future EV charging networks. These include access to all stations in places of public access, while simultaneously establishing different levels of service providers (e.g. electricity providers and charging station service providers) and a competitive environment in which all players bring value-added services to end-users: EV owners. The automatic financial settlement between all players is another Mobi.E feature that enhances the user's seamless experience.

Current EV features being developed for the MobicarInfo System include: EV station map with availability status; remote station reservation; access to information on all charging events, including date, start time, duration, station code and location, kWh, cost and CO2 impact; and real-time information on current charging events.

3.5 Integration with OSTInformation Platform

The One.Stop.Transport (OST) is an information platform that supports the development of mobility services. The platform is the aggregator of all mobility services/data. These concepts add value to the platform and provide a link between service/data providers and end-users supervised by mobility authorities.

The OST, from a high level perspective, has data providers of static and dynamic data from different sources and on the top the applications that interact with the user. The data providers will supply useful and important information to the platform, which will store, process and supply it for the mobility applications and services.

Through the MobicarInfo the user can access the platform, granting him different transport options and enhancing the whole mobility experience by providing useful information along the way. Points of Interest (POI) can be fetched, which contain, for example, bus stops, restaurants or museums associated to them. The system also provides Points of Advertisement (POA), inferring location-based advertisement services to the MobicarInfo, which will grant a whole new kind of advertisement related services, depending on the user's preferences. Other important information, such as charging stations location, can also be accessed. Schedules for different public transports or even updated traffic information are also available making the MobicarInfo an access point to an incredibly diverse transportation and information network, leading to an efficient and personalized urban mobility experience.

3.6 R&D Methodology Used by the Consortium

The MobicarInfo is a Research & Development (R&D) project following a prototyping lifecycle. This lifecycle is much more iterative than other software development processes because the project team will be searching for the best choice in terms of project architecture, technologies, goals, design, etc. Figure 2 presents the generic life cycle phases and identifies milestones for the prototype project.

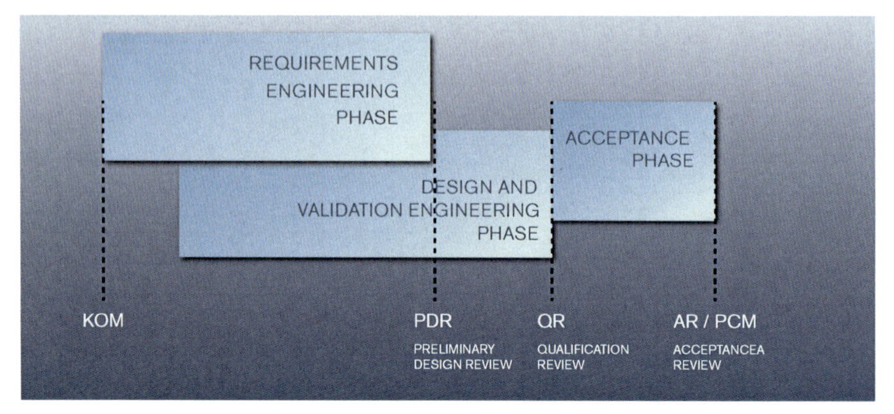

Fig. 2. Generic life cycle phases and milestones

The characteristics of this life cycle are:

- Requirements phase is longer and more iterative than usual because requirements take longer to consolidate. This phase includes System Engineering and Software Requirements Engineering.
- Design and validation starts earlier than usual because requirements need to be designed and validated earlier in order to be verified. These phases are conducted together because the validation of some requirements is performed just after its implementation when other requirements are still in design or implementation.
- The overlap between requirements and design phases is larger to allow definition, design and validation of requirements in parallel.
- The acceptance phase, if existent, is much shorter in relation to other phases.
- The Operations and Maintenance phase is typically not considered.

3.7 Partners and Partnerships

The MobicarInfotainment is being developed by a consortium of Portuguese companies with complementary capabilities. These capabilities include software development and deployment by Critical Software - software layer definition and development as well as the development of the interfaces with external devices and service providers -, vehicle integration by CEIIA access to vehicle data and control of specific vehicle features-, integration with EV charging network by INTELI - two-way exchange of EV-specific information between vehicle and the electric mobility management system and integration with public transport by IPN – two-way exchange of information between One. Stop.Transport and the infotainment system.

Finally Visteon has been a crucial external partner to the consortium by providing access to the Rapid platform (hardware and base software) on which the MobicarInfo system is based.

4 Remaining Steps

4.1 Further Developments – General

The integration of the navigation system will be one of the biggest challenges in the near future. The addition of new features regarding electric mobility to an existent navigation system will involve a deep understanding of navigation systems and other features.

Achieving a stable base system, with the correct power requirements, will also be a big challenge. A system with a touch screen, an onboard computer, audio amplifier and a 3G dongle will consume quite an amount of power; a few optimizations may be possible in order to establish a lower consumption of power.

4.2 Further Developments - Integration with the Vehicle

The car system development trend is to have a more distributed intelligence in order to reduce power cabling and also to have a more distributed computing. With CAN bus communications technology it is possible to have different systems on board the vehicle communicating with each other with reduced cabling.

Traditional vehicle control systems use a main ECU (Electronic Control Unit) to control most of the vehicle's systems. With this topology, it is necessary to have power wiring from a single source to each of the modules. With distributed intelligence and communications, it is possible to issue commands through the CAN bus and have a single power cable throughout the vehicle connected to each module.

4.3 Further Developments - Integration with Mobi.E Electric Mobility Management System

Future developments include Vehicle-to-Grid features that will provide grid managers additional control of EV charging processes by transferring electricity consumption from peak to low demand periods as well as the use of EVs as energy storage devices, hereby reducing the impact of peaks in consumption on electricity production.

4.4 Integration with OSTInformation Platform

The following features to be available in the MobicarInfo require the development and integration of new algorithms. Such features include the initially available data, making these suitable to be used for different purposes. Some of the possibilities are the integration of Intermodal Transportation Routing, Efficient Route Planning or Mobility Profile as described below.

The Intermodal Transportation Routing enables the user to plan and manage his mobility in an efficient way, combining EVs with different public transport systems or even car or bike sharing service. The MobicarInfo system will enable the user to plan a trip that contains various means of transportation, by accessing different information of public transports (bus, train, metro), as well as the transport network (roads, pedestrian trails), a mobility plan, tailored to the user's needs, can be produced.

With the Efficient Route Planning, an optimum route can be achieved by combining, for example, updated traffic information with the GPS navigation system. The user can plan a journey to visit museums and other cultural points based on the proximity of the charging stations to these spots and the autonomy of the vehicle. Moreover, other parameters like fast, economical or touristic trip, slope of the roads or time to destination can be defined by the user.

With a Mobility Profile the user can check his recent trips, ecological footprint, energy consumption and share this type of information with his network of friends. This feature applies the Personal Mobility Record (PMR) and the Vehicle Mobility Record (VMR) in order to create a complete profile for the user. The user can check his past activity, which includes the history of trips, journeys or energy saved under the VMR system. Whereas, the POI (point of interest) visited, different transportation methods, preferences and other characteristics are linked together in the PMR.

5 Contribution to Advancing the State of the Art in Infotainment Systems

The main factor that differentiates MobicarInfo from other systems is its capacity to communicate with cloud-based platforms (Mobi.E EV charging network platform and One.Stop.Transport – public transport information platform) and integrate the information exchanged with these platforms with vehicle data. From the point of view of the online platforms, probe data from the vehicle's system is a valuable asset that can increase the quality and reliability of other services provided by the same platform.

6 Conclusions

Infotainment systems are one major area of growth in the automotive industry with manufacturers and 1st tier suppliers competing for a position in a market that could reach $41.2 billion by 2016. New infotainment systems coming in to the market are moving beyond the features of previous systems by positioning themselves as extensions of the driver's personal laptop/tablet/smartphone. The challenge to the automotive industry is to differentiate itself from its ICT competitors by developing features that rely on the useful combination of vehicle data with information from online platforms. Mobicarinfo addresses that challenge. With a clear focus on electric mobility and the integration with public transport, the MobicarInfo system integrates vehicle data with information from public EV charging stations to produce information relevant to city commuters that combine the use of the private car with public transport.

References

[1] L. Ambroggi, Automotive Infotainment Electronics Market Set for Growth in 2012, IHS iSuppli Market Research, 2012.
[2] European Commission, Digital Agenda: Commission takes first step to ensure life-saving emergency call system for road accidents in place by 2015, 2011.

António Monteiro
INTELI
Inteligência em Inovação, Centro de Inovação
Avenida Conselheiro Fernando de Sousa n. 11 - 4°
1070-072 Lisboa
Portugal
amonteiro@inteli.pt

Rodrigo Maia
Critical Software
Parque Industrial de Taveiro, Lote 48
3045-504 Coimbra
Portugal
rtmaia@criticalsoftware.com

Pedro Serra
IPN
Instituto Pedro Nunes
Rua Pedro Nunes
3030-199 Coimbra
Portugal
pfserra@ipn.pt

Pedro Neves
CEIIA
Centro para a Excelência e Inovação na Indústria Automóvel
Rua Engenheiro Frederico Ulrich 2650
4470–605 Maia
Portugal
pedro.neves@ceiia.com

Keywords: infotainment, electric charging, public transport, range, vehicle, CAN, network, GUI, Mobi.E, OST, V2G, cloud

Estimation of In-Use Powertrain Parameters of Fully Electric Vehicle Using Advanced ARM Microcontrollers

R. Grzeszczyk, A. Hojka, Automex
J. Merkisz, M. Bajerlain, P. Fuc, P. Lijewski, Poznan University of Technology
P. Bogus, Rail Vehicles Institute - TABOR

Abstract

Electric vehicles are currently subject of many projects in the areas of research, design and production, but there is also an urgent need for new measurement tools and methods being developed to quantify their parameters and performance, and to assist researchers in their search of improved performance and optimal design decisions. The paper presents the design and implementation of such a measurement system which is devised to address these new demands. Subsequently the work is focused on example data logged from a fully electric vehicle, followed by evaluation of results and analysis.

1 Introduction

The first electric vehicle (EV) was built in the nineteenth century, and since then it is widely accepted that EV can be characterized by some superior qualities when comparing with internal combustion engine powered vehicles – they are relatively easy to build, operate quietly, no flammable or explosive liquid is stored on-board, lower vibration from the motor and no exhaust gas emission from the tailpipe. In addition, the use of the EV may reduce the dependence on fossil fuels and reduces energy costs by recuperation of braking energy.

The rapid development of internal combustion engines in the last century combined with the widespread availability and low price of crude oil caused that electric vehicles were not able to rival with vehicles powered by conventional combustion engines. The main problem was low energy storage density of the batteries, i.e. their low unit capacity resulting in excessive weight necessary to attain required energy supplies. This greatly limited the range of electric cars and thus prevented them from becoming more widespread. The oil crisis in 1973, resulting in introduction of first legislation limiting the emission from the vehicle caused car manufacturers to begin working on vehicles that use alternative energy sources and design principles.

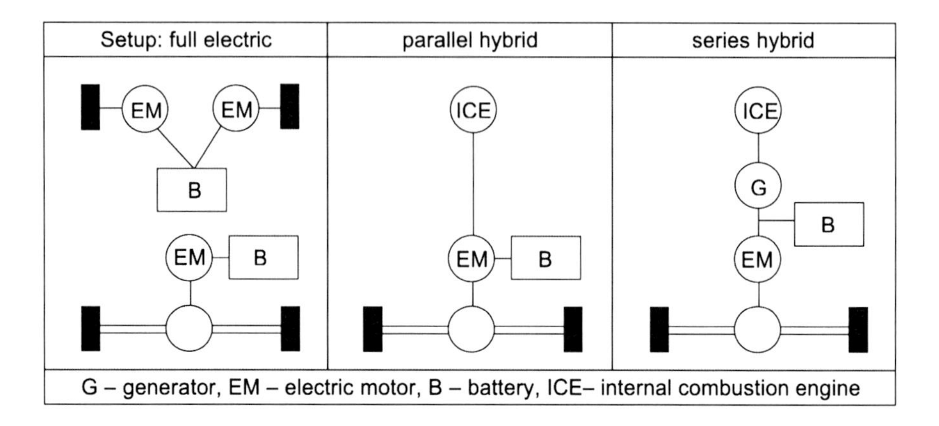

Fig. 1. Types of electric drive vehicles

One of many considered options was to return to vehicles that use electric drives, purely electric solutions (full electric vehicles) as well as solutions combining electric drives with the use of internal combustion engines (hybrids) (Fig. 1).

2 The System Design

Classical instrumentation currently used in the areas of research, design and production is not intended to focus on electrical properties of power train subsystems, and its practical use is limited due to the high value of electric power necessary to be transferred and controlled. To address this problem, a specialised, portable solution has been developed, to allow multi-channel measurement of electric quantities in power-electronic subsystem of the EV, as well as other traction-related data from the vehicle on-board communication networks, GNSS (GPS/Galileo) and inertial sensors.The proposed device allows to record up to 24 analogue channels to measure currents and voltages, as well as configurable digital data streams from double CAN port, position and speed data from GNSS sensor, and data from inertial sensor. Every single piece of data logged on SD card is precisely time-stamped to allow further processing and analysis. The numerical power of the used ARM microcontrollers is high enough to implement advanced data processing techniques in the devices. The ARM devices selected for this project are Texas Instruments ARM microprocessors: Stellaris ARM CORTEX M3 (LM3S9b96) for the front-end module, and OMAP4460 for post-processing and user interface module [1]. The OMAP4460 Dual-core ARM CORTEX-A9 MPCore with Symmetric Multiprocessing at 1.2 GHz each core allows running not only the GUI but also complex filtering for data fusion, processing and presentation. The firmware can be both implemented as applications written in designated C/C++ code, or even more interestingly, as these devices are devised to work also with ported Linux operating systems, it is possible then to use Scilab (free, open-source Matlab-like package) [2] which constitutes for a much more flexible tool for typical design and research activities in the area of electrical vehicle development.

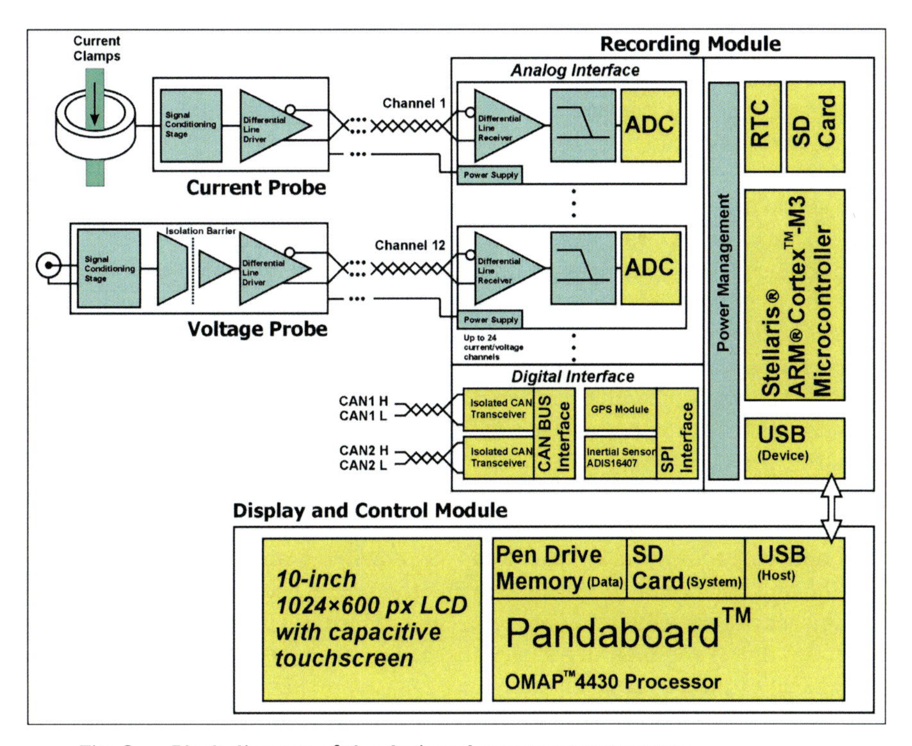

Fig. 2. Block diagram of the designed measurement system.

The designed device is composed of two basic modules (Fig. 2.) - a front-end data acquisition module which is responsible for real-time low-level processing and recording, and a GUI 'control and display module' which is used for communication with the user, test set-up configuration, visualization, and high-level processing.

The recording module contains 24 independent, identically analogue channels for measuring currents in the range of 0 - ± 500A and 0 - ± 1000 Amp, and voltages in the range of 0-± 1000 Volts. Software configuration determines the operating mode and number of active inputs. All probes are powered from the recording module which allows switching off unused channels, which is an essential feature for a battery-powered device. The determination of vehicle location and speed, and motion parameters is obtained from the GNSS module and 10-DOF (degrees of freedom) inertial sensor which provides gyroscopic data, accelerations, and electronic compass/magnetometer data. Additionally, acquisition of data streams from the vehicle on-board networks is made available by double CAN ports and a single Ethernet interface. The recording module is controlled by an efficient 32-bit ARM Cortex-M3 microcontroller manufactured by Texas Instruments, LM3S9b96 (Stellaris series). The recorded data is stored in the internal SD card. Fast timers and real time operating system provide precise and reliable time stamps. The unit after the initial setup can work autonomously (to save power), or can be configured to cooperate with control and visualization module (GUI and post-processing).

Solid state current sensors which operate either with alternating (AC) or direct (DC) current put in practise the principles of the Hall-effect wherein a Hall-effect device (HED) produces an output voltage linearly related to the amplitude and phase of a magnetic field applied to it. There are two types of current sensors based on HED elements: open and closed loop. In the open loop device the current sensing is performed by measuring the magnetic field surrounding a current-carrying conductor. The conductor is passed through the core flux collector which directs the magnetic field around the sensing HED element placed in the gap of this core. The magnetic field in the core gap is directly proportional to the current passing through the conductor, thus the waveform of this device output voltage will track the waveform of the measured current, both amplitude and phase. The closed loop device employs feedback system and the null balance or zero magnetic flux method to control the magnetic flux in the sensor core constantly at zero. The amount of current required to balance zero flux is the measure of the primary current. The through-hole design galvanically isolates the measurement device circuit from the vehicle circuit, and ensures overcurrent or high voltage transient protection. In general, the output of these devices can be adjusted by varying the supply voltage, varying the gap cut in the flux collector, or increasing the number of turns of the conductor passing through the centre of the flux collector. The main advantage of this technology is that it provides measurement of AC, DC and impulse currents at low power consumption, high level of electrical isolation between primary and secondary circuits, and that the diameter of primary current conductor is not limited to any certain size, with particular convenience of installation on existing vehicle harness in the case of the use of sensors with split core design [3]. A high voltage probe allows measurement of voltages in the range of 0-1000 V DC. The voltage probes were also designed to monitor lower voltages, such as control the status of the contactors (relays), where the operational input voltage range is of 0-40 V DC. Optical isolation applied in each interface provides safety and voids risks and problems related to ground loops and the possibility of electric shock. The isolation immunity is better than 5300 V_{rms}.

Probe inputs are also protected against damage in the case of application of reversed polarity or overvoltage.

The control and visualization module is based on the single-board computer PandaBoard [4] which is a low-power, low-cost development platform based on the Texas Instruments OMAP4430 Processor. The system runs a Ubuntu operational system, and ergonomic operation and intuitive graphical user interface is warranted by a LCD display with capacitive touchscreen. The control module allows the user to select the actual type of probes connected, activate and deactivate measurement channels, pre-configure CAN frame ID filtering, and set all the other measurement related settings. It is possible to perform digital filtering, processing and visualization of data in the place of registration, and to preview current and registered data in both graphical and numerical form. All logged data can be stored on the memory stick and easily transferred to a PC for further analysis.

3 Measurements

The developed system has been assembled and tested, all measurement channels calibrated and verified both in the laboratory and in field tests. The measurements included in this paper were made in a fully electric drive vehicle in real on-road operational conditions. The current and voltage probes were placed on power lines connecting all main power subsystems of the car, i.e. batteries, power inverter, driving motor, battery charger circuitry, EMC filters and power-dump resistors, so all vital parameters of the vehicle were observed. Additional signal inputs were wired to relay switches control terminals to log control signals as well. The vehicle CAN bus provided real-time data from the vehicle on-board modules such as velocity, accelerator and brake pedal positions, and control messages sent over the vehicle on-board network. The set of collected data was complimented by data stream from the GNSS receiver and inertial measurement module. Power conversion in car circuitry relies on high-frequency switching of high-power currents, so it makes up for a very noisy (in the sense of EMC) operational environment. Power circuits of the car should be designed in accordance with the best available engineering knowledge to date, but even then any precise measurement system utilised on-board has to be designed to withstand excessive electric distortions characteristic for controlling inductive loads with contradicting and dynamically changing demand values and control targets. During tests and trials it became known that a combination of different means to cope with distortions will be essential to ensure acceptable immunity against all unwelcome influence on measurement channels. The applied approach incorporate both hardware filtering at acquisition inputs (low-pass filters) as well as using relatively distortion-proof designs like differential lines for routing analogue signals over a-few-meter-long measurement cables. On the software side different available digital techniques were employed such as decimation, IIR/FIR filtering, outliers detection, as well as model-based methods at the layer of the whole vehicle parameter estimation (Kalman filtering).

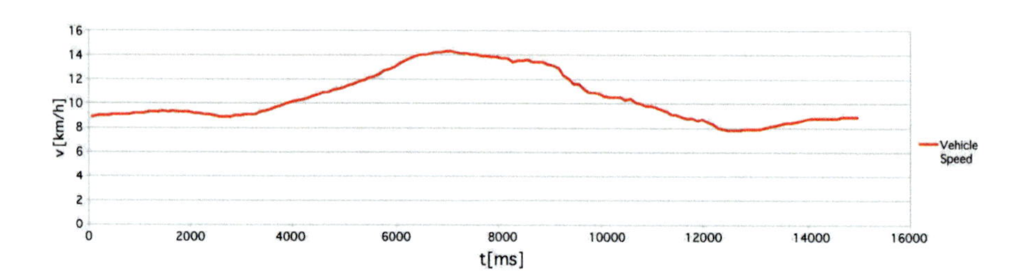

Fig. 3. Vehicle speed obtained from on-board CAN interface.

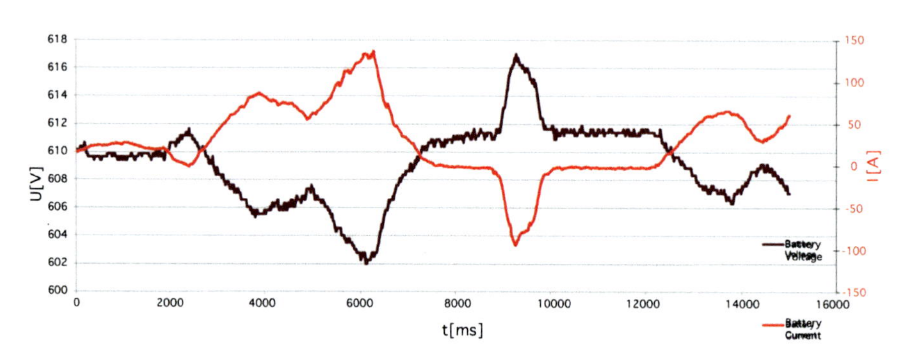

Fig. 4. The voltage and current for the vole vehicle powertrain registered for the velocity profile in the Fig. 3

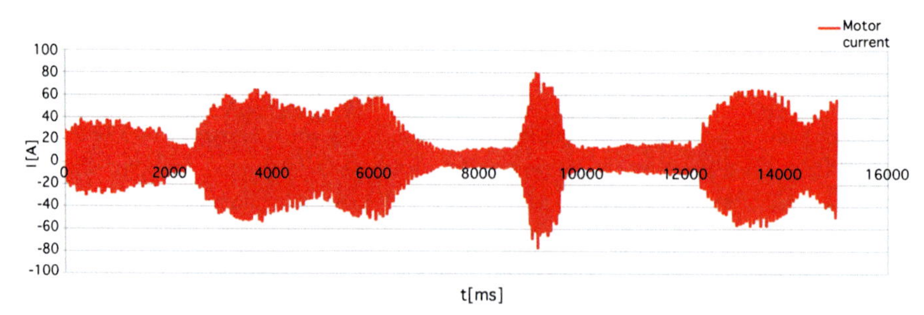

Fig. 5. The electric motor current (one phase) registered for the velocity profile in the Fig. 3

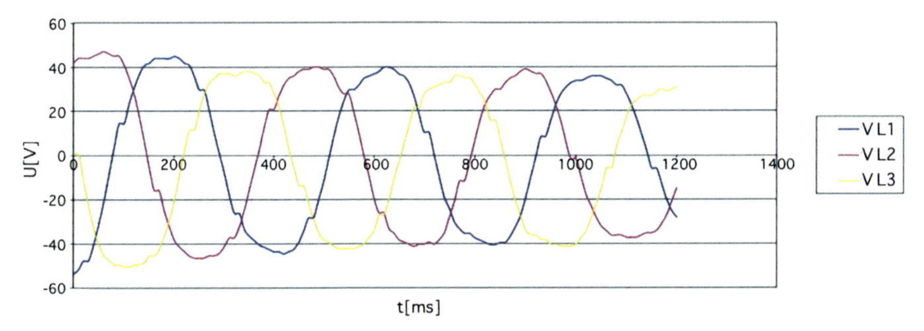

Fig. 6. Voltage waveform logged at the electric drive motor terminals.

The Figures 3 - 6 above present examples of measurements logged with the developed system. The vehicle speed (Fig. 3) was acquired from CAN messages routed between vehicle ECUs, while voltage and current in both figure no. 4 and figure no. 5 were measured as analogue signals converted in the device A/D converters. Figures 3, 4 and 5 represents measurements carried out over the same period of 16 seconds, and figure 6 is plotted over a different time scale (1.4s). Registered data is used for data fusion to estimate powertrain parameters of the vehicle, both electrical for drive sub-systems, as well as dynamical for the vehicle as a whole. Kalman filters allow to obtain better precision and accuracy by post-processing the data. Comparing vehicle performance (velocity, acceleration) with electrical parameters of the powertrain system allows to determine such parameters like e.g. efficiency of the electric drive, at the same time making it possible to analyse and diagnose elements and sub-subsystems which may need closer attention in the case of their underperformance.

4 Conclusions

Electric vehicles are currently object of many projects, in the areas of research, design and production, but there is also an urgent need for new measurement tools and methods being developed to quantify their parameters and performance, and to assist researchers in their search of improved performance and optimal design decisions. Classical instrumentation used to date is not intended to focus particularly on electrical properties of power train subsystems, and its practical use is limited especially due to the high value of electric power necessary of being transferred and controlled, which is a new requirement when comparing to vehicles equipped with internal combustion engine. EV is more an electric machine than a vehicle as we used to know it, so new approaches and instrumentations, as well as technical staff qualifications are now equally substantial. The system presented in this paper may fill the gap between existing instrumentation and the need created by the progress of technology in the automotive industry, especially in electric drive vehicles. The paper also presents sample results of recent measurements made on tested electric vehicle, followed by discussion and analyse of obtained data.

References

[1] http://www.ti.com/home_p_arm

[2] http://www.scilab.org/

[1] Solid State Sensors. Current Sensors. Sensing and Control Honeywell, www.honeywell.com/sensing, retrieved February 2012.

[4] http://www.pandaboard.org/

Rafał Grzeszczyk, Arkadiusz Hojka
ODIUT Automex
Marynarki Polskiej 55 D
80001 Gdansk
Poland
rafal.grzeszczyk@automex.pl
arkadiusz.hojka@automex.pl

Jerzy Merkisz, Maciej Bajerlein, Paweł Fuc, Piotr Lijewski
Poznan University of Technology
Piotrowo 3
61001 Poznan
Poland
jerzy.merkisz@put.poznan.pl
maciej.bajerlein@put.poznan.pl
pawel.fuc@put.poznan.pl
piotr.lijewski@put.poznan.pl

Piotr Bogus
Rail Vehicles Institute TABOR
Warszawska 181
61001 Poznan
Poland
piotr.bogus@gumed.edu.pl

Keywords: EV, fully electric vehicle, electric car, powertrain parameter estimation, HEV, hybrid electric vehicle

Safety & Driver Assistance

Laser-Based Hierarchical Grid Mapping for Detection and Tracking of Moving Objects

M. Schuetz, Y. Wiyogo, M. Schmid, J. Dickmann, Daimler AG

Abstract

Detection of moving objects is the fundamental component of both Advanced Driver Assistance Systems (ADAS) and autonomous driving systems in urban environments. This paper proposes a multi-layer laser scan projection approach for motion detection and tracking based on grid mapping. Consecutive grid maps are created to build a temporary dynamic occupancy grid map for each time step. The resulting dynamic grid map is used as an input to a particle filter track-before-detect (TBD) approach. This approach provides robust motion detection in highly dynamic environments. The proposed approach has been evaluated in a real-world scenario using a single moving object and multiple moving objects.

1 Introduction

Detection and tracking of moving objects (DATMO) are the crucial functions of autonomous driving systems. DATMO has motivated many researchers in the automotive field to build a generic platform that can support other ADAS functions. For example, detection of moving objects can be used to enhance collision avoidance and automatic parking systems. Moreover, such information can be used for free space analysis and improving the accuracy of the Simultaneous Localization and Mapping (SLAM) algorithm. However, the task of building a robust static map which includes dynamic objects is still challenging and not sufficiently solved.

The DATMO concept for SLAM in urban environments was first proposed by Wang [1]. He compares two scan segmentations using data association in order to extract moving objects. For the detection technique, Weiss [2] utilizes the trail or trajectory of moving objects in grid maps. Mori et al. [3] perform object detection with a grid classification method. Later, Konrad et al. [4] proposed the straightforward approach by computing the difference between two consecutive grid maps. This is followed by scan segmentation and grid-based Kalman filter tracking. Additionally, Bouzouraa [5] proposes a fusion interface between a grid map and object tracking, in order to update both components.

These approaches rely on the following processing steps: raw data segmentation, data association, motion extraction and tracking process. Each of these processes contains parameters that should be adapted or learned for different urban environments. Further, a dynamic environment may cause crucial data loss for each processing step, and therefore a trade-off between noisy data and data loss exists.

While those approaches use data segmentation and association, a grid-based approach for detection is currently preferred. Vu et al. [6] compare the acquired measurements with the occupancy values of the static grid map. A region in the static grid map may be occupied, free or unknown. Scan points resulting from moving objects are detected if a new measurement appears in a free region of the previous static grid map. However, this approach requires stable free-space modeling, and moving objects from the opposite direction of the ego-vehicle can be detected easily. In contrast, objects in front of the ego-vehicle which are moving in the same direction as the ego-vehicle require data association with previous inconsistencies. In a large free space area, such as in a large parking lot, registering a free space region requires a large number of memory allocation. Petrovskaya and Thrun [7] propose a model-based detection and tracking approach. Scan data is represented as a polar. Two consecutive virtual scans are compared for inconsistencies to detect moving objects. The paper is using a vehicle model-based range scan likelihood model to track vehicles. A 3-D lidar sensor is mounted on the top of the vehicle and combined with a 2-D lidar sensor at a lower position. Due to its high dimensionality, each box model with eight parameters is estimated by a Rao Blackwellized Particle Filter (RBPF). After performing RBPF update steps from three consecutive scans, moving objects can be detected. Instead using SLAM, a digital map is utilized to process measurements. In addition, Moras et al. [8] [9] proposes the Dempster-Shafer theory to detect moving objects. The conflicting information from the Dempster-Shafer theory can be analyzed to determine object motion. As stated in the paper, the main drawback of this approach is that very precise ego-motion estimation is needed. Most of these approaches work under optimal conditions with low measurement noise. Noise produced from dynamic environments and inaccuracy of ego-motion can lead to misinterpretation and false detection. This paper proposes a hybrid approach for DATMO in a real environment where unstructured objects (e.g. tree and bushes) and the weather influence the measurements. The self-localization and closing loop problems are not considered in this paper, assuming that several solutions exist [10]. Grid Mapping is already available as the basic platform for moving object detection. The paper is structured as follows: Section II introduces the basic frameworks. The grid-map-based track-before-detect approach is proposed in Section III. Section IV provides the evaluation results compared to other tracking algorithms. Finally, conclusions and outlooks are discussed.

2 Basics

2.1 Hierarchical Occupancy Grid Mapping

The proposed approach uses hierarchical grid mapping with octrees for automotive application, which Schmid et al. [11] proposed. This is an extension to occupancy grid mapping [12]. A 3-D grid map can be established depending on the incoming measurements. The map modeling is built as a concatenation of tree data structures (quadtrees or octrees). Each tree represents a region with equal size and is mapped in the global position. The map resolution is computed dynamically based on the tree level relative to a maximum level. Occupancy values are stored in the so called log odds form [13] (the log odds value of the ith cell of the map is later referred to as $L(m_i)$).

2.2 Particle Filter

A particle filter is used for the tracking of moving objects. The particle filter is a nonparametric approximation method, requiring no restriction of the process model (Gaussian and non-Gaussian models). A sampling approach is used to approximate the Bayes filter equations.

The Condensation algorithm [14] is used to implement the particle filter in this paper. It provides a state estimation using a sampling approach in three steps: resampling, reweighting and normalization. During the resampling step, higher weighted particles are sampled more frequently than the lower weighted particles. The reweighting step assigns a new weight based on a likelihood function of current measurements. Then, each weight is normalized from the sum of the assigned new weights.

3 Detection and Tracking

3.1 Dynamic Cells as Motion Features

A multi-layer or 3-D lidar sensor provides a large number of measurement points. A laser beam model that models free, occupied, and occlusion space in a grid map consumes much memory and many computational resources. Alternatively to the beam model, a point cloud can be projected into a 2-D hierarchical grid map. This means that the probability value of an occupied grid cell increases linearly proportional to the number of occupied cells along the projection. For example, a 3-D lidar sensor measures an obstacle and provides several measurements on 20 scan planes. The projection accumulates the log odds value of each plane.

Assume $M_{scan}[t]$ represents a grid map of the current scan at time t without ground measurements, $M_{stat}[t]$ is an accumulated static grid map and $M_{dyn}[t]$ is a dynamic grid map. M_{stat} is established by accumulating the static grid cells of different scans. M_{dyn} contains the nonstationary grid cells that change every time step.

By applying the intersection operator of two consecutive scans, the static grid map can be extended as a collection of temporary static cells:

$$M_{stat}[t] = M_{stat}[t-1] \cup (M_{scan}[t] \cap M_{stat}[t-1]) \quad (1)$$

where the intersection of two grid maps can be interpreted as:

$$\begin{aligned} M[t] \cap M[t-1] &= P(m_i[t] \cap m_i[t-1]) \\ &= \begin{cases} \max(L(m_i[t]), L(m_i[t-1])), \textit{if both} > 0.5 \\ \textit{do nothing} \qquad\qquad , \textit{else} \end{cases} \end{aligned} \quad (2)$$

The union operator is equal to addition of log odds ratios $L(m_i)$. Experiments showed that the intersection of two grid maps is not sufficient to capture all stationary objects. This is due to the susceptibility of the lidar sensor to a bad weather and short occlusion appearance in dynamic environments. Hence, static object measurements are not guaranteed to appear in every measurement cycle. In a static environment, this issue is not significant for the SLAM algorithm. In DATMO, since dynamic cells depend on the static cells, this phenomenon leads to false computation of dynamic grid cells. For this reason, the intersection operator is applied to three consecutive grid maps ($M[t]$, $M[t-1]$, $M[t-2]$). The recursive equation of the accumulated static grid map is proposed as:

$$M_{stat}[t] = M_{stat}[t-1] \cup (M_{scan}[t] \cap M_{stat}[t-1]) \cup (M_{scan}[t] \cap M_{stat}[t-2] \quad (3)$$

After the static grid map is obtained, the current dynamic grid map $M_{dyn}[t]$ can be computed by map subtraction between the previous scan and the accumulated static map. However, subtracting the grid map of the current scan from the static map does not generate the dynamic grid cells. Indeed, several static objects are also included in the dynamic grid map. This happens because subtracting the current scan and static grid map does not filter out new obstacles that are captured in the current scan for the first time.

To cope with this issue, $M_{scan}[t]$ is calculated from two map subtraction steps. The previous scan $M_{scan}[t-1]$ scan and the static map $M_{stat}[t]$ are subtracted first and second the result is then verified by the current scan using map subtraction. With this approach, dynamic grid cells can be computed robustly.

$$M_{dyn}[t-1] = (M_{scan}[t-1] - M_{stat}[t]) - M_{scan}[t] \tag{4}$$

where $M_{dyn}[t-1]$ represents to the dynamic grid map based on the previous scan. Fig. 1 depicts the map separation.

3.2 Grid Map Based Track-Before-Detect

Jung and Sukhatme [15] propose vision-based motion tracking from a mobile robot. They apply particle filters on a difference image frame. Using the similar concept, a track-before-detect (TBD) approach using particle filter is used to provide robust detection and tracking results. A constant velocity motion is used as the tracking model. We define the particle importance weight w with TBD as:

$$w^i[t] = p(M_{dyn}[t-1] \mid x^i[t]) = \sum_{n=1}^{N} P(z^n[t])\, p(z^n[t] \mid x^i[t]) \tag{5}$$

where $P(z^n[t])$ is the occupancy probability value of the nth dynamic grid cell and $x^i[t]$ is the ith particle. $z^n[t]$ corresponds to $M_{dyn}[t-1]$ as measurement input. $p(z^n[t] \mid x^i[t])$ refers to the likelihood function. During runtime, the covariance matrix Σ is adapted by the actual particle velocity variance. The weight function is proportional to the cells density and the cells probability value. Notice that this function can be applied optimally to multi-layer sensors (> 5 layers) since sensors with fewer layers do not provide occupancy values that enable distinguishing dynamic cells from noise. If all the current cells in a dynamic grid map have equal probability values, the particle filter cannot differentiate the densities between target cells and outliers. Detection of a moving object can be made after passing one to three particle update steps, depending on the self-localization error and the outliers. Based on the dynamic grid map

(Fig. 1d) the particle filter always follows the stable measurements. If the particles converge to a target simultaneously, the particle filter declares that a moving object is detected and an object track is confirmed. A track life time and the convergence of the particle filter characterize valid detection. For multiple target tracking, the initialization step is modified by performing a soft clustering before distributing particle filters. A single particle filter cannot track all objects because of following reasons: non-equal occupancy value of moving objects, no prior fix number of moving objects, different dynamic behaviors and limited number of particles. After clustering, each particle filter is assigned to each cluster. Note that this approach does not require optimal cluster parameters, since particle filters will converge to stable moving cells. The TBD approach can ignore noise segments and plays a significant role as an interface between the static and the dynamic map. For each iteration, it updates both maps simultaneously.

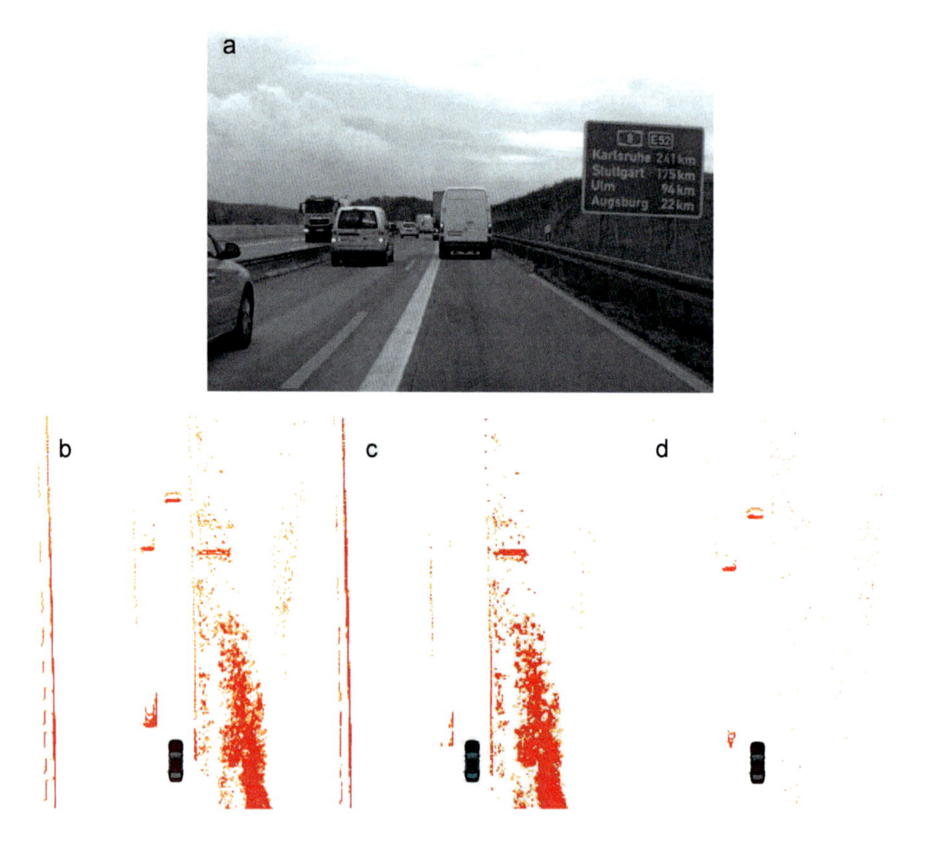

Fig. 1. Bird view of the resulting static (c) and dynamic (d) grid map corresponding to the real world (a) and the current scan (b).

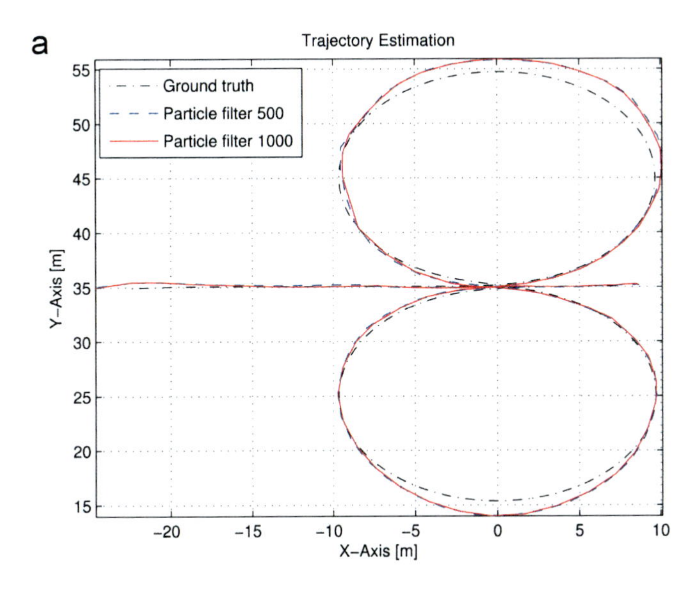

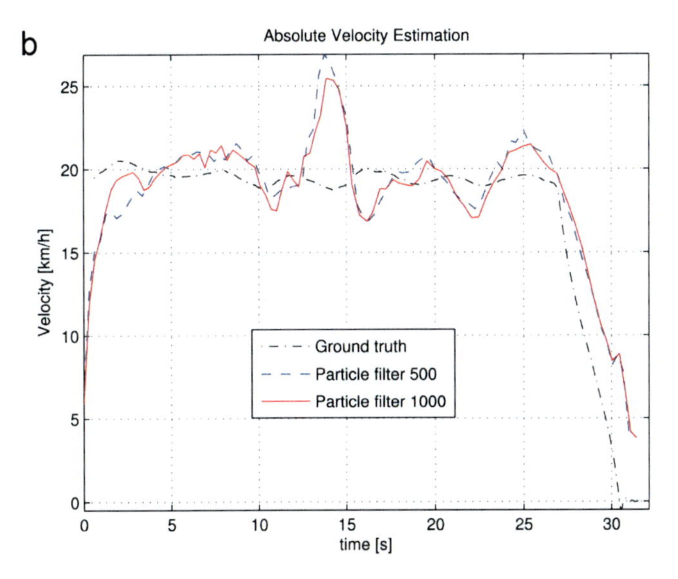

Fig. 2. Comparison of two trajectory (a) and velocity (b) estimations to the ground truth trajectory using 500 and 1000 particles.

4 Evaluation

4.1 Results

For evaluation purposes, an ego-vehicle is equipped with a camera and 3-D Velodyne HDL-64 S2 lidar sensor. This sensor is a 3-D lidar sensor that produces 64 scan layers with 360° horizontal and 26.8° vertical field of view. Each rotation scan is produced with a 10 Hz frame rate and over 1.8 million points per second output rate. The standard deviation of each point range measurement is less than 2 cm. Since more noise is captured due to the rain drops, a comparison with other segmentation and tracking algorithms is difficult to achieve. A robust detection can be achieved by sampling with a minimum of 200 particles. However, 500 particles lead to a better tracking result. Figs. 2a and 2b visualize the mean of the estimation conducted from ten simulations. We compare 500 and 1000 particles. The ground truth trajectory is obtained from the center of rear wheel axle. As shown, the proposed approach can estimate the trajectory smoothly. Regarding the absolute velocity estimation, it results in light peaks when the target turns from North to South. This effect will be considered in future works. The front side of the target suddenly has a higher probability density value and more grid cells than the rear side at that point. The velocity estimation jumps approximately to 26.5 km/h. The root mean square error for the velocity estimation is approximately 2.0 km/h, and 1.1 m for the trajectory estimation compared to the target's rear wheel axle. Fig. 3 shows a crossroad scenario with four tracked objects. Two cars, a truck and a child are tracked robustly. The right car is tracked even when it is temporarily occluded by the truck. The ego-vehicle is not moving, because robust self-localization is not implemented, yet.

Fig. 3. Multi-target track-before-detect from stationary ego-vehicle: Two cars, a truck and a pedestrian are detected. (a) actual scene, (b) extracted dynamic cells, (c) grid map with green particles, red and yellow specify static grid cells

5 Conclusions

We have presented a moving objects detection approach based on multi-layer lidar sensor and grid map separation. This paper presented a grid map separation technique using hierarchical data structure and the tree operator between consecutive grid maps. The TBD particle filter allows distinguish moving objects from measurement noise. This method relies on a multi-layer lidar sensor, since the TBD approach tracks the high occupancy values representing an actual object, rather than noise. Instead of detecting by directly extracting the motion grid cells, the proposed approach first tracks dynamic cells and detects after several iterative updates. Similar to the other grid-based approaches, the presented approach does not perform the data segmentation and association to detect moving objects. A segmentation algorithm is used for estimating and initializing new particle filters.

References

[1] C.-C. Wang, Simultaneous Localization, Mapping and Moving Object Tracking. PhD thesis, Robotics Institute, Carnegie Mellon University, Pittsburgh, PA, April 2004.
[2] T. Weiss, K. Dietmayer, Applications for Driver Assistant Systems Using Online Maps. In Proceedings of 5th International Workshop on Intelligent Transportation, 2008.
[3] T. Mori, T. Sato, H. Noguchi, M. Shimosaka, R. Fukui, T. Sato, Moving Objects Detection and Classification Based on Trajectories of LRF Scan Data on a Grid Map, In Intelligent Robots and Systems (IROS), 2010 IEEE/RSJ International Conference on, pages 2606 – 2611, 2010.
[4] M. Konrad, M. Fuchs, O. Lohlein, K. Dietmayer, Detektion und Tracking dynamischer Objekte in Occupancy Grids. In 7, Workshop Fahrerassistenzsysteme FAS 2011, pages 105–114, 2011.
[5] M.E. Bouzouraa and U. Hofmann, Fusion of Occupancy Grid Mapping and Model Based Object Tracking for Driver Assistance Systems Using Laser and Radar Sensors. In Intelligent Vehicles Symposium (IV), 2010 IEEE, pages 294 –300, 2010.
[6] T.-D. Vu, J. Burlet, O. Aycard, Grid-based Localization and Online Mapping with Moving Objects Detection and Tracking: new results, 2008 IEEE International Vehicles Symposium, pages 684 – 689, 2008.
[7] A. Petrovskaya, S. Thrun, Model Based Vehicle Detection and Tracking for Autonomous Urban Driving, Autonomous Robots Journal, 26(2-3):123 – 139, 2009.
[8] J. Moras, V. Cherfaoui, P. Bonnifait, A Lidar Perception Scheme for Intelligent Vehicle Navigation, In Control Automation Robotics Vision (ICARCV), 2010 11th International Conference on, pages 1809 –1814, dec. 2010.

[9] J. Moras, V. Cherfaoui, P. Bonnifait, Moving Objets Detection by Conflict Analysis in Evidential Grids, In 2011 IEEE Intelligent Vehicles symposium (IV), pages 1122–1127, Baden-Baden, Germany, 2011.
[10] S. Kohlbrecher, J. Meyer, O. von Stryk, U. Klingauf, A Flexible and Scalable SLAM System with Full 3D Motion Estimation, In Proc. IEEE International Symposium on Safety, Security and Rescue Robotics (SSRR). IEEE, 2011.
[11] M. R. Schmid, M. Maehlisch, J. Dickmann, H. J. Wuensche, Dynamic Level of Detail 3D Occupancy Grids for Automotive Use, In Proceedings of IEEE Intelligent Vehicles Symposium, pages 269–274, San Diego, CA, USA, 2010.
[12] S. Thrun and A. Buecken, Integrating Grid-Based and Topological Maps for Mobile Robot Navigation, In Proceedings of the AAAI Thirteenth National Conference on Artificial Intelligence, pages 944–950, 1996.
[13] S. Thrun, W. Burgard, D. Fox, Probabilistic Robotics, The MIT Press, 2005.
[14] M. Isard, A. Blake, CONDENSATION - Conditional Density Propagation for Visual Tracking, International Journal of Computer Vision, 29:5–28, 1998.
[15] B. Jung, G. Sukhatme, Real-time Motion Tracking from a Mobile Robot, International Journal of Social Robotics, 2:63–78, 2010.

Markus Schütz, Yonkie Wiyogo, Matthias Schmid, Jürgen Dickmann
Daimler AG
Wilhelm-Runge-Str. 11
89081 Ulm
Germany
markus.m.schuetz@daimler.com
yongkie.wiyogo@googlemail.com
mail@matthias-schmid.de
juergen.dickmann@daimler.com

Keywords: hierarchical grid mapping, particle filter, track-before-detect, moving objects detection, multiple layers lidar sensor

Lane Accurate Position Sensing of Vehicles for Cooperative Driver Assistance Systems

B. Schmid, M. Zalewski, U. Stählin, K. Rink, S. Günthner, Continental

Abstract

Since some years university and industry research in automotive systems have become strongly focused on cooperative driver assistance systems (Vehicle to X - V2X communication, consisting of Vehicle to Vehicle - V2V and to Infrastructure - V2I). Since the public funded project sim^{TD} [1] and the announcements of car manufacturers during the IAA 2011 in Frankfurt / Main it is obvious that such systems will strongly penetrate the automotive market within the next decade [2]. However, such systems with the aim to improve save driving have three basic challenges: First, they must secure a save and stable communication between the desired information providers, second, each partner needs to know his precise position with a time stamp on a global scale and third the provided information needs to be absolutely reliable. Continental provides such a dedicated solution for the sensory needs of V2X systems, the "Motion Information To X Provider" - M2XPro.

1 Introduction

M2XPro will determine the precise position of the vehicle by using all sensor information which are already available in the vehicle and combines them with GNSS data (Global Navigation Satellite System) in an intelligent fusion algorithm. Additionally this approach will improve the precision of the generic sensor information and provide lane accurate position information paired with a global time stamp. All the output information will be qualified by an integrity level, which informs the system about quality, reliability and availability of the output signals. This information will be made available to all known and future onboard safety and non-safety systems, like ESC, airbag systems, camera, radar, IR based assistance systems or navigation systems and the V2X systems. Even though M2XPro meets the needed safety requirements it is still an affordable solution. The development bases on the philosophy of using low cost and high volume standard components such as standard automotive gyros and accelerometers, wheel speed and steering angle sensors [3] and commercially available GNSS receivers. The architecture of M2XPro is built on a modular hardware and software concept and is independent of any user system. This makes the M2XPro concept easy to integrate into any vehicle architecture.

2 Technology

2.1 State-of-the-Art of Global Positioning Systems

Currently most position systems are using GNSS signals only to calculate the position of the signal receiver. According to this approach, at least 4 GNSS satellite signals are necessary for acceptable calculation. The precision of the calculation depends on the amount of satellites and the disturbances of the ionosphere [4]. In average such position data are less accurate then 5-10 m radius. In vehicle applications often no satellite signal can be received, e.g. when passing a tunnel or driving in cities. To encompass this "Dead Reckoning" was developed, which calculates the position while only insufficient satellite signals are available. The calculation is done by integrating the acceleration to recognize the speed changes and the second integration aims to determine the distance. In a second calculation the position is corrected by the integration of the yaw rate in case the vehicle changes the direction. As leading information the position is always aligned with the street map. In case no satellite positioning is possible as a function of time the calculation is getting more and more imprecise due to integration of the sensor errors.

2.2 M2XPro Hardware & Architecture

The main approach is to provide a precise position signal independent of any application feature. Therefore, in a first step 1-3D rate -, 3D acceleration MEMS sensors designed for ESC application are used to sense the inertial behaviour and steering angle, and wheel speed sensors are used to implement redundant information about speed and heading direction (odometry) (Fig. 1). The configuration with 1 or up to 3 rate sensors depends on the needed precision of the use cases.

2.3 M2XPro Fusion Algorithm

Many of the sensor information in modern vehicles are redundant, for example speed, taken course, acceleration, change of heading direction.

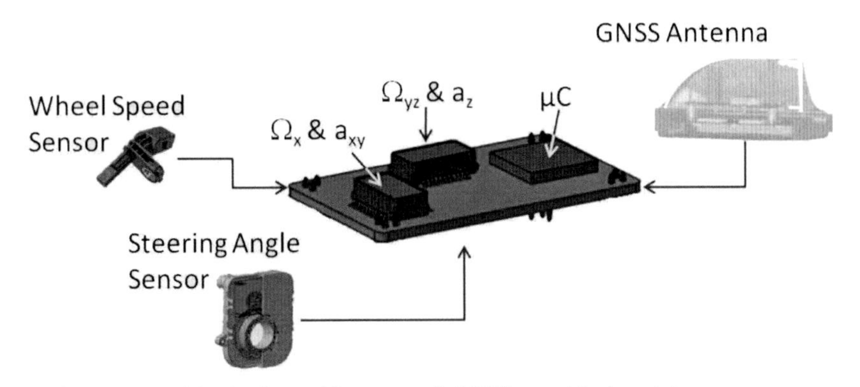

Fig. 1. Possible basis architecture of M2XPro with inertial sensors on board

But due to the fact that the sensor information is used in different independent systems, the benefits of combining them to more accurate, reliable and available information are not taken in advantage. The presented algorithm collects all the given sensor data and computes an improved motion. By exploiting the redundancy of the existing information the quality, availability and reliability of this data is improved. As a result you get more precise and dependable data under all circumstances even if one or more systems temporarily outages. Beyond that the M2XPro algorithm provides an integrity level of the fused information. With this information user systems can derive strategies to handle signals with less integrity and results in a higher failsafe performance for all user system.

Naturally not all sensors have the same influence. GNSS position computing is imprecise in short ranges but good and precise over long distance and then the absolute heading, velocity and position is very reliable and precise. Data from the odometry is highly precise in small distance, but heading errors or slippage leads to increasing position errors over the distance. The same applies for inertial sensors. They are good at high dynamic changes (acceleration, angular rates), but very inaccurate when moving constant, e.g. due to the offset drift behavior. By combining all sensors the advantages will compensate the weakness in a proper way, to get best and reproducible results under all circumstances and in all driving situations (Fig. 2).

2.4 Algorithm architecture

Handling the varying sensor data with independent sample times and fusing them to one comprehensive information is the challenging task of the algorithm. Because sensor data arrive sporadically and can also drop out, e.g. GNSS

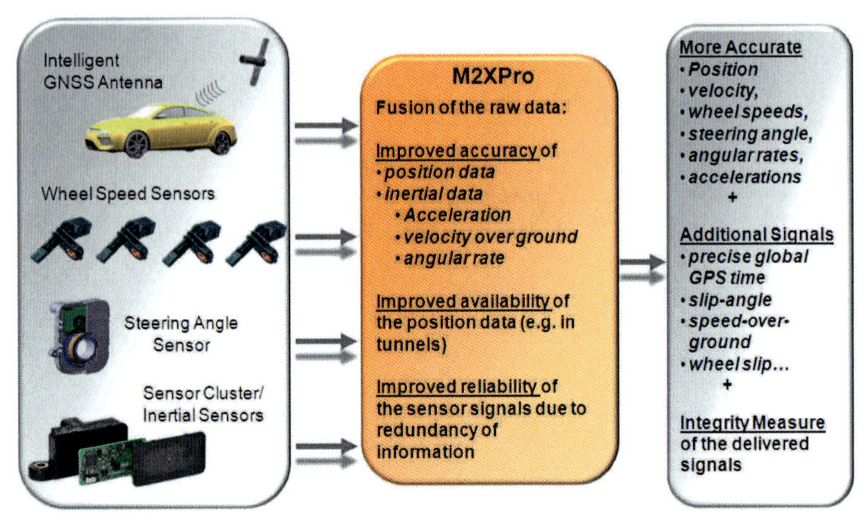

Fig. 2. M2XPro approach uses the raw sensor and satellite data for fusion algorithm

signal in tunnels, the algorithm output has to be extracted from other sensor data. A strapdown-algorithm computes the new position, using the fused data. Since it works independently from the filtering and fusion itself, new position data are available, directly from the input and free of delay (Fig. 3).

To fuse the different sensor data, an Error-State-Space-Extended-Sequential-Kalman-Filter is used (Fig. 4). The main advantage of the chosen filter is the sequential design, such that new sensor data can be applied if it is available. In this way the filter is flexible to add new sensor input data and also able to tackle sensors with different performance.

In contrast to the standard Kalman filter, this filter doesn't compute the new states but it does compute the errors of the system. The advantages are: First, the new errors are calculated by means of the given sensor input. If there is no input, there are no new errors to be calculated. Second, after the errors are calculated, they will be corrected by the algorithm. As a consequence the expected errors for the next Kalman filter cycle are zero, because all expectable errors have already been corrected, so the step of the Kalman prediction can be skipped. This saves computing time. The third advantage of this model is that it doesn't possess the typical self-reinforcing tendencies of the Kalman filter. This means to trust more and more in the model data and ignoring the sensor data, when the model works over a long time well. As a result the typical Kalman filter will ignore changes in sensor data when the movement is long time the same, for example driving straight on the highway.

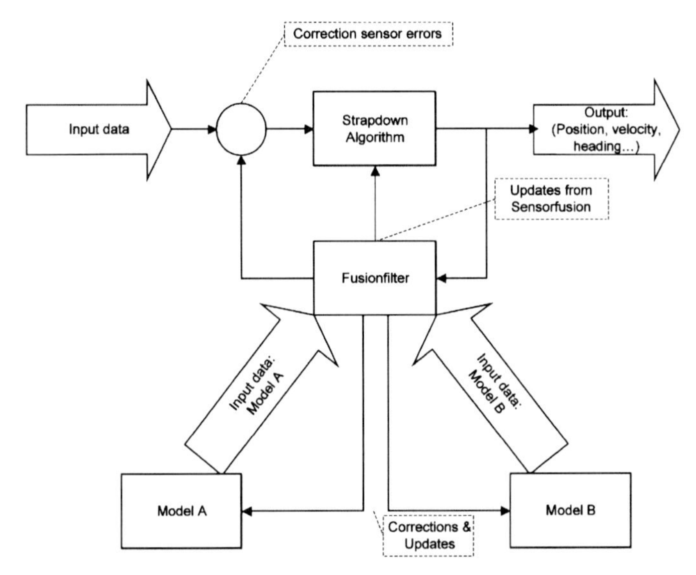

Fig. 3. Overview fusion algorithm

When the motion suddenly changes, this is interpreted as sensor faults and ignored. As a consequence the correct position overshoots until the error gets too big. The given filter doesn't have this problem and computes for every cycle new errors regardless from history trends.

The feedback from the Kalman filter can be used to correct all measurable sensor errors based on the fused information from redundant sensors. So even if a sensor fails the sensor values and errors can be appraised, based on the history and the other sensors. This brings more failsafe performance for the whole system.

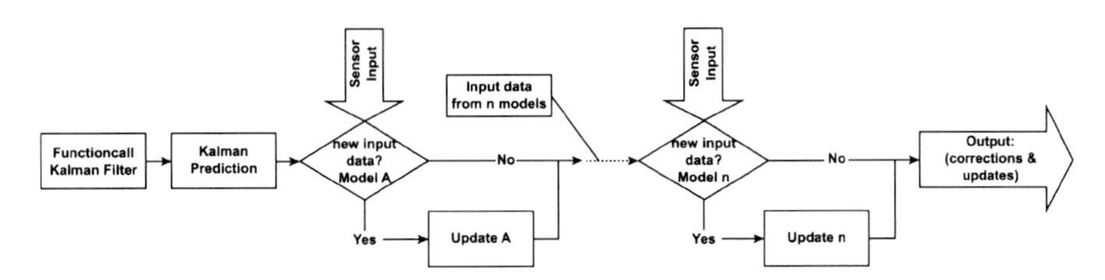

Fig. 4. Sequential design of the Kalman filter

Currently the algorithm development is focused on the elimination of some GNSS signal constellations, e.g. reflected but no direct GNSS signal (only pseudo signals). Further the plausibility check of all results have to be implemented and the rating of the algorithm output integrity as well as a self benchmarking.

3 Test Results

The algorithm is tested both offline in the development environment Matlab Simulink and online on a dSpace Autobox in the test vehicle with live visualization. In this way the direct influence of driving maneuvers and interactions with the environment can be seen and estimated live online. Later the recorded data can be analyzed and evaluated in detail in the offline simulation.

As can be seen in the left picture of Fig. 5 there are several situations typical for cities. First the start in the lower right corner where the GNSS raw data are massively disturbed because of reflections and multipath between buildings. In this case the GNSS position signal - which is calculated directly from the GNSS raw data without any corrections, filtering etc. – jumps about 15 m. In the following there is typical GNSS noise caused by smaller buildings, other vehicles and random noise of the sensor. The position output of the M2XPro algorithm is not influenced by this random noise. Later down the path there can be seen the turn in a backyard and the way back on the other lane which can easily be separated. Through time stamping, when the signal was computed (which is equivalent to the real-time position of the vehicle within the very small sampling grid of the fastest sensor), precise position data including heading and speed can be provided for applications using V2X communication.

Even while driving through the tunnel, where the GNSS signal fails, the algorithm continues calculating new position data based on the other sensors and renders the motion course. Of course the less sensor signals are available the less precise the position is getting over time. But knowing the exact sensor behavior and with a good sensor calibration through the algorithm, the output of the M2XPro algorithm can hold sufficient accuracy for some time, so that the GNSS loss can be compensated. In the right picture of Fig. 5 the tests on the Continental location Frankfurt / Main is shown. Due to the industrial environment (metal covered buildings, strong radio sources etc.) the GNSS signals are heavily disturbed but as in the city test drive the M2XPro algorithm stays on track.

The test results in Fig. 5 show the achievable accuracy and repeatability. Even on bad cobblestone ways where the odometry and inertial sensors have massive noise, the position of M2XPro stays within the required values.

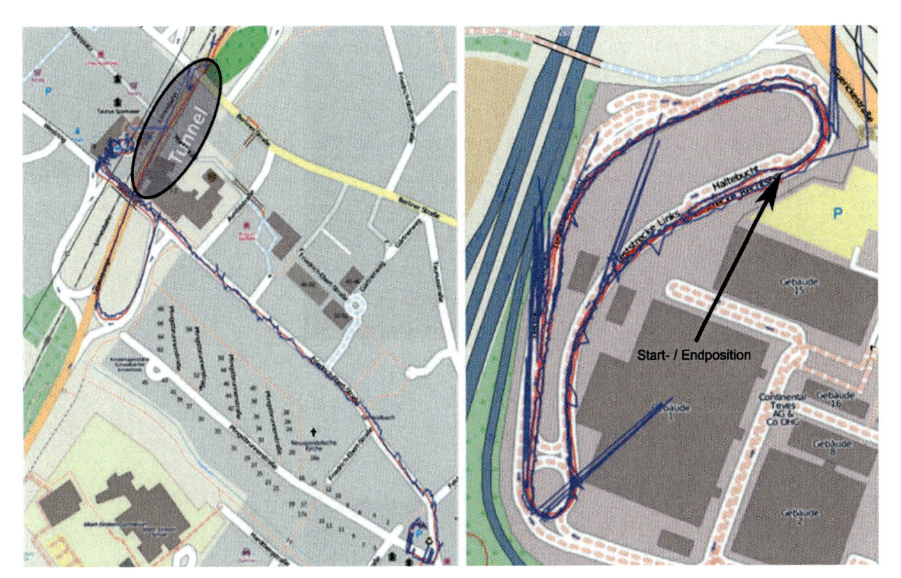

Fig. 5. Left: Test result driving trough city tunnel (map layer OpenStreetMap, blue = GNSS raw data, red = computed position M2XPro algorithm) - Right: Test result driving several times with GNSS noise around the Continental test track [50...130 km/h] (map layer OpenStreetMap, blue = GNSS raw data, red = computed position M2XPro algorithm)

4 Use Cases

In Europe, the automotive industry has gathered in the Car2Car Communication Consortium (C2C-CC) [5] to contribute to standardization of cooperative systems. One of the major goals is the development of realistic deployment strategies for V2X systems such that the introduction of these systems becomes reality. The right choice of use cases plays a major role for the introduction scenarios. The use cases can be grouped mainly in the areas of traffic safety and efficiency.

4.1 Green Light Optimal Speed Advisory (GLOSA)

Being a major use case of V2X systems, this is also the best example of cooperation not only between vehicles, but also between vehicles and the infrastructure. To support the function, the participating traffic light has to be equipped with a so called Road-Side-Unit (RSU). The RSU broadcasts the traffic light information to vehicles in the surrounding. The information consists of the position of the traffic light, the topology of the intersection and the phase schedule of each individual signal. Based on this information, each vehicle can calculate the optimal speed to approach the traffic light such that the vehicle can pass at green. In order to do so it is very important to know exactly which lane the vehicle is driving at and the distance to the traffic light. Typically traffic light arrays have more than one signal that could be the right one for the approaching vehicle, e.g. for driving straight or taking a left/right turn. Only if the vehicle knows which lane it is driving at, the application can provide the correct information to the driver. Otherwise the information for all possible manoeuvres has to be given to the driver and can cause more distraction from the traffic in a complex situation (intersection).

4.2 Stationary Vehicle Warning

One of the benefits of V2X Communication is its ability to distribute information about invisible obstacles. This possibility to "look around the corner" is also one of the major reasons for trying to deploy such systems. A typical example enabled by this ability is the stationary vehicle warning. A vehicle in a blind spot broadcasts the information that it has stopped and that it could therefore be dangerous for other approaching vehicles. A typical example is a traffic jam behind a curve. Again, it is evident that a precise positioning is very beneficial for the quality of this use case. Looking at streets with more than one lane per direction, like highways, it makes a major difference whether the resting vehicle is positioned on the ego-vehicle's lane or on another lane. Only with lane-accurate positioning the distinction between a highly dangerous situation and a normal passing is foreseen.

4.3 Intersection Assistance

Looking more into the future, one of the key features of V2X systems is the intersection safety. Since the time window separating a dangerous approach of two vehicles from a normal crossing of these two vehicles is very narrow, a very precise positioning is crucial. Only if the position is available with high accuracy, the system can assess such situations correctly. The results of several projects, like the German project AKTIV [6] and the European project INTERSAFE-2 [7] have shown clearly the importance of precise positioning, especially for the feature "Intersection Assist". Today typically differential GNSS is used for this use case, but research is directed towards the intelligent fusion of "normal" GNSS and vehicle dynamics sensors. Based on the results of Continental's participation in AKTIV, the M2XPro also follows the latter approach.

5 Architecture Scenarios

Today it remains a huge challenge to cope with all possible vehicle system configurations. All car manufacturers follow more or less their individual strategy to implement vehicle systems. E.g. inertial sensors for ESC are nowadays integrated in the ESC ECU or the airbag ECU. This fact has to be considered from the beginning of the architectural analysis. On the other hand V2X features are computing resource consuming and OEMs ask for implement their proper applications (Apps) in the hosting ECU. To cope with this constrains in total more than 50 integration scenarios have been analyzed. Out of them less then 10 scenarios make currently sense to evaluate. However, the selection of best fit M2XPro architecture depends on use cases and their system requirements and therefore, on the feature and architecture strategy of the OEM. From the perspective of Continental the first step of implementing M2XPro is the usage of a separate M2XPro Cluster integrated in the existing bus system of the vehicle (Fig. 6). This cluster offers a configurable µC concept which is at least able to run M2XPro fusion algorithm up to a powerful µC for the implementation of V2X functions, customer App's as well as the fusion algorithm of M2XPro. Steering angle signal (SAS) and wheel speed signal (WSS) signals are received via the CAN of FlexRay© bus. The yaw-, roll-, pitch-rate sensors (YRS, RRS, PRS) as well as the acceleration sensors (ax, ay, az) are integrated in the cluster. The GNSS receiver is integrated in the so called intelligent antenna, where also the first filtering of the incoming V2X data happens. The output of M2XPro then is provided via the bus system, e.g. for the HMI (Human Machine Interface) and other systems.

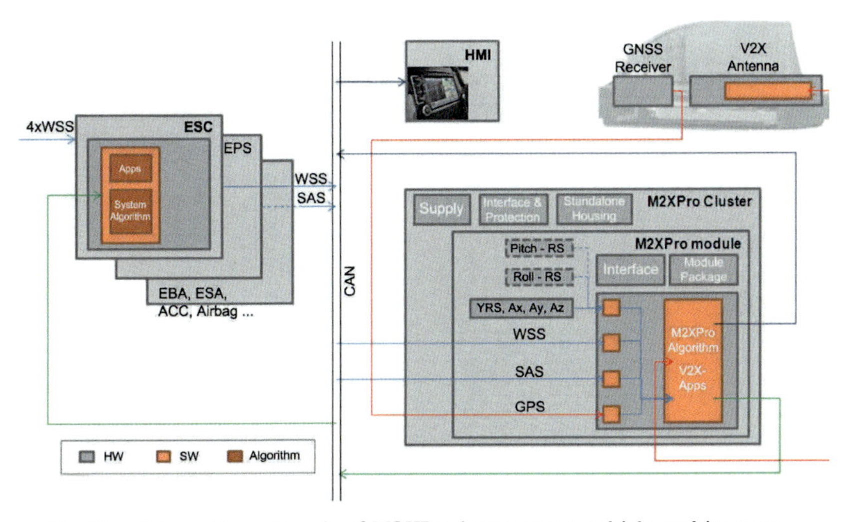

Fig. 6. Integration scenario of M2XPro into current vehicle architecture

6 Conclusions

Emerging cooperative vehicle systems are requesting a higher precision and reliability of vehicle positioning then current GNSS systems can provide. The presented approach of the vehicle onboard sensors and GNSS input signal fusion by an intelligent algorithm cope these requirements independently of any application and provides integrity information about the signal condition. This makes M2XPro easy to integrate in any vehicle architecture, is scalable towards performance and function, depending on the user system requirements, and is safe to apply. The given hardware architecture bases on commercially available sensors and GNSS receivers but still further development work is necessary to optimize the fusion algorithm.

References

[1] simTD, http://www.simtd.org/index.dhtml/154f1040d8ee825502dp/-/enEN/-/CS/-/

[2] Agenda2020 - Beiträge der Automobilindustrie zu einer nachhaltigen Verkehrspolitik, VDA, 27.11.2008, http://www.vda.de/de/downloads/616

[3] N. Dziubek, et. al., Fahrstreifengenaue Ortung von Kraftfahrzeugen durch Datenfusion und Fehlerkompensation von Standard-Seriensensoren", DGON Posnav 2011, Darmstadt, 2011

[4] Globales Navigationssatellitensystem, Wikipedia, 2011, http://de.wikipedia.org/wiki/Globales_Navigationssatellitensystem

[5] C2C-CC, http://car-to-car.org/

[6] AKTIV Kreuzungsassistent, http://www.aktiv-online.org/deutsch/aktiv-as.html#KAS

[7] INTERSAFE 2, http://www.intersafe-2.eu/public/

Bernhard Schmid, Michael Zalewski, Ulrich Stählin, Klaus Rink, Stefan Günthner
Continental
Guerickestr. 7
60488 Frankfurt / Main
Germany
bernhard.schmid@continental-corporation.com
michael.zalewski@continental-corporation.com
ulrich.staehlin@continental-corporation.com
klaus.rink@continental-corporation.com
stefan.guenthner@continental-corporation.com

Keywords: GNSS, GPS, dead reckoning, M2XPro, inertial sensor, V2X, V2I, V2V, cooperative vehicle systems, vehicle positioning

On the Design of Performance Testing Methods for Active Safety Systems

H. Eriksson, J. Jacobson, J. Hérard, SP Electronics
M. Lesemann, RWTH Aachen University
A. Aparicio, IDIADA Automotive Technology SA

Abstract

Performance testing provides an unbiased way of presenting the benefits of active safety systems. However, some issues must be solved before performance testing can be executed for a system targeting a specific traffic scenario. First of all, the selected scenarios must be motivated; second, a suitable driver model shall be selected; and finally, a generic test target shall be defined. All these issues will be addressed in this paper.

1 Introduction

Although the impact of active safety systems on reduction of accidents is hard to quantify, an IIHS study [1] indicates that Volvo cars equipped with City Safety, which allows fully-automatic braking up to 30 km/h, are involved in fewer accidents than other cars of the same type. Meanwhile, there is a problem inherent to the quantification issue; a fully functional active safety system may help to completely avoid an accident, and thus its benefit never shows up in accident statistics. Therefore, performance testing is needed, which has a couple of advantages. First, it provides easy-to-understand results, and shows the potential in equipping active safety systems in a graspable manner for the consumer. Second, it can be used as leverage in providing an incentive for the car manufactures to put active safety systems as standard in their vehicles. A good rating in a performance test is a strong sales pitch, proven for years by Euro NCAP ratings for passive safety. Three critical components in the design of methods for performance testing are: test scenarios, driver models, and test targets. These will be discussed in the next sections.

2 Test Scenarios

For cost and efficiency reasons, the number of test scenarios used during performance testing must be small. As a consequence, it is important to carefully select those. Accident databases are used for the selection. However, the level of detail and accident classification may differ; both between databases and individual records within a database. The extracted scenarios are ranked based on frequency, seriousness of injuries, or a combination of these two, as e.g. done in the ASSESS project [2]. When the feasibility with respect to proving ground and target limitations has been analyzed, the remaining top ranked scenarios are chosen. The main parameters of a scenario are test vehicle and target object directions, velocities, and accelerations, respectively. Other important parameters are road topology and friction, neighboring infrastructure, light conditions, and possible precipitation. Since braking is expected it is important to specify tires and road friction as well. The used driver model in the scenario is also important. Driver models will be the topic of the next section.

As was explained in the introduction, there are several on-going or recently finished initiatives and research projects devoted to defining performance testing methods for longitudinal active safety systems, e.g. forward collision warning (FCW) and autonomous emergency braking (AEB) systems. To the initiatives belong ADAC [3], AEB [4], and vFSS [5]. ASSESS [2] and eVALUE [6] are two research projects which has/had performance testing methods for longitudinal active safety systems as parts of their scopes. Additionally, standardization organizations such as ISO and SAE have released standards for performance testing of FCW systems [7,8], and an ISO standard for AEB is under development. NHTSA has defined three scenarios for FCW systems in their NCAP confirmation test [9].

Examples of proposed scenarios are shown in Fig. 1. Regardless of the databases which have been used to guide the initiatives and projects, most of them end up with a similar set. All of them have scenarios where the vehicle in front is braking, travelling at constant speed, or is stationary. Besides those, some of them also specify cut-in and junction scenarios as well as scenarios with vulnerable road users such as pedestrians and (motor) cyclists.

The common scenarios are summarized in Table 1 for comparison. Test vehicle (TV) speed and lead vehicle (LV) speed and deceleration are presented. The offset is the lateral mismatch between the centerlines of the two vehicles. Curvature tells the direction of the road. A curve is defined by its radius. The reaction, or driver model, specifies if the driver is passive or alert/reactive, and how hard and fast (s)he presses the brake pedal.

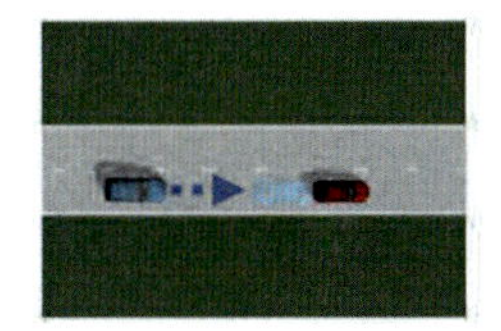

A1A: Urban scenario
- Lead vehicle speed: 10 km/h
- Subject vehicle speed: 50 km/h
- Initial distance based on TTC >> 3 s
- No lateral offset
- Driver reactions: A1A2:slow, A1A3:fast, A1A1: no

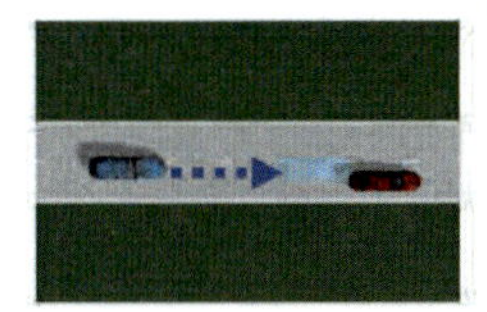

A1B: Urban scenario
- Lead vehicle speed: 10 km/h
- subject vehicle speed: 50 km/h
- Initial distance based on TTC >> 3 s
- 50% lateral offset
- Driver reactions: A1B2: slow,A1B3: fast,A1B1: no

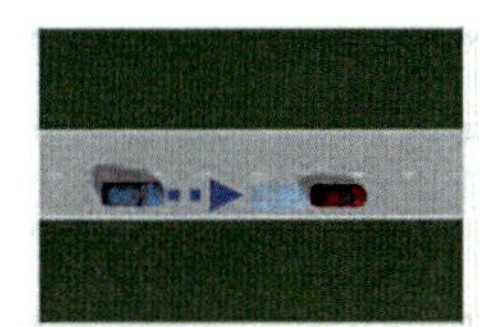

A1C: Motorway scenario (traffic jam)
- Lead vehicle speed: 20 km/h
- Subject vehicle speed: 100 km/h
- Initial distance based on TTC >> 3 s
- No lateral offset
- Driver reactions: A1C2: slow, A1C3: fast, A1C1: no

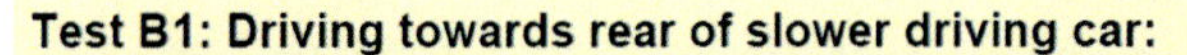

Test B1: Driving towards rear of slower driving car:

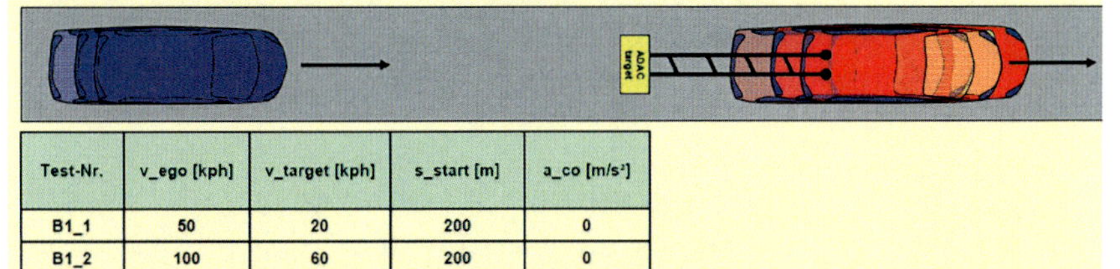

Test-Nr.	v_ego [kph]	v_target [kph]	s_start [m]	a_co [m/s²]
B1_1	50	20	200	0
B1_2	100	60	200	0

Test B2: Driving towards rear of braking car:

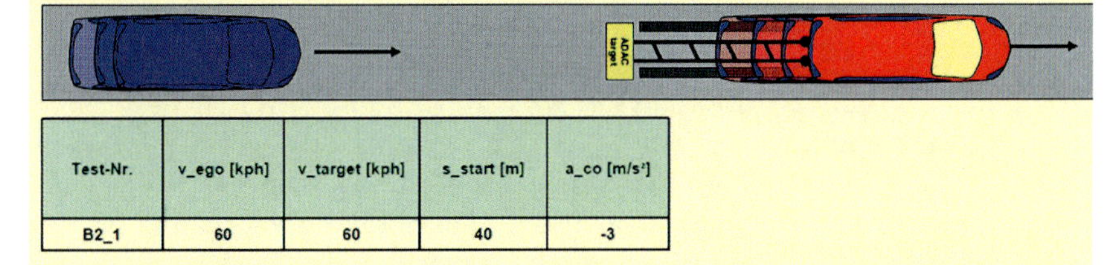

Test-Nr.	v_ego [kph]	v_target [kph]	s_start [m]	a_co [m/s²]
B2_1	60	60	40	-3

Fig. 1. Examples of car-to-car scenarios proposed by ASSESS (top) and ADAC (bottom)

Scenario	Parameter	ADAC	AEB	ASSESS	eVALUE	vFSS	NHTSA	SAE
Stationary LV (LV speed = 0) (LV decel. = 0)	TV speed [km/h]	20 30 40 70	10+10n[a] (n=0 ... 5) 10	50 80	50 70	72[c] 25[d] 50[d]	72[e]	100[f]
	Offset	0	0 0 ... 45°	0 50% 0	0	0	0	0
	Curvature	Straight	Straight	Straight	Straight Curve	Straight	Straight	Straight
	Reaction	No	No	No Slow Fast	No Slow Fast	No	No	No
Slower LV (LV decel. = 0)	TV speed [km/h]	50 100	10+10n[a] (n=0 ... 5)	50 50 100	70 70	72[c] 90[d]	72[e]	60[f] 50[f]
	LV speed [km/h]	20 60	20	10 10 20	30 50	32[c] 50[d]	32[e]	10[f] 30[f]
	Offset	0	0	0 50% 0	0	0	0	
	Curvature	Straight	Straight	Straight	Straight Curve	Straight	Straight	Curve Straight
	Reaction	No	No	No Slow Fast	No Slow Fast	No	No	
Braking LV	TV speed [km/h]	60	50[b]	50 80	70 70	72[c] 50[d]	72[e]	100[f] 100[f]
	LV speed [km/h]	60	50	50 80	70 70	72[c] 50[d]	72[e]	100[f] 100[f]
	LV decel. [m/s²]	-3	-2 -6	-4 -7	-3 -5	-2.9[c] -6.2[d]	-2.9[e]	-1.5[f] -3.4[f]
	Offset	0	0	0	0	0	0	0
	Curvature	Straight	Straight	Straight	Straight Curve	Straight	Straight	Straight
	Reaction	No	No	No Slow Fast	No Slow Fast	No	No	No

Tab. 1. Scenarios and parameters proposed by different initiatives, standards, and research projects; a: step increment until collision occurs b: at two different headways c,e,f: only for FCW testing d: only for AEB testing

One precondition, which is not listed in the table but common in many of the test scenario specifications, is the headway (time gap) which needs to be established between the TV and LV before the test sequence is initiated. As can be seen in the table there are commonalities and differences. For example, ASSESS scenarios are the only ones covering different offsets, whereas eVALUE and SAE are the ones considering curves. Curves put more demands on the proving grounds since all of them have straight roads (naturally), but do not have many different radii for curves. NHTSA, SAE and some of the vFSS scenarios are only applicable for testing of FCW systems. FCW systems are easier to test: if the minimum required time-to-collision has been passed without a warning is issued, the test can be aborted. Thus, a real vehicle can in principle be used as target (lead) vehicle. However, this way of testing may be a bit too simplified. First of all, in a situation where the FCW should help to avoid or mitigate a collision, the driver is in the loop. Therefore the HMI becomes very important: the warning modalities and their design. Also the driver reaction becomes important. An alert driver might be the difference between a collision and not. Additionally, support systems, such as brake assist and well-designed brake/stability systems, are not awarded.

3 Driver Models

Driver models can be used for several purposes. A common use is to analyze the benefits that possibly could be expected by introducing a new active safety system which has the driver as part of the loop. In this case, the model can be used to tune e.g. the timing of the warning signals of the active safety systems in order to optimize the trade-off between useful and nuisance alarms. A somewhat different use for driver models is to mimic driver behavior in critical situations. The model is implemented in a robot which controls the steering wheel as well as the brake pedal and throttle. The robot can be programmed to react to a specific warning, wait a delay and then perform a number of actions.

In the rear-end scenarios, it is usually assumed that the driver cannot perform an evasive maneuver and thus has to stay in the same lane as the obstacle, possibly with a lateral offset. Then the important driver reaction parameters become: reaction time, brake pedal application time, and brake pedal force. Both the eVALUE and the ASSESS project propose three models:

- No reaction
- Slow reaction
- Fast reaction

No reaction is self-descriptive. The driver does not react to the warning, and consequently no brake action is performed. In e.g. eVALUE, a slow reaction is defined as a brake force of 350 N that is applied with a ramp up time of 0.4 s after a "reaction time" delay of 1.5 s. A fast reaction, on the other hand, is defined as a brake force of 700 N that is applied with a ramp up time of 0.2 s after a "reaction time" delay of 1.0 s.

Georgi et al. [10] have defined three classes of driver behavior with respect to braking action: best, realistic, and lethargic. The best driver reacts to an acoustic warning after 0.7 s, and uses the brake capability fully. A realistic driver has a reaction time of 1.0 s, and uses 80 % of the brake capability. Finally the lethargic driver has a reaction time of 2.0 s and only uses 60 % of the brake capability.

In Annex A of ISO 15623 [7], typical brake reaction times and decelerations are presented. An average reaction time of roughly 0.7 s and a 70 % use of brake capability are concluded. The average brake reaction time seems a bit too fast, but in the referenced study the drivers were probably told to expect a warning.

An interesting aspect when it comes to performance testing is a possible need for system conditioning. That is, the test vehicle might need to be driven according to a well-specified driving profile prior to the test to put the active safety system into a specific state. The reason is that some active safety systems adjust e.g. the warning timing after the driving behavior during the last time period. A driver who seems active and alert will experience later warnings than an unfocused driver.

4 Test Targets

To protect test vehicles and test drivers from harm, it is in many cases necessary to use test targets during performance testing of active safety functions. The target shall resemble a real vehicle with respect to radar, lidar, and/or IR signature, and visual appearance. At the same time, it shall be possible to collide with the target a number of times without harming the test vehicle or wearing the target too much. In many scenarios, the target has to be synchronously positioned and moved with respect to the test vehicle to be able to create well-defined and repeatable tests.

In Fig. 2 - Fig. 5 four different concepts for test targets are shown. Fig. 2 shows the ASSESSOR, which is being developed in the ASSESS project [2]. Special care has been taken to ensure that the ASSESSOR has a true radar signature from any viewing angle. The physical characteristics of an Opel Astra-like car have been the goal. Air filled tubes are used as frame to which vented boxes are mounted to cushion the impact. Without a propulsion system, the ASSESSOR weighs 90 kg. It is intended to be self-propelled, and different propulsion systems can be used. Fig. 3 shows the second generation of the target developed by ADAC [3]. The target is towed by a support vehicle, which also can be used to transport the target to the test site. To be detected by different sensors, the

Fig. 2. Example of test target ASSESSOR

target's visual appearance is photo realistic and lidar as well as radars reflectors are part of the design. To further enhance the detection possibility, the target is equipped with a real license plate, a 3D bumper, and a realistic shadow under the car. The maximum impact speed is 50 km/h. The EVITA target [11] is shown in Fig. 4. A cable and a winch are the link between the target and the towing car. The maximum towing speed is 100 km/h, and the deceleration is adjustable up to $9 m/s^2$. No impact should occur between the test vehicle and the EVITA target. Therefore, the target is equipped with a radar distance sensor, and if the test vehicle gets too close, the target is pulled away using the winch. The EVITA target has real brake lights as well as distance indication lights to guide the driver of the test vehicle to the correct relative position. Fig. 5 shows the Guided Soft Target (GST) from Dynamic Research [12]. It consists of two parts: one soft car foam body and a satellite guided self-propelled low profile robot vehicle as carrier. The low profile carrier can be overrun without being damaged or causing damage to the test vehicle.

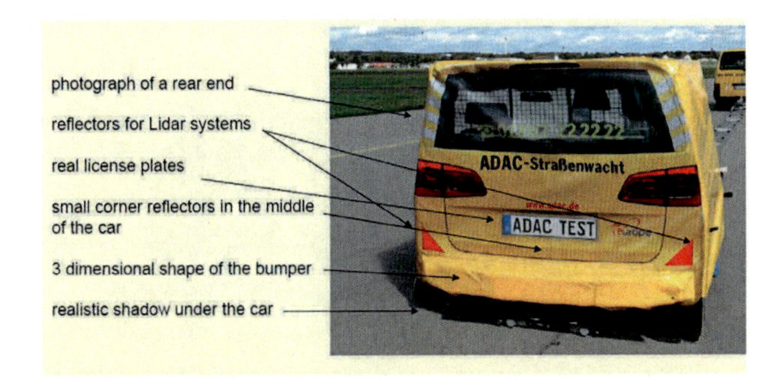

Fig. 3. Example of test target ADAC

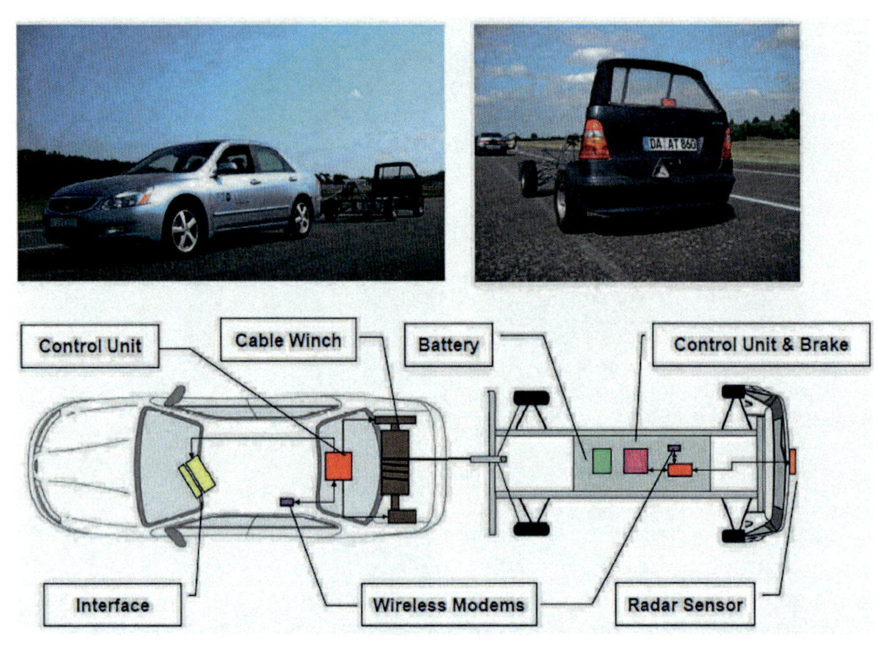

Fig. 4. Example of test target EVITA

Fig. 5. Example of test target GST

5 Conclusions

Proposed test scenarios, driver models, and test targets have been presented in this paper. Despite many similarities, there is a need for harmonization and standardization within these areas. This fact was observed by the ASSESS project as well as the vFSS and AEB initiatives. This group has formed three harmonization platforms, of which two are targeting scenario definition and target development. The ActiveTest project [13] also provides an arena where these issues are brought up and discussed.

Acknowledgement

The research leading to these results has received funding from the European Community's Seventh Framework Programme (FP7/2007-2013) under grant agreements n° 269904. This publication solely reflects the authors' views. The European Community is not liable for any use that may be made of the information contained herein.

References

[1] IIHS, Status Report, Vol. 46, No. 6, 2010.

[2] ASSESS project, http://www.assess-project.eu, 2012.

[3] C. Gauss, D. Silvestro, ADAC Test Procedure for Advanced Emergency Brake Systems, presentation at the first ActiveTest workshop, 2011.

[4] M. Avery, AEB - Autonomous Emergency Braking, presentation at the first ActiveTest workshop, 2011.

[5] H Schebdat, Advanced Forward-Looking Safety Systems – Working Group – Introduction and Status Update, presentation at the first ActiveTest workshop, 2011.

[6] eVALUE project, http://www.evalue-project.eu, 2012.

[7] ISO, ISO 15623:2002 Transport Information and Control Systems – Forward Collision Warning Systems – Performance Requirements and Test Procedures, 2002.

[8] SAE, SAE J2400 Human Factors in Forward Collision Warning Systems: Operating Characteristics and User Interface Requirements, 2003.

[9] NHTSA, Forward Collision Warning System NCAP Confirmation Test, 2010.

[10] A. Georgi, et al., New Approach of Accident Benefit Analysis for Rear End Collision Avoidance and Mitigation Systems, in Proceedings of 21st International Technical Conference on the Enhanced Safety of Vehicles, 2009.

[11] A. Weitzel, Darmstädter Method with EVITA, presentation at the first ActiveTest workshop, 2011.
[12] Dynamic Research, Inc, http://www.dynres.com, 2012.
[13] ActiveTest project, http://www.activetest.eu, 2012.

Henrik Eriksson, Jan Jacobson, Jacques Hérard
SP Electronics
Box 857
50115 Borås
Sweden
henrik.eriksson@sp.se
jan.jacobson@sp.se
jacques.herard@sp.se

Micha Lesemann
RWTH Aachen University
Institut für Kraftfahrzeuge (ika)
Steinbachstraße 7
52074 Aachen
Germany
lesemann@ika.rwth-aachen.de

Andrés Aparicio
IDIADA Automotive Technology SA
Integrated Safety
L'Albornar
43710 Santa Oliva (Tarragona)
Spain
aaparicio@idiada.com

Keywords: performance testing methods, active safety systems, test scenarios, driver models, test targets, advanced driver assistance systems

Advanced Driver Assistance System Supporting Routing and Navigation for Fully Electric Vehicles

K. Demestichas, E. Adamopoulou, M. Masikos, NTUA
T. Benz, W. Kipp, PTV AG
F. Cappadona, PININFARINA S.P.A.

Abstract

The emergence of Fully Electric Vehicles has sparkled visions of pollution- and noise free cities. However, towards this challenging end, a lot has yet to be accomplished. One of the first priorities should be placed on improving the reliability and energy efficiency of the fully electric vehicles. This paper presents a new Advanced Driver Assistance System that has been implemented, which automatically helps the driver to save more energy while on-trip, by choosing the most energy efficient routes and by providing recommendations whenever necessary. This advanced functionality is based on the collection and exploitation of experiences – through machine learning.

1 Introduction

In the years to come, Fully Electric Vehicles (FEVs) [1] will be emerging as a high-tech alternative road transportation means that can potentially solve the pertinent problems of vehicle-originating air pollution in urban areas. However, despite this potential, whether FEVs will prevail in the future is still unclear, due to the fact that their market penetration and commercial viability will strongly depend on their range of autonomy and their energy efficiency. FEVs' reliability in terms of energy autonomy and efficiency is influenced by a number of factors, all of which contribute to a relevant degree of uncertainty. Such factors are the road characteristics and conditions that are met while on-trip; the traffic conditions that are encountered on the route to the destination; the availability or not of surrounding recharging points; and the impact of FEV-specific characteristics, such as regenerative braking [2].

In order to raise the energy efficiency and range autonomy of FEVs, this paper introduces the notion of autonomous optimized route planning, through means of continuous consumption monitoring and advanced machine learning techniques [3]. A new Advanced Driver Assistance System (ADAS) suitable for FEVs is designed and implemented, aiming at utilizing past knowledge and experience, and applying suitable machine learning processes. The ADAS is empowered with traffic estimation and optimal route selection capabilities, which help the driver take the right routing decisions to save energy and increase residual range.

A number of industrial initiatives attempting to address the electric vehicles' routing problem have recently taken place, including the fruitful joint-venture between TomTom and Renault, which launched their first in-dash navigation system, Carminat TomTom Z.E. LIVE [4], whose key features involve electric vehicle routing functionality (including routing to and via charge stations) and integration with battery management. Another noteworthy initiative within the context of routing optimisation for electric vehicles is the partnership established between Hyundai and NAVTEQ: the latter will provide its Advanced Driver Assistance System (ADAS) for Hyundai's latest navigation platform, which includes a "green routing" option, beside the traditional "fastest" and "shortest" routes [5]. The NAVTEQ-provided content includes additional information on altitude, slopes and curves, which allows more precise calculations according to map terrain: together these attributes enable the navigation system to find routes which can minimise fuel consumption (in case of Internal Combustion Engine vehicles) and energy consumption (in case of FEVs).

At present time, only few navigation devices include an eco-driving or eco-routing functionality, such as: (i) Vexia's ecoNav solution [6], which provides advice to the driver and monitors fuel consumption using vehicle characteristics and acceleration, (ii) Garmin's ecoRoute software [7], which calculates a fuel-efficient route based on road speed information and vehicle acceleration data, and (iii) The Freightliner Predictive Cruise Control (developed by DAIMLER) [8], which is integrated in the truck's cruise control system and exploits road topology data from maps, in order to fine-tune the target speed to reduce fuel consumption.

Nonetheless, these systems are not seamlessly integrated with the vehicle and make limited use of map data content or traffic information. Most importantly, they lack the ability to monitor the vehicle's consumption-related experience along routes, and to learn according to this accumulated experience, which constitutes the main innovation of the proposed ADAS. The remaining sections provide details of the proposed ADAS from business, architectural, functional and physical implementation perspectives.

2 Business-Layer Architecture

2.1 Business Processes and Events

The proposed ADAS incorporates two main services that are of fundamental importance for enhancing the energy efficiency of FEVs, namely:

(On-board) Data collection and storage. This service collects and processes a wide range of context and environmental data, accumulating a specific knowledge about the travelled routes, with special attention to all aspects concerning energy efficiency. Such knowledge is shared with the community of EcoGem-enabled vehicles (via the V2V/V2I interface), thus integrating different travelling experiences into a global database. With reference to EcoGem concept architecture, it maps to the following in-vehicle functionalities: On-going measurement collection, Secure Measurement Storage.

Energy-driven routing and navigation, his service encapsulates all basic functionalities concerning the energy-efficient routing and navigation. More specifically:

- It provides the driver with energy-efficient route planning functions. By means of a machine-learning based engine, it integrates static map information with experience database, and calculates the most efficient route to destination, according to current context.

- It guides the driver on the way to destination, providing standard turn-by-turn navigation functions.

- It continuously monitors the surrounding traffic conditions on the way to destination (as propagated by the central platform, following the integration of EcoGem-generated real-time data and traffic events gathered from external providers), and dynamically calculates alternative routes, in case such conditions affect the efficiency of remaining part of the route.

- It continuously monitors the level of the battery and provides range estimation. By means of a machine-learning based engine, it integrates actual measurements (battery level, current consumption, active auxiliary services, etc.) with the accumulated experience (behaviour of the system battery-vehicle in different contexts), and provides a reliable estimation of the residual range, thus ensuring that the selected destination is safely reachable.

- In case the destination is not reachable, it calculates the most efficient route to a recharging point, taking into account the final destination, residual range, possible vehicle/battery constraints, availability of recharging points and suggested recharging options provided by battery recharge management service.

Fig. 1. presents the business processes and events that determine the behaviour of the two aforementioned services.

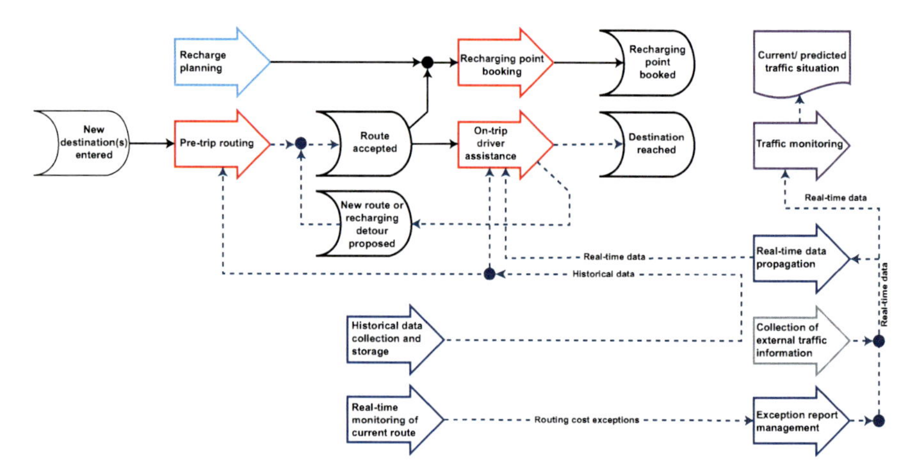

Fig. 1. Business processes and events of the proposed system

2.2 Pre-Trip Routing and On-Trip Driver Assistance

According to the specified business behaviour, the two main business processes oriented to driver interaction are the Pre-trip routing and the On-trip driver assistance. The Pre-trip routing is triggered by entering the address of a new destination; such action may occur immediately before leaving for the journey, or some hours in advance (early planning), and provides the driver with a list of optimized routes to choose from, according to the desired optimization criterion (most efficient route, fastest route). Subsequently, the On-trip driver assistance guides the driver to his/her destination by means of regular turn-by-turn navigation functions (graphical map indications, text-to-speech suggestions, etc.), plus a range of specific features, such as real-time traffic monitoring and destination reachability assessment, that may lead to the calculation and proposal of a new route or recharging detour.

Since Pre-trip routing and On-trip driver assistance are rather complex and present some distinctive features, they will be further analyzed in the following Figure (Fig. 2), which provides more information and operating details. As may be observed, the two main business processes are further broken down into eight sub-processes.

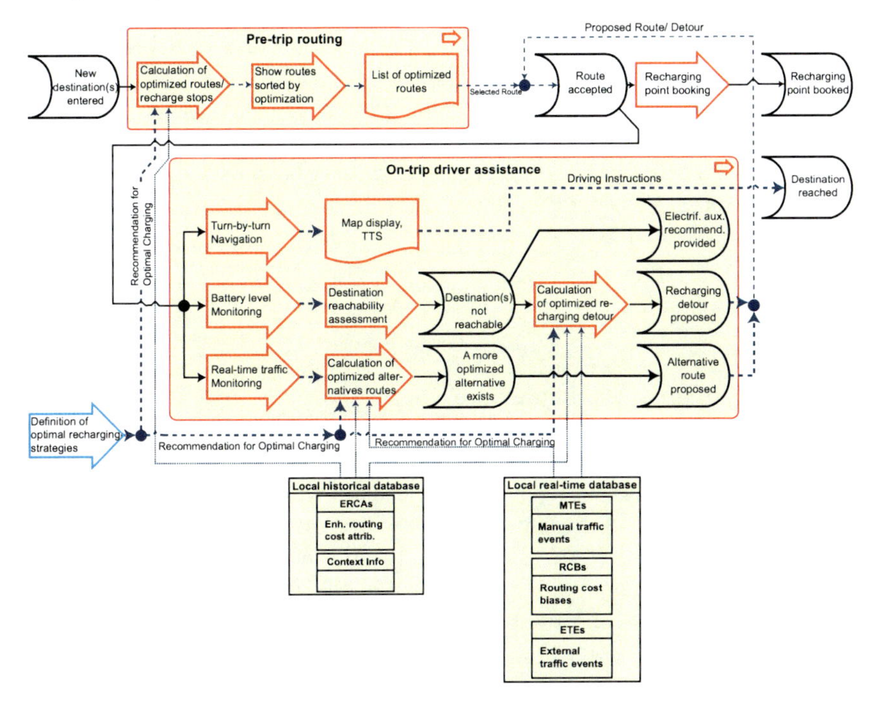

Fig. 2. Detailed business specifications of the proposed Pre-trip routing and On-trip driver assistance

In Fig. 2., it is particularly interesting to describe what runs behind the On-trip driver assistance process:

- Turn-by-turn navigation: this process guides the driver to destination, by means of graphical (current position on map, 2D/3D visualization, graphical indication of next turn, lane selection, points of interest, etc.) and vocal indications (text-to-speech suggestions).

- Battery level monitoring, Destination reachability assessment, Calculation of optimized recharging detour: these processes work together, in order to monitor current battery level and to evaluate the expected energy consumption to destination. If the final destination is not safely reachable, the system warns the driver and proactively suggests recovery actions, which may include the calculation of optimized detours to reach a recharging point, as well as providing recommendations for electrified on-board auxiliaries (e.g., to switch off the air conditioner).

- Real-time traffic monitoring, Calculation of optimized alternative routes: these processes work together, in order to ensure that the remaining section of the current route is not negatively affected by current traffic conditions. If this occurs, the system calculates a set of possible alternative routes (from current position to final destination, including remaining intermediate stops) and compares their costs with the current route. If, according to real-time traffic conditions, any alternative route becomes more efficient than the current one, the system proposes a route change.

3 Application-Layer Architecture

Fig. 3 depicts the aforementioned eight business processes supporting the energy-driven routing and navigation service, which have been defined in Fig. 2., and illustrates the main application services that are used by such processes. It should be noted that not all application services associated to energy-driven routing and navigation are included in this Figure, only those are most closely and directly associated to the business processes.

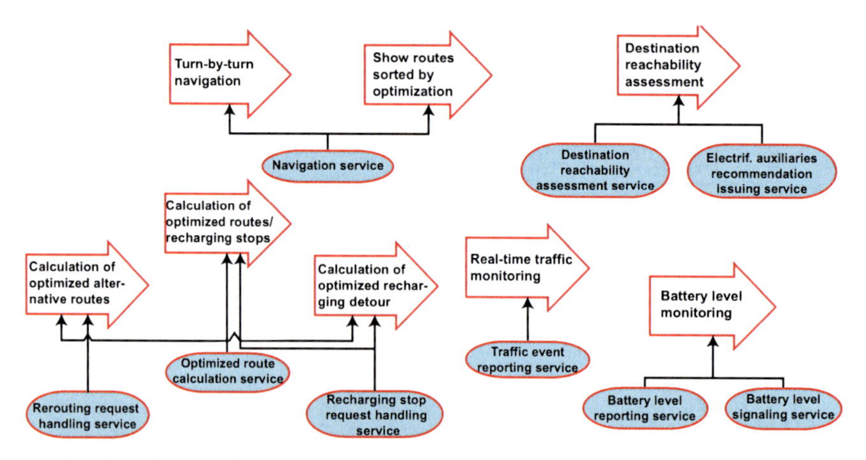

Fig. 3. Energy-driven routing and navigation – Mapping of application services to business processes

Fig. 4. goes one step further, and provides the complete overview of the application landscape, containing both the main applications (application components) of the functional architecture, as well as the application services that each one uses and realizes, thus effectively illustrating the interfaces between applications. It is common for each application to "realize" one or more application services (that it exposes to other applications) and to "make use of" one or more other application services (typically realized by other applications). According to Fig. 4, the main application components of the proposed system are grouped under four different products or categories:

- Routing and navigation application components: This category contains components that are relevant to the calculation of energy-efficient routes, both pre-trip and on-trip, and the provision of navigation instructions to the end-user.
- Machine learning application components: This category contains all components whose functionality is heavily based on machine learning usage. Machine learning is used by the ADAS for efficient and reliable predictions of energy consumption (primarily) and travel time (secondarily).
- Destination reachability application components: This category contains the components that are concerned with the continuous assessment of the ability of the vehicle to reach the desired destination, as well as with the suggestion of proper recommendations for (energy consumption minimization), as far as the usage of the electrified auxiliaries is concerned.
- Awareness application components: This category contains the components that render the other applications aware of traffic events and of the battery status.

4 Conclusion & Outlooks

4.1 Physical Implementation and Concluding Remarks

The functionality of the proposed system, as described in the previous sections, is implemented in C++ using NAVTEQ's ADAS-RP fast prototyping platform (Fig. 5). The ADAS's User interface consists of a map display window and three splitter-window areas where Plug-ins can be placed. Plug-in windows can also be free floating or undocked. This allows placing the window on a display like a touch-screen connected to an on-board small factor PC. One of the PC's USB ports is physically connected to the FEV through the vehicle's Controller Area Network (CAN) bus, so as to read and record measurements (e.g., battery levels, auxiliaries' status, etc.). As the implemented ADAS follows the standard user interface guidelines set for Windows GUI applications, any graphical layout can be defined for the User Interface Plug-in. First experimentations show that there is an improvement in consumption estimation accuracy by a factor of two. Immediate future work includes relevant trials both in test tracks as well as in urban environments.

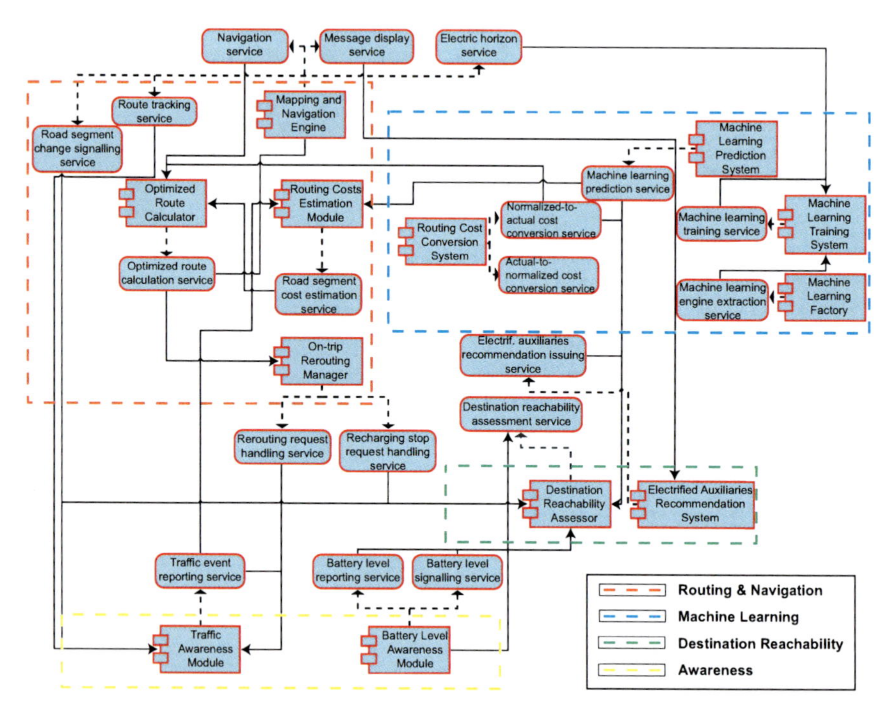

Fig. 4. Complete application-layer architecture of the proposed system

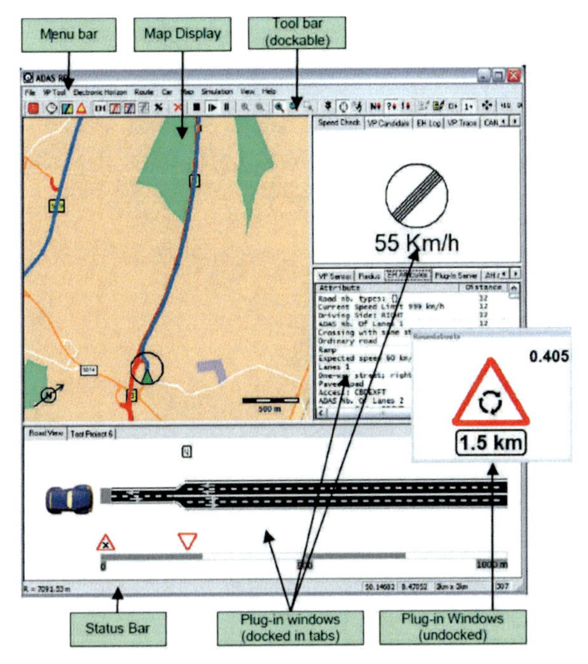

Fig. 5. User interface with docked and floating plug-in windows

Acknowledgments

This work has been performed under project EcoGem, which has received research funding from the European Community's Seventh Framework Programme. This paper reflects only the authors' views, and the Community is not liable for any use that may be made of the information contained therein.

References

[1] Hybrid cars website retrieved in Feb. 2012 under http://www.hybridcars.com/electric-car.

[2] EcoGem Project Official website retrieved in Feb. 2012 under http://www.ecogem.eu.

[3] K. Demestichas, E. Adamopoulou, M. Masikos, et al., EcoGem: An Intelligent Advanced Driver Assistance System for Fully Electric Vehicles, 18th World Congress on Intelligent Transport Systems, Orlando, Florida, U.S.A., 16-20 October 2011.

[4] Carminat Tom Tom Navigation System, website retrieved in Feb. 2012 under http://www.tomtom.com/en_gb/products/built-in-car-navigation/carminat-tom-tom/.

[5] NAVTEQ ADAS for Hyundai, website retrieved in Feb. 2012 under http://press.navteq.com/NAVTEQ-Map-Powers-New-Navigation-System-For-Hyundai-Solaris.

[6] Vexia's ecoNav solution, website retrieved in Feb. 2012 under http://www.vexia.es/en/econav_technology.htm.

[7] Garmin ecoRoute software, website retrieved in Feb. 2012 under http://www8.garmin.com/buzz/ecoroute/.

[8] The Freightliner Predictive Cruise Control, website retrieved in Feb. 2012 under http://www.freightlinertrucks.com/Multimedia/MediaLibrary/Innovation?AssetId=15990456-b11e-4966-a5be-c70e22cafcca.

Konstantinos Demestichas, Evgenia Adamopoulou, Michalis Masikos
National Technical University of Athens
Institute of Communication and Computer Systems
Heroon Polytechneiou, 9
15773 Zografou, Attiki
Greece
cdemest@cn.ntua.gr
eadam@cn.ntua.gr
mmasik@telecom.ntua.gr

Thomas Benz, Wolfgang Kipp
PTV AG
Stumpfstr. 1
76131 Karlsruhe
Germany
thomas.benz@ptv.de
wolfgang.kipp@ptv.de

Filippo Cappadona
PININFARINA S.P.A.
Via Nazionale, 30
10020 Cambiano, Torino
Italy
f.cappadona@pininfarina.it

Keywords: fully electric vehicle, advanced driver assistance system, energy efficiency, routing and navigation, recharging, machine learning

Slippery Road Detection by Using Different Methods of Polarised Light

J. Casselgren, Luleå University of Technology
M. Jokela, M. Kutila, VTT Technical Research Centre of Finland

Abstract

Road friction measurement is an important issue for active safety systems on vehicles; hence knowledge of this key parameter can significantly improve the interventions on vehicle dynamics. This study compares two different on-board sensors for the classification of road conditions with polarised infrared light. Several tests are performed on a dedicated track, with focus on detection of dry or wet surfaces, and the presence of ice or snow. The work shows the capability of both sensors to provide a correct classification. In particular, results indicate how the monitored area, the presence of active illumination and the mounting position influence measurements and response times. It is concluded that both systems classify different road conditions in all cases. Performance of the Road eye system varied from 80 to 90 % whereas the camera based IcOR achieved 70-80 % accuracy level. Since these are being prototype sensors more development is needed before implemented into advanced safety applications.

1 Introduction

Friction estimation has been the subject of intense research for some years. Hence, good friction estimation is of great importance for safety and handling applications in vehicles. Statistics regarding road accidents show that slippery road conditions are often the cause of severe crashes or road departures. [1] Incorporating a device that estimates the friction in front of the vehicle could enable more effective interventions aiming to avoid or at least mitigate accidents. Gustafsson [2] has reviewed the existing automotive safety systems where the tire-based friction monitoring technology is reviewed. The conclusion of the article is that tire monitoring technology alone is not sufficient for reliable road state estimation and hence, sensor fusion with alternative sensors such as GPS and optical devices should be more exhaustively investigated.

Today there are many prototypes of optical devices using optical characteristics to make a classification of different road conditions. Yamada et al. [3] pioneered the work in the area of polarization-based detection using standard CCTV cameras. Their work also includes texture and coarseness analysis in order to improve the classification performance. The drawback of this method is that a multivariable approach is needed requiring a lot of computational power. Another sensor technique employs laser diodes of two wave lengths and a photo detector [4]. The two wave lengths are chosen because of the differences in spectral response between water, ice and snow [5]; moreover they are cheap off-the-shelf laser diodes.

Within this paper we have chosen to investigate two optical sensors, Road eye and the IcOR system. The idea is to compare the two vision systems to investigate the pros and cons. The Road eye is a spectral based sensor employing a laser diode technique and the IcOR sensor is based on the work of Yamada et al. [3] and Yusheng and Higgins-Luthman [6] using polarisation differences for black ice detection. However, our experiences have indicated that polarisation difference features do not provide reliable estimation alone and therefore the graininess analysis is required especially for wet and snowy road detection. The paper is divided into sections, were section 2 describes the measuring technique and algorithms of the two sensors. In Section 3 the test scenario is explained and the results are presented. The paper is ended with some discussion and conclusions.

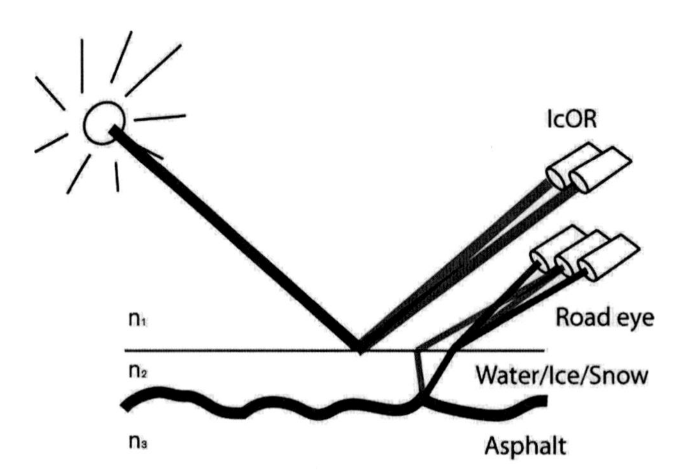

Fig. 1. Measuring principles.

2 Measuring Principles

The benefit of an optical sensor compared with other non tire-based sensors is the fast response time and the preview mode, meaning that the classification of road conditions can be carried out well ahead of the vehicle. The main difference between the two sensors used in this investigation is the illumination principle where the IcOR is dependent on surrounding illumination while the Road eye has laser diodes as illumination source and the monitored area.

Because of the different illumination techniques, shown in Fig. 1 as black lines, the propagation paths of the measured light will be different as well, shown in Fig. 1 as grey lines. However the propagation is described by the same physical Fresnel formula [7, 8] for both methods. The Fresnel formulae describes how a plane electromagnetic (light) wave that falls on to a boundary between two mediums of different optical properties gets split into two waves: a transmitted wave proceeding into the second medium and a reflected wave propagating back into the first medium.

The formulas are given by:

$$T_V = \frac{2n_1(\lambda)\cos\theta_i}{n_2(\lambda)\cos\theta_i + n_1(\lambda)\cos\theta_t} A_V, \quad T_H = \frac{2n_1(\lambda)\cos\theta_i}{n_1(\lambda)\cos\theta_i + n_2(\lambda)\cos\theta_t} A_H, \tag{1}$$

$$R_V = \frac{n_2(\lambda)\cos\theta_i - n_1(\lambda)\cos\theta_t}{n_2(\lambda)\cos\theta_i + n_1(\lambda)\cos\theta_t} A_V, \quad R_H = \frac{n_1(\lambda)\cos\theta_i - n_2(\lambda)\cos\theta_t}{n_1(\lambda)\cos\theta_i + n_2(\lambda)\cos\theta_t} A_H. \tag{2}$$

Where T_V, T_H, R_V, R_H, A_V and A_H are the complex amplitudes of the transmitted (T), reflected (R) and the incident (A) waves for vertically (V) and horizontal (H) polarisation. The $n(\lambda)$:s are the wave length dependent refractive indexes of the different road surfaces and θ_t and θ_i are transmitted and incident angles of the waves.

In the case of the IcOR sensor measuring the direct reflex (see Fig. 1) Eq. 2 are used to classify the road conditions. While the Road eye measuring the back-scattered light exploits Eq. 1 in the first reflection and then more or less all four equations when the light travels back trough the medium. In the case of the IcOR only one wave length is measured, while for the Road eye the wave length dependence of the refractive index [9] is exploited measuring two wave lengths. It is the difference in these physical properties for the different road conditions that enables a classification by measuring the reflected intensities.

3 Monitoring Area and Mounting

3.1 Road Eye

The environmental sensor Road eye provides a classification of road conditions at short distance (0.5-1.5m) and has a Short Wave InfraRed (SWIR) active illumination consisting of two laser diodes emitting at wave lengths $\lambda_1 = 1323$ nm and $\lambda_2 = 1566$ nm. As laser diodes emit linearly polarized light it is possible to control the polarization of the illuminated light by mounting the diodes in a specific manner. Fig. 1 shows that the light measured by the Road eye will propagate into the medium. For a sensor it is important to get a signal that is as strong as possible. To ensure this the laser diodes were mounted such that the emitted light is vertically polarized. For angles around 50° all vertical polarized light will propagate down into the medium [7, 8] resulting in a reflection that is as strong as possible. The Road eye's focusing optics gives an illuminated spot with a radius of 10 mm on the road surface at a distance of 0.8 m. In order to acquire data from the reflected light, a lens focuses the reflected light on a photodiode. The amplitude-modulated signals are sampled at 20 Hz. The output signal consists of two voltages (mV) representing the reflected intensity of the two wave lengths, respectively. The active amplitude modulated illumination ensures insensitivity to disturbances, such as other vehicle's headlights or daylight.

For this investigation the Road eye sensor was mounted in a tube in front of the right front wheel of a truck. The tube was used to keep the sensor clear from splash and pollutions. The mounting angle of the tube was around 45°, with a height of 0.8 m from the ground, resulting in a measuring distance in front of the wheel of 0.8 m. Due to the sampling rate of the Road eye sensor and the simple classification algorithm the response time of the system ensures a preview measurement even when measuring only 0.8 m in front of the wheel. However, the response time is too short for preview information to the driver but systems as the ESC, ABS and TCS could benefit from the information.

3.2 IcOR

The environmental system IcOR provides a classification of road conditions over the whole road surface at distance of 0-40 m in front of sensor, in this case a vehicle as shown at the top of Fig. 2. Since the IcOR system first was developed as an on-board driver support system, it was obvious that illumination in the visible band was sufficient when driving. Therefore the camera system uses the visible and near-infrared spectrum to which many automotive camera systems are adapted. In order to keep the hardware costs reasonable and keeping a real passenger car product in mind (see Fig. 2) the IcOR system does not include any dedicated illumination; it relies on ambient illumination from external light sources as the sun, headlights, etc. As a detector the IcOR system uses a stereo camera pair by Videre Design [10]. Each camera acquires monochrome images with a size of 640 x 480 pixels. They have a global shutter that exposes all the pixels at the same time. This prevents some artifacts that arise with a rolling shutter (e.g. skew). The camera pair is connected to the computer with a FireWire cable through which it also is powered. The system uses external synchronization i.e. all DSCG devices on the same FireWire bus capturing the images at the same time. This is convenient because then there is no need to synchronize the images afterwards. A polarization filter is mounted on each camera in such a way that one camera measures the incoming horizontally polarized light and the other the vertically polarized light.

The camera pair of the IcOR system was installed inside the cabin of the vehicle behind of the windscreen and therefore, the vehicle wipers maintains the optical path clear. The installation corresponds with the typical lane tracker cameras since they are often mounted behind the interior mirror together with the rain detector. Another benefit concerning the installation location is that expensive heated optics is not needed during wintertime. The evaluation of the IcOR system was first done in a passenger car (see Fig. 2). To make results fully comparable between the IcOR and the Road eye systems, both systems were instead installed into a Volvo tractor. IcOR also requires a computing unit running with the Microsoft Windows XP operating system. The actual version runs in a special vibration proof laptop, which is practical for evaluating the performance of the system. However, the software is portable to any MS Windows XP computer if there is a PCI slot for the Firewire data acquisition board available.

4 Algorithm Description

4.1 Road Eye

The two intensities outputs from the Road eye sensor, here after named λ_1 and λ_2, represent the reflected light from the road surface. These two quantities are implemented in the classification algorithm by computing the two magnitudes s and q as:

$$s = \lambda_1 + \lambda_2, \quad q = \frac{\lambda_2}{\lambda_1} \tag{3}$$

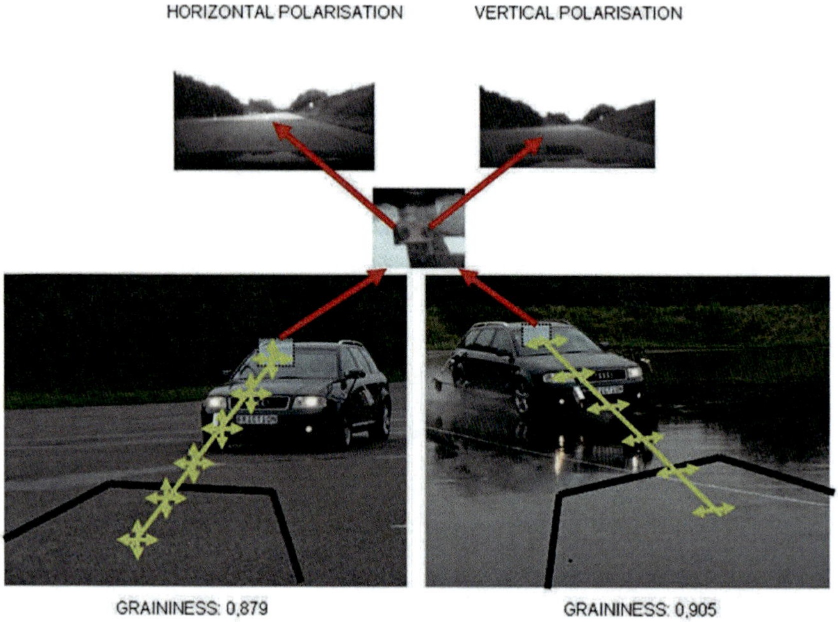

Fig. 2. The IcOR system measurement principle bases on suppression of vertical light polarization when reflecting from mirror like road surface. In addition, the graininess analysis is implemented to distinguish icy and wet road.

Where s is the total reflected intensity and q is the ratio of absorption. The s magnitude will explore the differences in the surface structure meaning if the surface is rough (Dry asphalt and snow) more light will be reflected back compared with if the surface is smooth (Water and Ice). For the q magnitude the differences in $n(\lambda)$ for different road conditions will be explored [7, 8]. For example dry asphalt will have a value of 1 for q as the absorption is almost equal for the two wave lengths, while for snow it will be close to 0 as almost all light for λ_2 will be absorbed.

The two magnitudes s and q is then used to draw up a plane where q represents the x-axis and s the y-axis, by plotting the measured values in this plane each road surface will create a separate cluster. By entering starting values based on previous measurements for the four different road conditions dry asphalt, water, ice and snow covered asphalt and implementing a K-mean [11] clustering algorithm, the measurements will be affiliated to a certain cluster, i.e. classifying different road conditions. The output from the algorithm is one number representing which cluster the measure belongs to and one Euclidean length, i.e. the distance from the measured point to the centre of the cluster.

The distance is then used to calculate a validity of each classification. The limits of the distances are calculated with a 90% confidence interval. This limitation is set to disregard outliers; hence if the distance is too large the classification can't be "trusted". If the distance is outside the confidence interval the validity is set to 0 otherwise it is 1.

4.2 IcOR

The IcOR cameras with polarization filters are utilized to measure intensity difference. In fact, the value is equal with difference between horizontal and vertical reflection indexes (R_H and R_V from Eq. 2). The polarization differences measured are used to distinguish wet and icy road surfaces from dry asphalt. The difference tends to increase between horizontal and vertical polarization intensity for dry asphalt compared with wet and icy asphalt. The polarization difference is for each pixel defined as:

$$I_{diff} = I_{horizontal} - I_{vertical} \tag{4}$$

Moreover, dry and snowy road surfaces have unique responses due to the fact that the brightness of white snow is higher than black asphalt and therefore also the difference between the measured polarization planes becomes stronger. In addition there is a road graininess calculation method implemented in the IcOR system to distinguish whether the reflected surface is due to ice or water on a road. The graininess analysis is based on a similar algorithm that was originally proposed to analyse paintings. The graininess is a measure related to the high frequency content in the image. The analysis begins by performing low-pass filtering to an image, which makes it blurrier. Then calculating the contrast difference of the original and low-pass filtered images provides information on "the small particles" in the picture. Using the wiener filter, which assumes that the additive white noise has a Gaussian distribution, performs the low-pass filtering. The wiener filter is selected in this case since typical road surface graininess statistics can be better modelled in time-domain than in frequency space. The wiener filter is mathematically expressed as:

$$G = \frac{H^*}{|H|^2 + \frac{P_n}{P_g}} \tag{5}$$

where H^* is the complex conjugate of the Fourier transform of the point-spread function. P_n is the noise power and P_g is the power spectrum of the model image, which is calculated by taking the Fourier transform of the signal auto-correlation. Blurriness is measured by estimating the total amount of contrast in the image. The contrast (C) is defined as the difference between adjacent pixels aligned horizontally or vertically (see Fig. 3). Therefore, the contrast value is determined by firstly summing all pixel differences between two neighbourhood pixels (p_{row_i}) vertically (C_{row}) and then doing same horizontally (C_{column}).

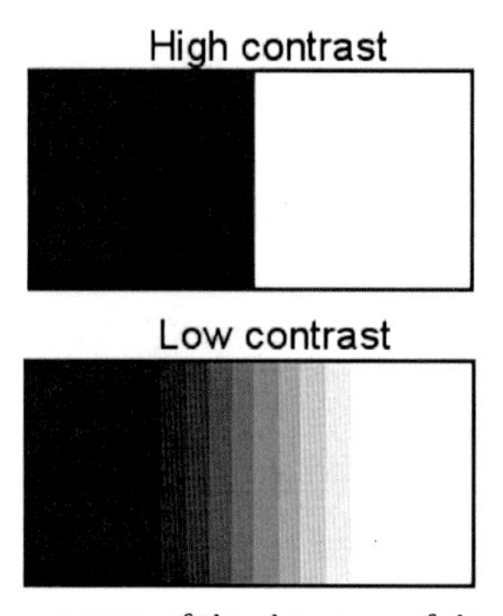

Fig. 3. Contrast is a measure of the sharpness of the changes

The total contrast is here defined as the sum of row and column differences that is sufficient when considering practical road surface images [12]. In order to compare the contrasts in different images the relative change is calculated, which also represents the graininess (S) of the image, defined as:

$$C_{total} = C_{row} + C_{column}, \quad S = \frac{C_{original} - C_{filtered}}{C_{original}} \tag{6}$$

5 Evaluation

The two sensors Road eye and IcOR are prototype sensors and the purpose of this investigation is to compare the functionality and precision of the two sensors. Therefore, the measurements of interest were carried out on a test track with distinct road conditions as dry, wet, icy and snowy asphalt. The weather

Fig. 4. The test track situated in Arjeplog: 150 m asphalt, 50 m ice and 50-100 m snow.

during the test was cloudy with some elements of snow and a temperature between 0-3 °C. Hence the snow that fell on the asphalt melted and became water. The test track was around 250-300 m with 150 m heated asphalt, 50 m ice and a 50-100 m snow surfaces. The test track is shown in Fig. 4. The left corner of the picture shows a small part of the track where the asphalt stayed dry due to extra heating, making it possible to measure all four road conditions in one run. The track layout is depicted as fields in Fig. 5 and Fig. 6 as a reference for the classifications.

Experimental results have been synchronously recorded for the two sensors Road eye and IcOR. Translating the individual results from IcOR to a continuous measure of the road condition makes it possible to compare the results with the Road eye measure. The results are depicted in Fig. 5 and Fig. 6. Due to the different measuring distances the IcOR signal is time shifted to be synchronised with the Road eye signal. Units and absolute values of the measures have been ignored since the graphs are intended to visualise the behaviour of the tested sensors. Both sensors give a correct and similar classification of the road conditions along the track. The different road parts are fairly visible in the graphs of figures Fig. 5 and Fig. 6.

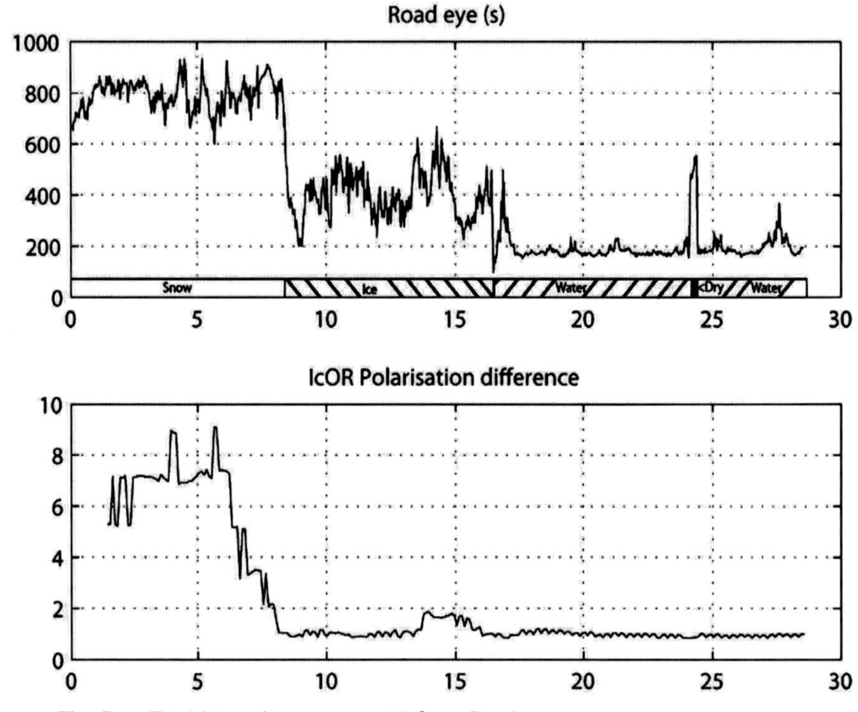

Fig. 5. Total intensity measure (s) from Road eye

Especially in the signals from Road eye the shifts between road types are steep and clear. The dry part around 24 s causes problems for the IcOR because of the available illumination and shortness of the dry asphalt. This is not so much a problem for the Road eye as the dry road part can be seen as a clear peak in Fig. 5. The results show that both sensors can classify water from ice and ice from snow. In Fig. 5 for the Road eye measurements it can be seen that the response for snow, ice and water are distinct for s, but for dry asphalt and ice the value is the same. Shifting to Fig. 6 and q the difference for ice and dry asphalt is clearly separated making it possible to classify all four road conditions. In the case of the IcOR the S value is the same for ice and water, however incorporating the Granularity there is a clear difference between ice and water.

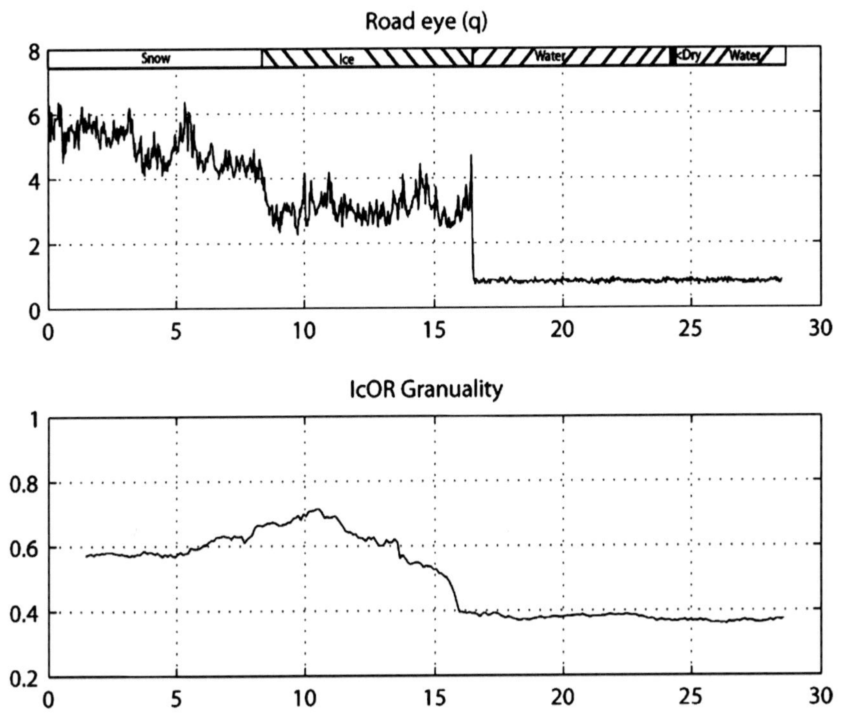

Fig. 6. Ratio of absorption (q) from the Road eye and Polarisation difference (S) from the IcOR Granuality (G) measure from IcOR

6 Conclusions

Here, two different indirect methods for monitoring road conditions called Road eye and IcOR have been compared. Both methods are non-contact methods for mounting on a vehicle. In a test performed on a dedicated track both sensors has shown with good results that it is possible to classify different road conditions such as a dry surface, and the presence of water, snow or ice with polarised near infrared light. Due to their strengths and weaknesses the two techniques can be used for different applications as road maintenance supervision or driver alert, etc. In particular, IcOR gives an overall evaluation of the road conditions without specific details, while Road eye provides data referred to a well defined spot in front of the vehicle. It is concluded that the Road eye system is better for detecting ice/water in a small area in front of the tire with 80 - 95 % performance rate whereas the performance of IcOR is limited to 70 – 80 %. But on the other hand a larger area is "scanned" instantly.

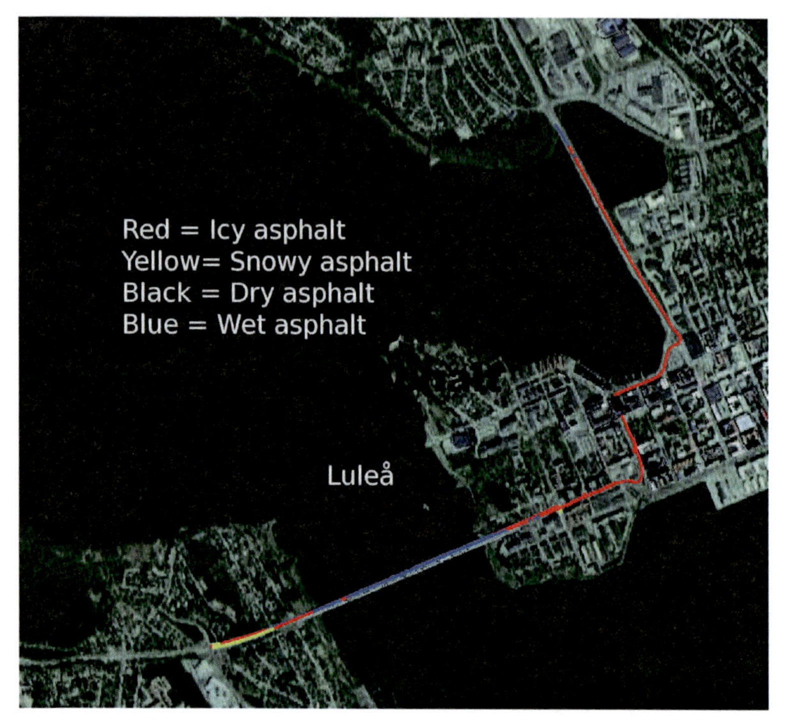

Fig. 7. Positioned Road eye measure making it possible to plot different road conditions on a map

These conclusions suggest that due to the larger monitoring area and the need for active illumination the IcOR could perform better in stationary measurements as for example intersection monitoring. Where it is possible to add an active illumination source to improve the classification even further. In the case of the Road eye the fast response time and the short computation time of the classification makes it suitable for road condition monitoring in real time by plotting road conditions on a map as shown in Fig. 7. This information could be used by other vehicles or maintenance vehicles for example to save salt and in the next step money.

Due to the fact that both sensors make a classification of road conditions the methods could be improved and made more valid in combination with direct methods estimating the actual friction between surface and tire contact patch.

References

[1] C-G. Wallman, H. Åström, Tema vintermodell – olycksrisker vid olika vinterväglag, Technical Report N60-2001, VTI, 2001.

[2] F. Gustafsson, Automotive Safety Systems. IEEE Signal Processing Magazine. p. 32-47, July 2009.

[3] M. Yamada, K. Ueda, I. Horiba, and N. Sugie, Discrimination of the Road Condition Toward Understanding of Vehicle Driving Environments. IEEE Transaction on Intelligent Transportation Systems. Vol. 2. No. 1 pp. 26-31, 2001.

[4] F. Holzwarth and U. Eichhorn, Noncontact Sensors for Road Conditions. Sensors and Actuators a-Physical, 37-8: p. 121-127, 1993.

[5] J. Casselgren, Road surface classification using near infrared spectroscopy". Licentiate thesis. Dept. Appl. Physics and Mech. Eng., Luleå Univ. of Tech, Luleå, Sweden, 2007.

[6] L. Yuesheng, & M. Higgins-Luthman, Black ice detection and warning system. Patent application. Application number: US2007948086A. 2007.

[7] M. Born, E. Wolf, Principles of Optics, 7th edition, Cambridge University Press, 2003.

[8] E. Hecht, Optics, 3rd edition ed.Addison Wesley Publishing Company,, pp. 694, 1998.

[9] J. Casselgren, M. Sjödahl and J. LeBlanc, Angular spectral response from covered asphalt. Applied Optics, Vol. 46, No 20, pp 4277-4288, 2007.

[10] Videre Design, STH-DCSG-VAR/-C Stereo Head - User's manual, www.videredesign.com, June 2005.

[11] K. Wagstaff, C. Cardie, S. Rogers and S. Schroedl, Constrained K-means Clustering with Background Knowledge, Proceedings of the Eighteenth International Conference on Machine Learning, Williamstown,MA, USA, 2001.

[12] ART SPY. Graininess Detection. Available in [www.owlnet.rice.edu/~elec301/Projects02/artSpy]. Cited on 27th Nov 2007.

Johan Casselgren
Luleå University of Technology
971 87 Luleå
Sweden
johan.casselgren@ltu.se

Matti Kutila, Maria Jokela
VTT Technical Research Centre of Finland
Tekniikankatu 1
33101 Tampere
Finland
matti.kutila@vtt.fi
maria.jokela@vtt.fi

Keywords: friction, road conditions, optical reflection, camera, IcOR, polarization, Ice, snow, machine vision, texture

A Centralized Real-Time Driver Assistance System for Road Safety Based on Smartphone

A. Corti, V. Manzoni, S. M. Savaresi, Politecnico di Milano
M. D. Santucci, O. Di Tanna, Piaggio Spa

Abstract

This work proposes a smart system to increase road safety. The main idea is to exploit existing technologies and devices, as the smartphone and the existing cellular network, to implement a centralized, real-time, Advanced Driver Assistance System: Each vehicle is equipped with a smartphone that periodically computes and sends the vehicle's global position and speed through GPRS or UMTS network. A centralized server snaps all running vehicles into digital maps and implements safety algorithms to detect potential dangerous situations. The prediction of a prospective crash generates a warning on the smartphone to alert the driver. This work presents the general hardware and software architecture and the implementation of two use cases. Finally, an evaluation of the critical aspects of the system is provided.

1 Introduction

The economic and social costs of road accidents have encouraged car manufacturers to equip modern vehicles with driver assistance systems. It is well known that many dangerous situations cannot be detected and avoided by using only the sensors available on the vehicle. For this reason, a massive effort in the automotive research field has been devoted at developing safety applications based on vehicle communications.

Modern Intelligent Transportation System approaches are based on vehicles' communication, i.e. vehicles are interconnected using wireless connection called Vehicular Ad-Hoc Networks (VANET) [4]. Each vehicle is equipped with an accurate positioning system and is able to communicate with all other neighbours. Therefore, each vehicle knows data from all the others traveling in the same road. Such data are, for example, position, speed and warning status. The communication extends sensor ranges and increases neighbours' awareness [1]. Unfortunately, it has strictly requirements: it must be reliable enough to develop safety application and it requires customized additional electronics. Therefore, this is usually not considered a cost effective solution. Furthermore, the efficiency in detecting dangerous scenario is strictly dependent on the market penetration: if only few vehicles are adopting the system it is ineffective.

This work proposes an alternative approach that leverages the pervasiveness of the smartphone to implement a real-time, centralized Advanced Driver Assistance System. Nowadays the market penetration of these ubiquitous devices is constantly increasing. In 2011 Nielsen expects that smartphones will overtake feature phones penetration in the US market. Therefore, in the future every driver will be able to adopt the system. In literature the first attempt to increase driver safety using cellular phone has been done by [5]. The authors propose to increase the interaction between a motorbike and a rider through a smartphone. Also the cellular network has already been proposed as an alternative solution to interconnect vehicles [6], [7]. Experimental results proved its reliability and its latency time regularity [2]. However, to the best of our knowledge, no one has ever proposed a centralized real-time ADAS, based on smartphones.

In the proposed system each vehicle is equipped with a smartphone that periodically computes and sends the vehicle's global position and speed using the existing cellular network. A centralized cluster of servers gathers all the data from different vehicles, snaps them into a digital map, performs safety algorithms to detect dangerous situations and sends back warnings, all in a 1-second timeframe. A preliminary prototype of the system has been developed and tested using two cars and a motor vehicle in a real city scenario. Two typical ADAS use cases have been considered: the safe-overtaking and the obstructed view at intersection. Finally, the critical aspects of the proposed system have been analysed and a detailed feasibility analysis is provided. The intent is to demonstrate how the proposed concept can be used to improve road safety

The remainder of this work is organized as follow: in Section 2, the architecture, the data flow and the safety algorithms are explained. The implementation of the whole system and of two interesting use cases is presented in Section 3. In Section 4 numerical analysis demonstrate the feasibility of the proposed solution. Finally, Section 5 summarizes the presented content and gives suggestions for further works.

2 System Architecture and Implementation

The architecture of the proposed system comprises the following elements (see Fig. 1.):

- **Smartphone** equipped with a UMTS/GPRS communication system and a multitasking operating system (iOS, Android). It also includes an advanced Human Machine Interface (HMI) that gives the possibility to show pictures, play audio messages and vibrate.

- **GPS Receiver** providing the vehicle's global position with a refresh rate of at least 1 Hz and an accuracy of at least 10 m CEP (Circular Error Probable). Modern smartphones already integrate a GPS receiver. Alternatively, an external one can be connected through Bluetooth or Wi-Fi.

- **Database** with GIS extension (Geographic Information System). It is the core of the centralized system. It periodically runs safety applications to detect potentially dangerous situations. It is aware of all the vehicles' motions that are real-time monitored using digital maps of the urban area.

- **Web Server** is the interface between smartphones and the database. It receives vehicle's data and it propagates warning to the drivers.

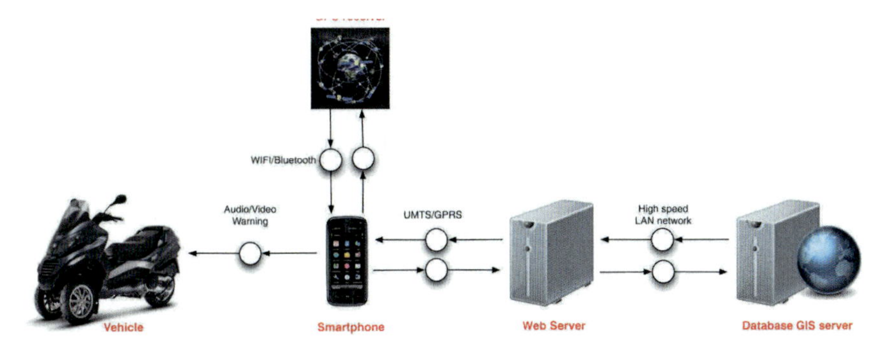

Fig. 1. System architecture: components and connections

Every second the smartphone computes the vehicle's position. It processes the information and sends data to the server through an HTTP request exploiting the cellular network connection. The web server receives requests and is responsible for storing the information in the database. Whenever a new position is inserted a trigger is activated and the safety applications are executed.

The road safety algorithms proposed in this work are composed by three consecutive phases:

- **Position Prediction** compensates the transmission/reception delay, due to the cellular network. Using linear regression it predicts the vehicle's next location and speed, given the previous ones.

- **Point Snapping** analyzes the vehicles' predicted positions and it snaps points into digital maps. We decided to implement a snapping algorithm widely used in literature [8]. It grants good performances and is relatively simple.

- **Scenario Analysis** is the core of the ADAS algorithm. Each safety application is a single thread. Every thread performs spatial queries analyzing all neighbors in the road, searching for potentially dangerous situations.

The whole system has been implemented with open source software on a home server (AMD Athlon 64 bit 2.2 GHz, 2 GB RAM) with Ubuntu Server 9.04 as operating system. The server has been connected to a home DSL network (348 Kbps uplink, 7 Mbps downlink). The database and the Web server have been integrated into the same machine. We choose PostgreSQL8.3 as database. It is an open source solution which guarantees high flexibility and native GIS support with the PostGIS extension. Digital maps of the test site have been taken from Open Street Map. They are digital maps created by an open source community. Furthermore, these maps are royalty free and they can be easily imported in the PostgreSQL database using an appropriate tool. Generally, they are accurate; however, we have manually verified the coherence of every street of the test site. All the vehicles have been equipped with a smartphone Nokia 5800, running Symbian OS 5th edition and a SIM card with GPRS/UMTS connection. Two external GPS Bluetooth antennas have been used. They grant a position accuracy of 10 m CEP, and a refresh rate of 1 Hz.

3 Implementation and Case Studies

To demonstrate the feasibility of the proposed ADAS system two different manoeuvres have been considered:

- **Safe Overtaking** Fig. 2a shows a motorbike while overtaking a car (or queue) on the left lane. The cars are notified of the dangerous situation, so they avoid changing the lane;

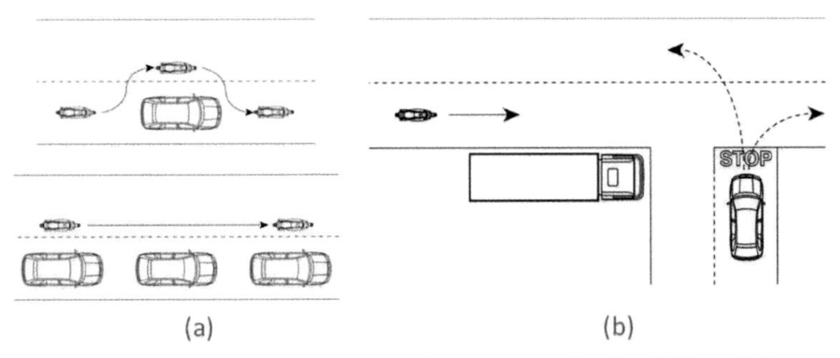

Fig. 2. The use cases considered to demonstrate the feasibility of the system: on the left side two typical overtaking scenarios; on the right side, an example of obstructed view at intersection.

- **Obstructed View at Intersection** Fig. 2b shows a motorcycle while moving on the main street; meanwhile a car is approaching an intersection. The car is warned and stops its motion even if there is a vehicle occluding is visual field.

We have considered two very common use cases in urban area. In many situations, they can become fatal especially when lightweight vehicles, like bikes or motorcycles, are involved. As an example, in Fig.3 the flowchart of the algorithm implemented for the safe-overtaking manoeuvre is depicted. Every time a vehicle sends its data to the server geographical queries are executed to isolate the vehicles running in the same road and direction. Then, the so called time to overtake is computed and is compared with thresholds in order to identify dangerous situations.

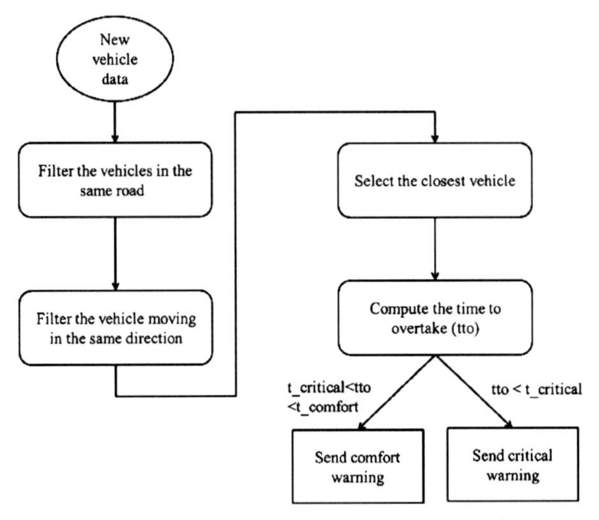

Fig. 3. Flowchart of the safe overtaking scenario analysis.

The correct implementation of the system has been first tested with a micro-traffic simulator. Finally, real trials have been carried out using three vehicles: a Piaggio MP3 and two cars (see Fig. 4).

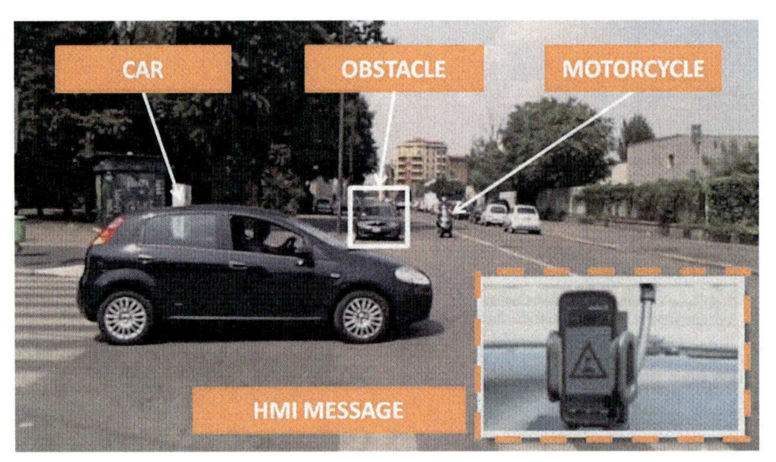

Fig. 4. Vehicles during a demonstration of the manoeuvre of Fig. 2b in the test site. The Car driver is warned by a message on the smart phone that a motorbike is coming at intersection.

4 Performance Analysis

In order to demonstrate the practical feasibility of the proposed system two critical aspects have been further investigated. The real-time ADAS must guarantee a response within 1 second to prevent crashes. From this point of view, the main challenges are the cellular network delay and the computational time required to execute centralized algorithms.

4.1 Cellular Network Delay

The GPRS/UMTS cellular network is the medium chosen for transmitting vehicles' data and delivering notifications. Nowadays, this communication channel is usually reliable but for road safety applications a response time within one second has to be guaranteed. To this aim, a specific evaluation of cellular network performance has been carried out using two SIM cards of different Italian operators. We designed 10 trials of almost 1000 seconds for each SIM card: 5 stationary and 5 performed while the vehicle was moving in the urban area.

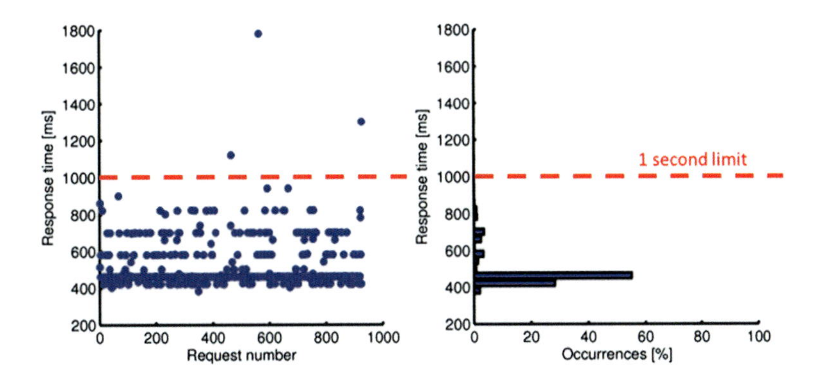

Fig. 5. Cellular network delay measured during a trial with the operator B in Table 1. On the left side the measured response time for each request is shown. On the right side the percentage of occurrences for response time intervals.

The smartphone application sends location data to the web server and measures the round-trip time. In order to evaluate the network delay only, we have disabled all database computations described in the previous sections. Fig. 5 reports the results obtained for the second operator while the vehicle is traveling. It can be noticed that 95% of the requests are managed within 0.5 s. Moreover, the network delay increases when the vehicle is moving. As reported in Tab. 1, the measured performances are sufficient to implement safety application within 1 second timeframe. The results obtained are similar to the one presented in [2].

	Network Operator A	Network Operator B
Stationary	156.71 ms	399.08 ms
Moving	216.40 ms	490.45 ms

Tab. 1. Average cellular network delay [ms]

4.2 Server Performances

The overall centralized architecture performance has been explicitly considered. As the number of vehicles traveling in the area increases more conditions must be checked every second by safety applications. The bottleneck is clearly the scenario analysis composed by many geographical queries on digital maps. For this reason, it is crucial to understand the number of vehicles to be simultaneously supported by a single machine.

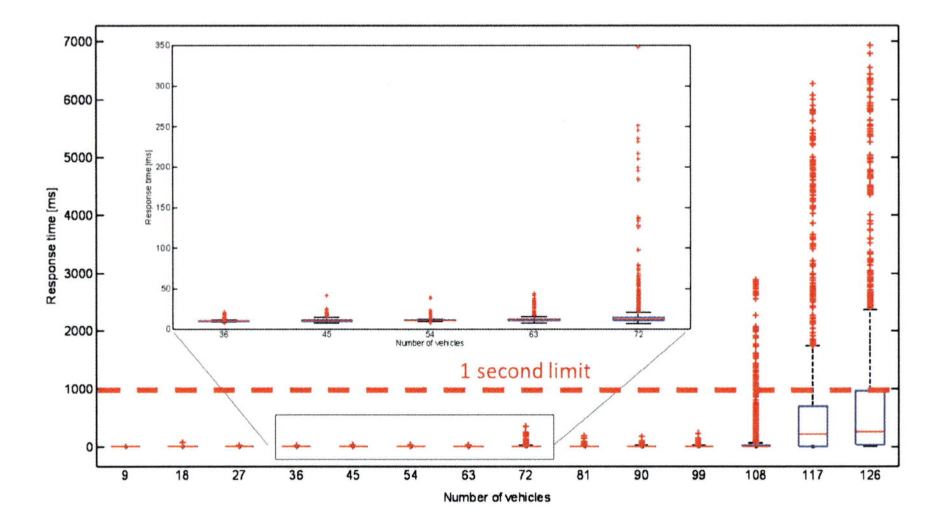

Fig. 6. Box plot of the response time obtained simulating an increasing number of vehicles traveling in the urban area. On each box, the central mark is the median, the edges of the box are the 25th and 75th percentiles, the whiskers extend to the most extreme data points not considered outliers ($\pm 2.7\ \sigma$), and outliers are plotted individually.

To estimate this performance index, we used a micro-traffic simulator directly connected to the centralized system, so as to minimize network delays. In the server all the algorithm phases have been enabled. We have simulated in the same urban area an increasing number of vehicles traveling for 100 seconds and we analysed the elaboration time of each request. As shown in Fig. 6, experimental results reports that the server used can support up to 108 vehicles with 95% of the responses being less than 15 ms. Fig. 6 shows that with 117 vehicles the server performances diverge and the time requirements are violated.

5 Conclusions

This paper presents an innovative approach based on smartphones and a centralized server to implement a real-time Advanced Driver Assistance System. The system is devoted to increase the road safety level: it provides an audio/video interaction with the driver trough the smartphone HMI capabilities. Two use cases have been implemented: the safe overtaking and the obstructed view at intersection. They have been tested using a simulator and with real vehicles.

A detailed analysis evaluates critical aspects and demonstrates the practical adoption of the system, even with many vehicles traveling at the same time.

Starting from the proposed architecture and exploiting vehicles' interaction, several other road safety applications and value-added services can be defined. Moreover, it is important to point out that both the localization technology and the communication technology used in this work in the next few years will dramatically increase their performance with the introduction of the 4G cellular networks and of the Galileo global navigation system.

References

[1] L. Andreone, M. Provera, Inter-vehicle communication and cooperative systems: local dynamic safety information distributed among the infrastructure and the vehicle as virtual sensors to enhance road safety. ITS Hannover, 2005.

[2] J. Cano-Garcia, et.al, Experimental analysis and characterization of packet delay in UMTS networks, Next Generation Teletraffic and Wired/Wireless Advanced Networking, pp. 396–407, 2006.

[3] J. Chen, et. al, Modelling and predicting future trajectories of moving objects in a constrained network, IEEE Proceedings of the 7th International Conference on Mobile Data Management, pp. 156-162, 2006.

[4] J. Jakubiak, Y. Koucheryavy, State of the art and research challenges for VANETs, IEEE Proceding on Consumer Communications and Networking Conference, pp. 912–916, 2008.

[5] C. Spelta, et. al., Smartphone-Based Vehicle-to-Driver/Environment Interaction System for Motorcycles, IEEE Embedded Systems Letters, vol. 2, no. 2, pp. 39–42, 2010.

[6] J. Santa, et. al, Architecture and evaluation of a unified V2V and V2I communication system based on cellular networks, Computer Communications, vol. 31, no. 12, pp. 2850–2861, 2008.

[7] J. Santa, et. al, Experimental evaluation of a novel vehicular communication paradigm based on cellular networks, IEEE Intelligent Vehicles Symposium 2008, pp. 198–203, 2008.

[8] H. Yin, O. Wolfson, A weight-based map matching method in moving objects databases, IEEE Proceedings of the 16th International conference on Scientific and Statistical Database Management, pp. 437-438, 2004.

Andrea Corti, Vincenzo Manzoni, Sergio M. Savaresi
Politecnico di Milano
Piazza Leonardo Da Vinci, 32
20133 Milan
Italy
corti@elet.polimi.it
manzoni @elet.polimi.it
savaresi@elet.polimi.it

Mario D. Santucci, Onorino Di Tanna
Piaggio Spa
Via Rinaldo Piaggio, 25
56025 Pontedera
Italy
mario.santucci@piaggio.com
onorino.ditanna@piaggio.com

Keywords: road safety, real-time, advanced driver assistance system, intelligent transportation systems, road safety application, smartphone, ADAS, ITS

Simulation of Advanced Lateral Safety Systems as a Cost Effective Tool to Estimate Potential Success

E. Cañibano Álvarez, J. Romo García, B. Araujo Pérez, C. Maestro Martín,
J.C. Merino Senovilla, Fundación CIDAUT

Abstract

The need of promoting lateral vehicle safety is leading to more and more complex systems. As complexity increases, the evaluation of the potential success of new alternatives becomes a key point in reaching a cost-effective development. A system of four cameras mounted on the side of the vehicle has been suggested. By means of artificial vision, the control unit will be able to monitor the impacting car, and thus take effective measures to prepare the vehicle for the impact, changing the seat set up to aid the passenger and improve passive safety. By means of a simulation model different cases have been simulated varying speeds, locations of impact and studying the period of time on each camera between detection and impact. This simulation model has enabled evaluating the success of the proposed cameras set up.

1 Simulation Model and Input Data

The simulation model proposed in this research has been developed within the frame of national Spanish project OPTIVE (CCT/10/VA/0002). In order to foresee the lateral impact of another vehicle a four camera system on each side of the vehicle has been proposed. The set up of these cameras can be observed in Fig. 1.

The two cars involved in the simulation will be named from now as impacted vehicle and impacting vehicle, where the former refers to the car which has the cameras and will be impacted by the latter. Speed range for these two cars will vary between 20 and 56 kmph, and angle of their trajectories will range between 5° and 175°. Lateral impacts have been proposed on three vehicle locations, which are depicted in Fig. 2.

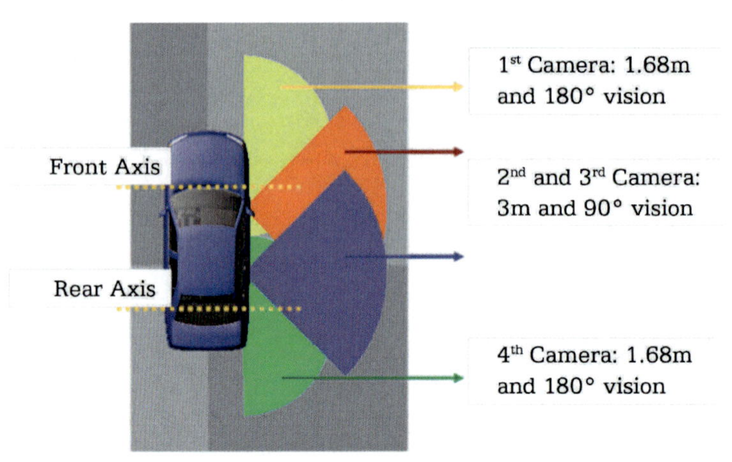

Fig. 1. Location and set-up of cameras system.

The implemented simulation model has been developed with CarSim. CarSim is a dynamic vehicle simulation software that predicts the behaviour of the vehicle in response to the driver inputs (throttle, brake, clutch, steering) in a given environment (roads, friction, wind). The simulation of different manoeuvres allows studying the influences of many factors on vehicle handling and viewing results that are comparable to road tests. In this case, a sedan car from

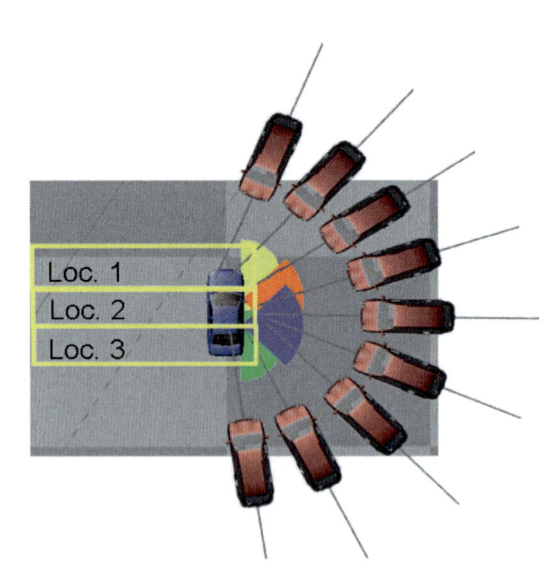

Fig. 2. Impact Locations on the impacted vehicle and trajectory angle variation range

the CarSim database has been selected as representative geometry for the study. The main objective of this simulation model is to study the number of cameras that will see the impacting vehicle and the period of time going from detection to impact in the range of situations mentioned.

CarSim functionality has been extended by means of a Simulink model. The inputs for this simulation model are the position of each vehicle and the speeds, with which the model will compute the moment when each of the cameras detects the impacting vehicle and the moment when the impact takes place, thus obtaining the span of time during which the safety system can take effective measures to prevent or prepare the driver and the vehicle for the upcoming impact. The Simulink simulation model is presented in Fig. 3.

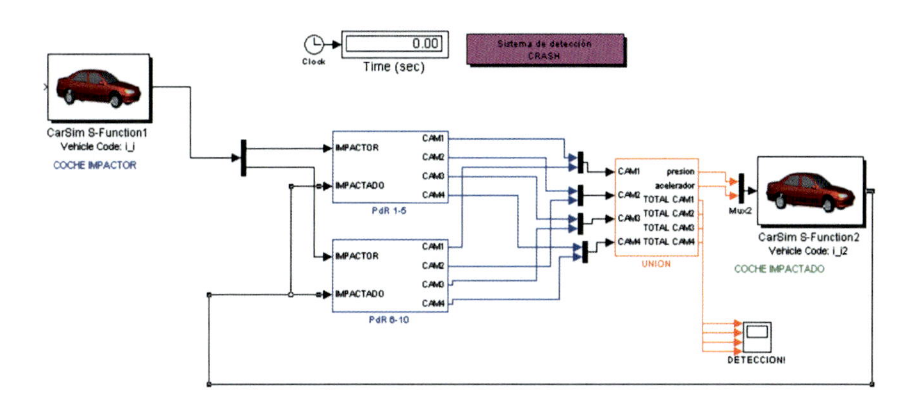

Fig. 3. Simulink & Carsim Simulation Model

2 Study Cases Selection

Given the ranges of speeds and angles of the two vehicles, one of the challenges this project faced was how to discretize a limited amount of study cases that was nonetheless representative of a bigger number of situations. The solution achieved within this research involves the use of relative speeds. If relative speeds are computed it can be seen that the three input variables, speed of impacted vehicle, speed of impacting vehicle and angle sustained by both trajectories, are interrelated and can be reduced to a system of two independent variables: the relative speed on X axis and the relative speed on Y axis (as stated in Fig. 4).

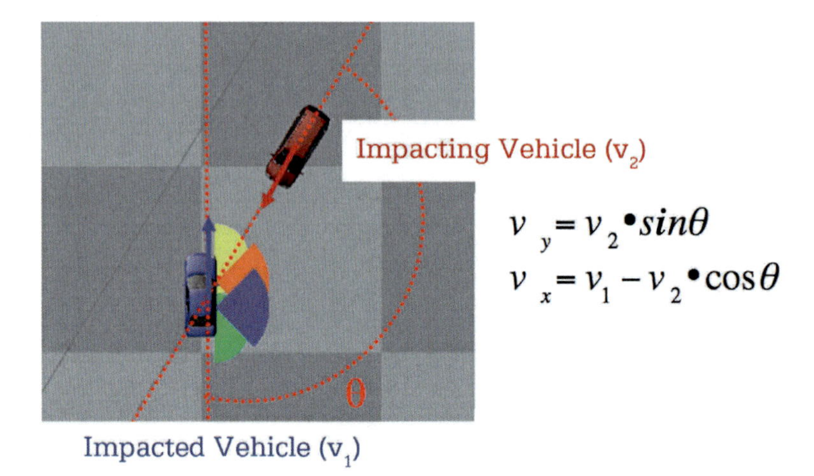

Fig. 4. Relation between absolute speeds and trajectory angle and variables selected for the analysis (relative speeds)

This means that by giving values to these two variables a bigger range of real situations is represented. As an example if the relative speed on X axis is 35kmph and the relative speed of Y axis is 40kmph it can be deduced that there are 23 equivalent cases if angle is discretized every 2.5 degrees, or 56 equivalent load cases if angle is discretized every degree. The impact of studying relative speeds instead of absolute speeds and trajectory angle not only implies that the study will cover a huge number of real situations, but also, that selected study cases cover all the input parameter ranges. Having used absolute speeds and trajectory angle as input parameters for the study cases selection could have derived in the selection of equivalent load cases for the analysis and a loss of information.

The possible envelope of study cases is delimited by the cases when both vehicles run at maximum speed and by the cases when the impacted car is kept to the maximum and the impacting car is at its lowest. Out of this area 54 study cases have been selected, and out of this study cases, one of the equivalent absolute speeds – angle case has been chosen for simulation. In Fig. 5 the envelope of possible cases is presented, along with the selected 54 study cases.

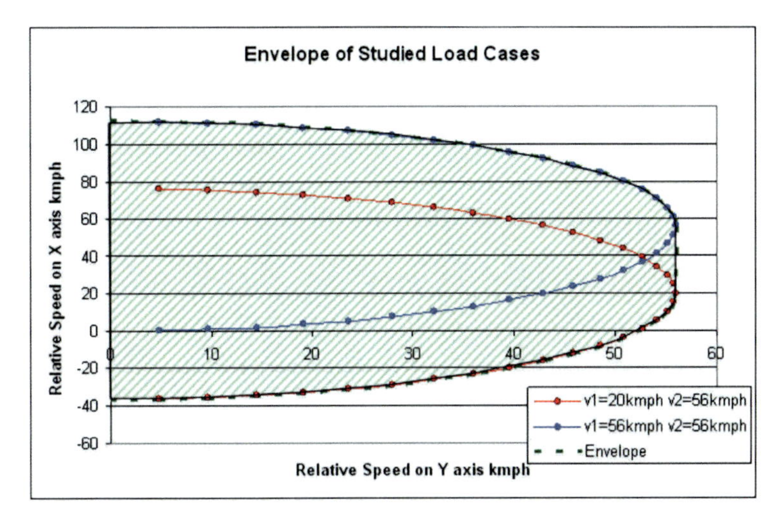

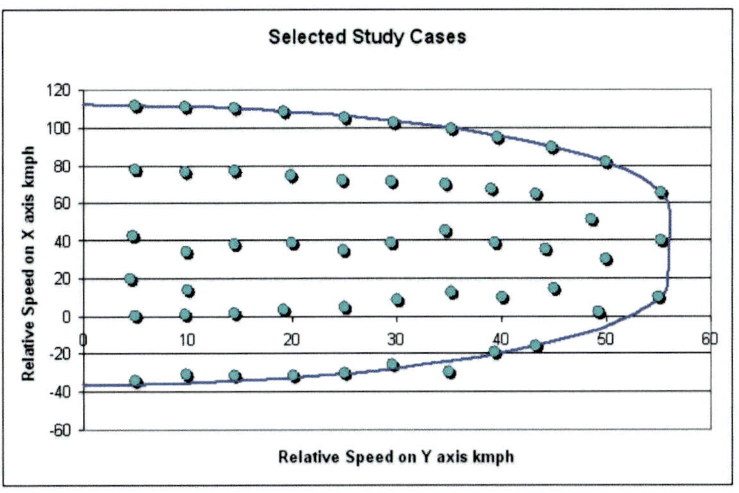

Fig. 5. Envelope of possible cases (top) and selected study cases (bottom)

3 Results

A bubble graph has been selected to show the results of each camera, displaying the relative speed on X axis on the abscissa axis and the relative speed on Y axis on the ordinate axis (inputs) and the size of the bubble accounts for the span of time during which each camera detects the impacting Car. The results of impacting location 1 are shown in Fig. 6.

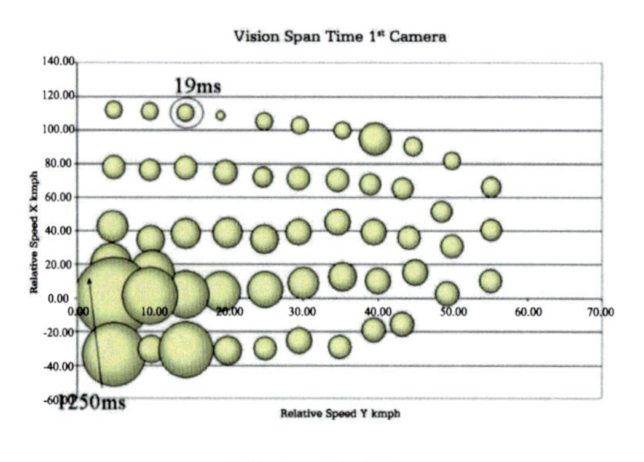

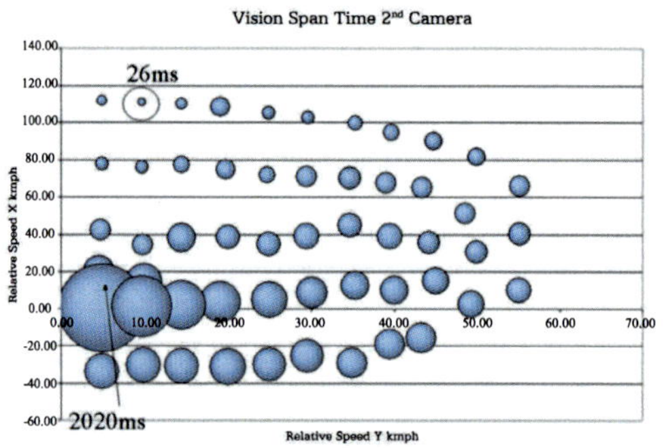

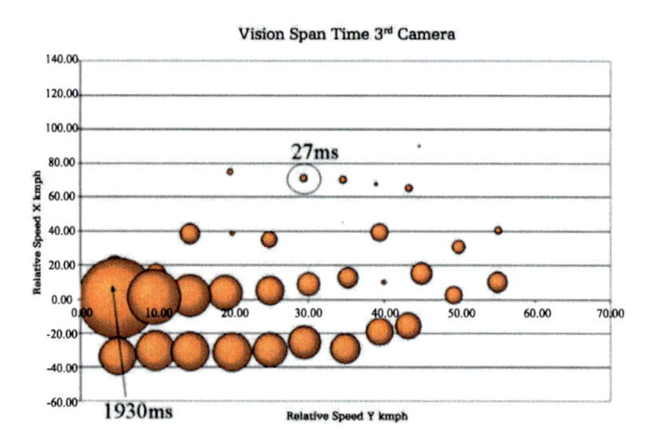

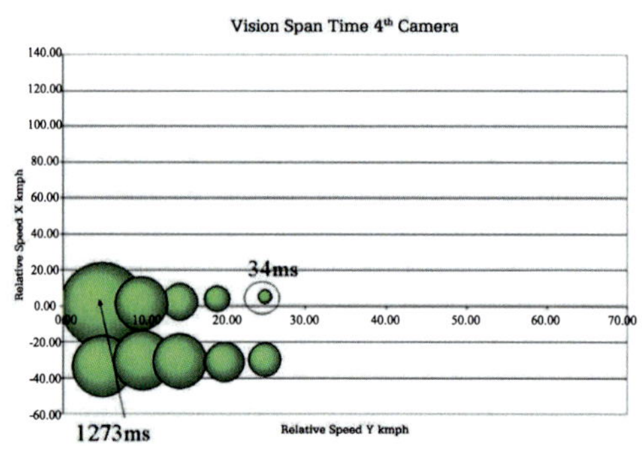

Fig. 6. Time Span detected by the different cameras for impacts against location 1

Regarding the number of cameras that detect each study case, and considering that each camera needs at least ,120ms to react, results of impacting location 1 are the depicted in Fig. 7.

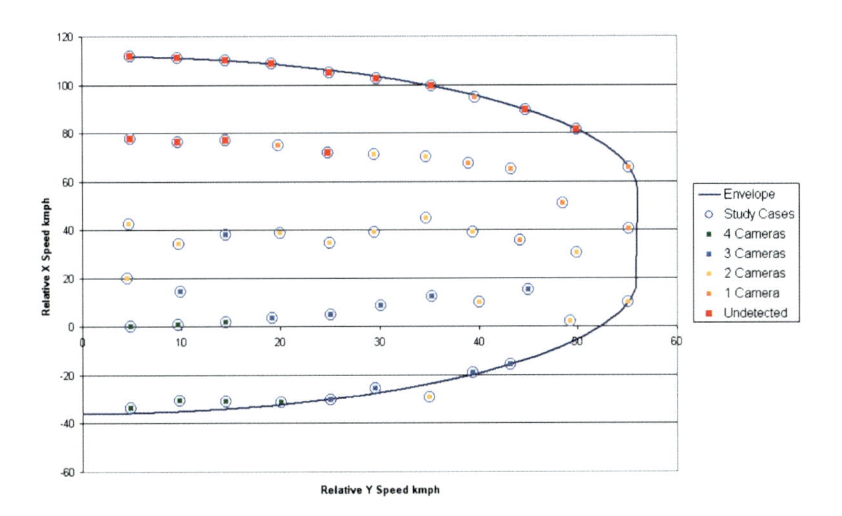

Fig. 7. Number of cameras detecting each Study Case

The same analysis has been conducted for locations 2 and 3, analyzing a total of 162 study cases. Achieved results indicate that 82.10% of the cases would be detected by two or more cameras, 7.41% of the cases would only be detected by one camera, and a 10.49% of the cases would remain undetected. This means that the range within which the proposed lateral system will be successful is very large. This evaluation is key to the cost effective development of this safety system, since a wide number of tests have been conducted on a virtual basis, allowing effectively reaching a global conclusion.

4 Conclusions

A simulation tool to predict the success of a passive safety system has been presented in this article. In this research it has been found that using relative vehicle speeds is a robust way to filter study cases and cover a wide range of situations effectively.

The analysis conducted has allowed an estimation of the number of situations in which the proposed lateral safety system would operate successfully, achieving a success rate of over 82% with two cameras used and almost 90% with a single camera being used. With the gained insights and valuable data about detection times and camera's vision the system's design team surely will be able to improve the further development in a cost-effective way.

References

[1] C. Erbsmehl, Simulation of real crashes as a method for estimating the potential benefits of advanced safety technologies, 21st International Technical Conference On the Enhanced Safety of Vehicles, 2009

[2] E. Cañibano, et al., Development of mathematical models for an electric vehicle with 4 in-wheel electric motors, In: G. Meyer, J. Valldorf [Eds.], Advanced Microsystems for Automotive Applications 2011, Smart Systems for Electric, Safe and Networked Mobility, Springer, Berlin, 2011.

[3] E. Cañibano, et al., Control Algorithm Development for Independent Wheel Torque Distribution with 4 In-Wheel Electric Motors, IEEE UKSim 5th European Modelling Symposium on Mathematical modelling and Computer Simulation, 2011.

Blanca Araujo Pérez, Cesar Maestro Martín, Esteban Cañibano Álvarez, Javier Romo García, Juan Carlos Merino Senovilla
Fundación CIDAUT
Parque Tecnológico de Boecillo p.209
42151 Boecillo
Valladolid
Spain
blaara@cidaut.es
cesmae@cidaut.es
estcan@cidaut.es
javrom@cidaut.es
juamer@cidaut.es

Keywords: vehicle safety, vehicle dynamics, enhanced safety systems, vehicle dynamic simulation, artificial vision

Networked Vehicles

Traffic Jam Warning Messages from Measured Vehicle Data with the Use of Three-Phase Traffic Theory

H. Rehborn, B. S. Kerner, Daimler AG
R.-P. Schäfer, TomTom

Abstract

Based on Kerner's three-phase theory, we study an algorithm for the generation of traffic jam warming messages from measured GPS and GSM probe vehicle data that have been collected in TomTom's HD-traffic service both from nomadic devices and vehicle's embedded systems. We find that the data allows us to reconstruct the structure of congested traffic patterns with a much greater quality of spatiotemporal resolution than has been possible before. It occurs that congested traffic in measured traffic patterns consists of the two traffic phases of Kerner's three-phase theory, synchronized flow and wide moving jams. The application method distinguishes between the fronts of the congested traffic phases, wide moving jam and synchronized flow. It will be shown that a penetration of about 2% of the total traffic flow is enough to implement a precise traffic jam warning message for navigation systems.

1 Introduction

The efficient operation of most intelligent transportation systems (ITS) including traffic control and management methods needs traffic information both in time and space covering a traffic network. Because the road detector technology as well as other stationary road installations required for such spatiotemporal traffic measurements in the whole road network are relatively expensive, probe vehicle data is considered as a future traffic measurement technology that will determine the quality of future ITS. This is because of two well-known features of this technology:

- Probe vehicle data can provide spatiotemporal traffic information covering the traffic network.
- Probe vehicle data can automatically be produced by vehicles moving within the network and they are not necessarily associated with any road installations.

However, probe vehicle data is "raw data" that requires subsequently processing to identify spatiotemporal traffic features. In this article, we consider a method for the reconstruction of spatiotemporal traffic dynamics in probe vehicle data measured on highways with the aim of generating jam warning messages. This method is based on three-phase traffic theory introduced by Kerner [1],[2]. In this theory there are three traffic phases: free flow, synchronized flow (see Fig. 1.), wide moving jam.

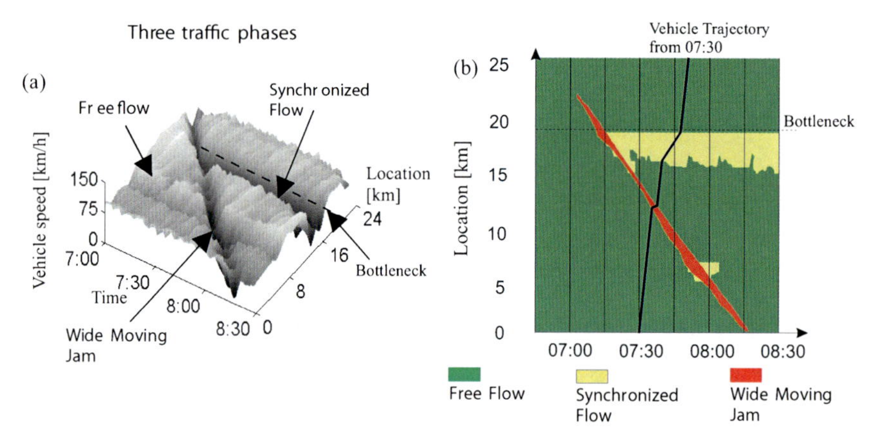

Fig. 1. Explanation of the traffic phase definitions in three-phase theory based on measured data of speed in time and space (a). (b) Representation of speed data in (a) on the time-space plane [1],[2].

We will use the phase designations F, S, and J for free flow, synchronized flow, and wide moving jam, respectively. The synchronized flow and wide moving jam traffic phases belong to congested traffic. The empirical criteria *[J]* and *[S]* that define the wide moving jam and synchronized flow phases are as follows: A wide moving jam is a moving traffic jam, i.e., a localized structure of congested traffic with great vehicle density and low speed, spatially limited by two jam fronts, which exhibit the characteristic jam feature *[J]* to propagate through bottlenecks while maintaining the mean velocity of the downstream jam front. Note, that at the downstream jam front vehicles accelerate from a low speed state within the jam to either free flow or synchronized flow downstream of the jam.

Synchronized flow is defined as congested traffic that does not exhibit the jam feature *[J]*; in particular, the downstream front of synchronized flow is often fixed at the bottleneck. Note, that at the downstream front of synchronized flow vehicles accelerate from a lower speed within the synchronized flow to a higher speed in the free flow phase. The use of three-phase traffic theory for the understanding of spatiotemporal traffic features in TomTom probe vehicle data is explained as follows. Traffic is a complex dynamic process that occurs in space and time. Empirical spatiotemporal features of traffic have been understood in 1998-2002 (see [1]). Therefore, it is not surprising that classical traffic flow theories based on fundamental diagram hypothesis (see e.g., [3]-[6] and references therein) cannot show the fundamental empirical features of traffic breakdown at a highway bottleneck (see [2] for detail).

2 Methodology

For a correct evaluation of any method for traffic phase reconstruction one needs single-vehicle data with the vehicle speed $v(t)$ and location $x(t)$ as time-functions measured along a vehicle trajectory while the vehicle moves though all traffic phases F, S, and J. However, currently such precise single vehicle data measured in real traffic are almost not available. For this reason, we describe the following methodology (Fig. 2.) (see [7]):

1. We use averaged measured traffic data (1-min data). The data are measured through the use of road detectors installed on more than 30 km long freeway section of A5 in Germany (Fig. 2.(a)). The analysis of the database from 1996-2012 shows characteristic congested traffic patterns whose qualitative spatiotemporal features are reproducible and predictable (see Sect. 2.4 of [1]).

2. We choose a representative empirical congested pattern from the data. We choose a representative data set from this empirical database and then process the data with the ASDA and FOTO models [1],[2],[8] (Fig. 2.(b)). The data set exhibits spatiotemporal structures of congested traffic that are common for all data measured on different highways in various countries (Fig. 2.(b)).

3. We simulate this empirical pattern with a three-phase traffic flow model. We simulate the representative pattern with a stochastic microscopic three-phase traffic flow model [9] (Fig. 2.(c)).

4. We make a random choice of single vehicle trajectories within the simulated pattern. For different percentages of stochastic distributions of vehicles moving through the congested pattern single vehicle trajectories are found (Fig. 2.(d)).

5. Traffic phases along each of the vehicle trajectories are identified (Fig. 2.(e)).

6. Jam warning messages by using the traffic phase identification are generated (Fig. 2.(f)).

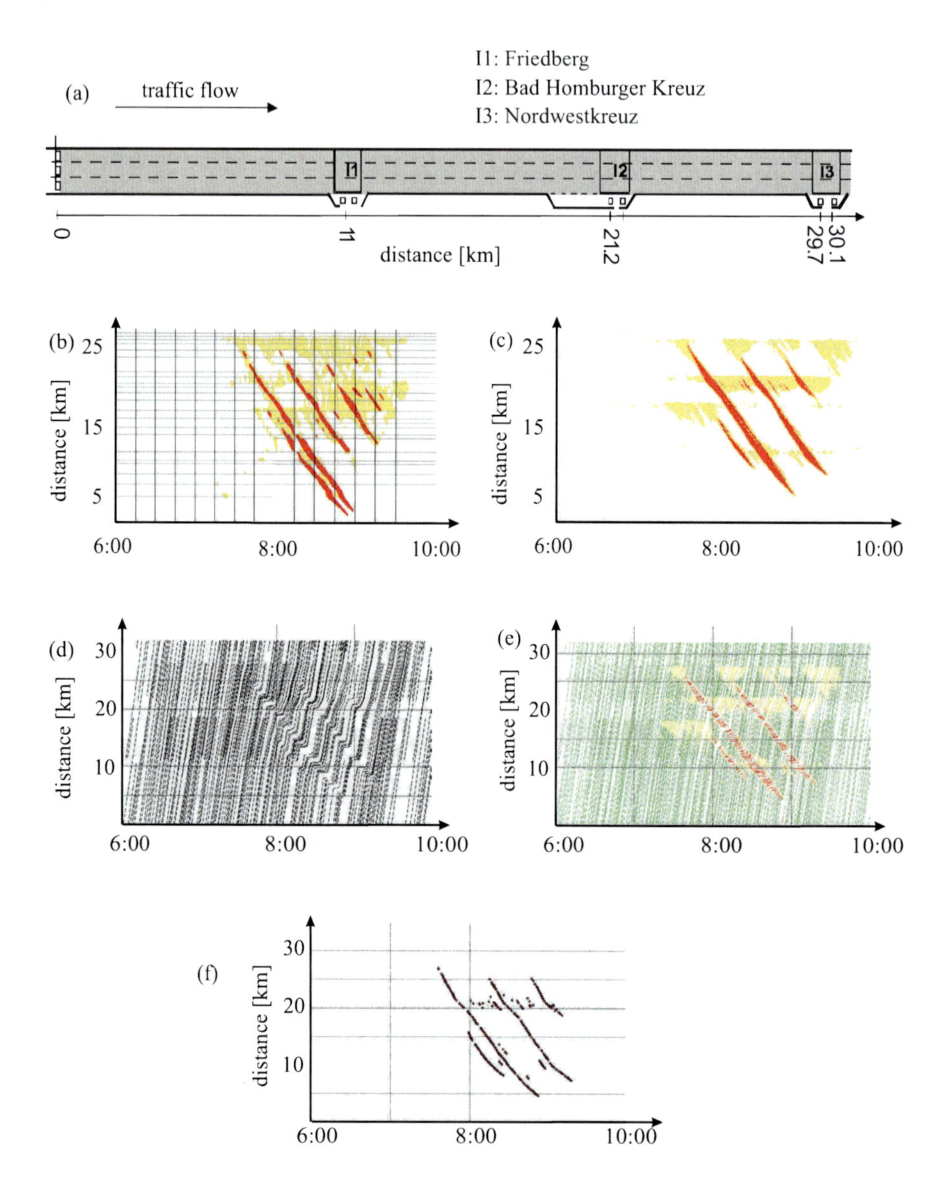

Fig. 2. Methodology of traffic reconstruction [7]: (a) A5 highway stretch. (b) ASDA/FOTO pattern reconstruction based on detector data on 10th December, 2009. (c) Simulations of the pattern in (b) with Kerner-Klenov stochastic microscopic three-phase traffic flow model. (d) Simulated vehicle trajectories of random distributed 2% probe vehicles within the pattern in (c). (e) Traffic state detection for probe vehicles in (d). (f) Jam warning messages by probe vehicles in (d).

3 Identification of Traffic Phases

After the congested pattern has been found in the microscopic simulations (Fig. 2.(c)), we have the information about vehicle trajectories of all vehicles moving through the pattern. Now, we apply a model for an approximate identification of traffic phases in a probe vehicle (Fig. 3.). The model for traffic phase identification is as follows: The recognition of a new traffic phase by the vehicle is made based on conditions that during a time interval T that is longer than a given one the vehicle speed v is lower or higher than a given value [2],[10]:

(i) The probe vehicle detects synchronized flow, when

$$v_{jam} < v \leq v_{syn} \text{ during time interval } T > T_{syn}. \tag{1}$$

(ii) The probe vehicle detects a wide moving jam, when

$$v \leq v_{jam} \text{ during time interval } T > T_{jam}. \tag{2}$$

(iii) The probe vehicle propagating initially within synchronized flow or within a wide moving jam detects free flow, when

$$v > v_{free} \text{ during time interval } T > T_{free}. \tag{3}$$

In (1)-(3), T_{free}, T_{syn}, T_{jam}, v_{free}, v_{syn} and v_{jam} are model parameters.

The results of the application of this model to trajectories of probe vehicles at η=2% (Fig. 4.(a)) and η=0.5% (Fig. 4.(b)) and η=0.25% (Fig. 4.(c)) show that probe vehicles reconstruct many important features of spatiotemporal congested patterns already at a penetration rate of about η=0.5%. Based on 2% probe vehicles the congested pattern from Fig. 2. (c) can be reconstructed with very high quality: we would call this 2% probe vehicle penetration as "premium" quality. Traffic phase identification along a vehicle trajectory allows us to generate jam warnings as follows: a jam warning message is generated by a probe vehicle when either a $F \rightarrow J$ transition (transition from free flow to wide moving jam) or a $S \rightarrow J$ transition (transition from synchronized flow to wide moving jam) is identified along the vehicle trajectory.

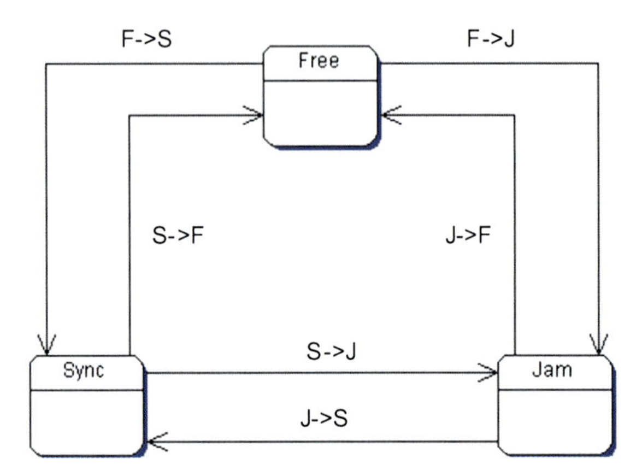

Fig. 3. State diagram for the detection of three traffic phases: "Sync" - synchronized flow. "Jam" - wide moving jam [10].

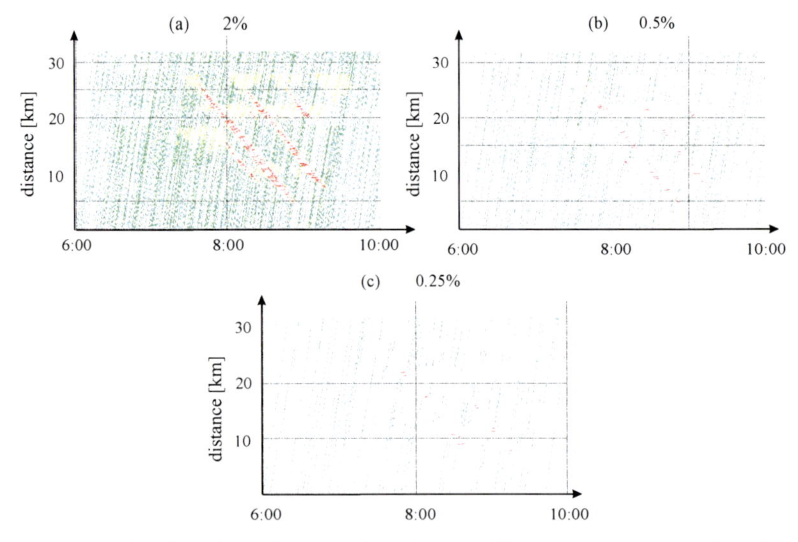

Fig. 4. Results of traffic state detection at different percentages of probe vehicles: (a) η=2%, (b) η=5%, (c) η=0.25% [11]

Then a sequence of these jam warning messages generated by all probe vehicles shows a spatiotemporal behaviour of the propagation of the jam upstream front on the road (Fig. 5). This jam front propagation can be used for the warning of all other vehicles moving upstream of the jam front.

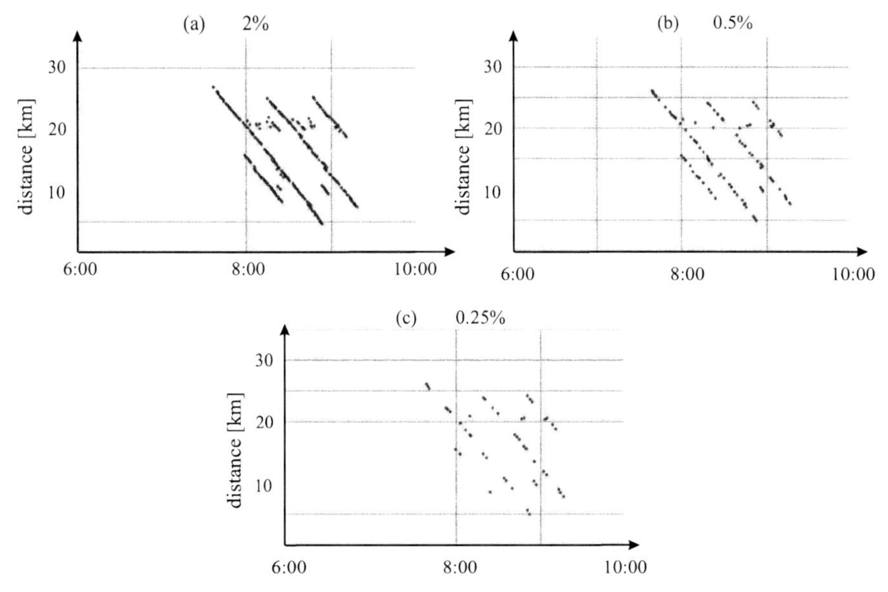

Fig. 5. Detection of upstream jam fronts ("jam warning messages") with different percentage of probe vehicles: (a) η=2%, (b) η=0.5%, (c) η=0.25% [7]

Figure 5. shows that 2% probe vehicle penetration allow a precise spatial resolution of the jam upstream fronts and their related propagation over time.

4 Traffic Patterns in TomTom's Probe and Phone Data

Let us compare spatiotemporal structures of traffic patterns reconstructed by the ASDA/FOTO models from traffic data measured by road detectors (Fig. 6.(a)) with spatiotemporal structures of traffic patterns found in on-line TomTom's GPS probe vehicle data (Fig. 6.(b)) and TomTom's GSM probe vehicle data (Fig. 6(c)). In particular, the ASDA/FOTO models show usual spatiotemporal structures of traffic patterns consisting of synchronized flow and wide moving jams that have initially emerged within this synchronized flow and then propagate through the synchronized flow while maintaining the mean velocity of the downstream jam fronts. Such congested patterns are called general congested patterns (GP) in three-phase theory, because each of the patterns consists of both traffic phases of congested traffic, synchronized flow and wide moving jams. Considering spatiotemporal structures of traffic patterns in both in TomTom's GPS probe vehicle data (Fig. 6(b)) and TomTom's GSM probe vehicle data (Fig. 6(c)), one can make the conclusion that each of the associated

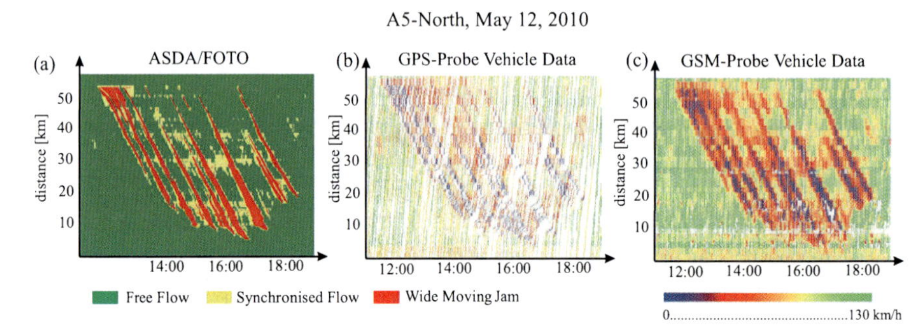

Fig. 6. Comparison of congested traffic pattern measured on May 12, 2010 with detector data processed with ASDA/FOTO models (a) with on-line TomTom's GPS (b) and GSM (c) probe vehicle data for this day. A5-North, Germany [12].

patterns for the related day and freeway coincides almost exactly with the associated pattern reconstructed with the ASDA/FOTO models that have used road detector mea-surements [12].

5 Conclusions

Three-phase traffic theory allows us to explain, reconstruct, and predict the spatiotemporal picture of traffic breakdown and resulting congested patterns on multi-lane roads almost as precisely as observed in real measured data. For this reason, methods and models based on three-phase traffic theory can be effectively used for the generation of danger and other messages that can improve vehicle safety and increase driving comfort. In particular, we have shown that a model of traffic phase identification along a vehicle trajectory based on three-phase theory allows us to make a reliable generation of jam warning messages already at the penetration rate 0.5% of probe vehicles randomly distributed in traffic flow (see for detail [7],[11],[12]). Penetration rates of 2% probe vehicles might be a boundary condition for a "premium" quality for traffic service messages.

TomTom's probe vehicle data allows us to reconstruct structure of traffic patterns with a much greater quality of spatiotemporal resolution than has been possible before. It occurs that congested traffic in measured traffic patterns consists of two traffic phases of Kerner's three-phase theory, synchronized flow and wide moving jams. Therefore, TomTom's probe vehicle data opens qualitative new perspectives for the creation of many new ITS applications like traffic jam warning, input data for traffic control, hybrid vehicle strategies, fuel consumption reduction, etc..

References

[1] B.S. Kerner, The Physics of Traffic, Springer, Berlin / New York, 2004.

[2] B.S. Kerner, Introduction to Modern Traffic Flow Theory and Control: The Long Road to Three-Phase Traffic Theory, Springer, Berlin / New York, 2009.

[3] A.D. May, Traffic Flow Fundamentals, Prentice-Hall, Inc., New Jersey, 1990.

[4] C.F. Daganzo, Fundamentals of Transportation and Traffic Operations, Elsevier Science, Oxford, 1997.

[5] G.B. Whitham, Linear and Nonlinear Waves, Wiley, New York, 1974.

[6] Highway Capacity Manual 2000, National Research Council, Transportation Research Board, Washington D.C., 2000.

[7] B.S. Kerner, H. Rehborn, J. Palmer, Klenov, S.L., Using probe vehicle data to generate jam warning messages Traffic Engineering & Control, Hemming, 3:141–148, 2011.

[8] B.S. Kerner, H. Rehborn, H. Kirschfink, Verfahren zur automatischen Verkehrsüberwachung mit Staudynamikanalyse, German patent DE19647127C2, 1996; US-patent US5861820, 1999; Kerner, B.S., Rehborn, H., Verfahren zur Verkehrszustandsüberwachung und Fahrzeugzuflußsteuerung in einem Straßenverkehrsnetz, German patent publication DE19835979A1, 1998; Kerner, B.S., Aleksic´, M., Denneler, U., Verfahren und Vorrichtung zur Verkehrszustandsüberwachung, German patent DE19944077 C2, 1999; Kerner, B.S., Verfahren zur Verkehrszustandsüberwachung für ein Verkehrsnetz mit effektiven Engstellen, German patent DE19944075 C2, 1999.

[9] B.S. Kerner, S.L. Klenov, Microscopic theory of spatial-temporal congested traffic patterns at highway bottlenecks, Phys. Rev. E 68 036130, 2003.

[10] B.S. Kerner, H. Rehborn, S.L. Klenov, M. Prinn, J. Palmer, Verfahren zur Verkehrszustandsbestimmung in einem Fahrzeug, German patent publication DE 102008003039, 2009.

[11] J. Palmer, H. Rehborn, B.S. Kerner, ASDA and FOTO Models based on Probe Vehicle Data, Traffic Engineering & Control, Hemming, 4:183–191, 2011.

[12] R.P. Schäfer, S. Lorkowski, N. Witte, J. Palmer, H. Rehborn, B.S. Kerner, A study of TomTom's probe vehicle data with three phase traffic theory, Traffic Engineering & Control, Hemming, 5:222-230, 2011.

Hubert Rehborn, Boris S. Kerner
Daimler AG
HPC: 050-G021
71059 Sindelfingen

Germany
hubert.rehborn@daimler.com
boris.kerner@daimler.com

Ralf-Peter Schäfer
TomTom
Content Production Unit Berlin
An den Treptowers 1
12435 Berlin
Germany
ralf-peter.schaefer@tomtom.com

Keywords: networked vehicles, nomadic devices, road network monitoring, probe vehicle data (FCD/FPD from GPS and GSM), Kerner's Three-Phase traffic theory

Realtime Roadboundary Detection for Urban Areas

S. Hegemann, S. Lueke, Continental
C. Nilles, Altran

Abstract

In the context of driver assistance systems lane detection systems are used for lane departure warning and lane centering functions. In some cases, especially in urban scenarios, no lane marks are available and for some circumstances lane marks are not sufficient to assist the driver. For novel urban functions it is necessary to know in which areas the car can operate. This problem requires detecting the free space as a traversable area for vehicles and has to take into account small objects such as curbstones. In this paper we present a real-time road boundary detection for urban areas to realize an easy lane recognition algorithm by using a series hardware stereo camera system. The algorithm is based on dense stereo information.

1 Introduction

In the last years more and more driver assistant systems are developed to provide the driver with increased comfort and safety e.g. Adaptive Cruise Control and Lane Recognition [1]. These systems support the driver by longitudinal control, warning and or Emergency Braking [2]. The research results in recent years show the potential of functions based on environment perception by radar, laser and camera sensors to prevent accidents [3]. European research projects such as HAVEit [4] demonstrate impressively the possibilities with current sensor systems, for example Highly Automated Driving in a construction site [5]. The systems were primarily developed for non-urban areas. The focus for new functions is changing and so it is necessary to find solutions to percept the environment in urban areas for braking and steering assistance functions.

The problem of navigation in urban scenarios requires detecting the free space as a traversable area for vehicles. Most of the existing algorithms are designed to detect large objects and have their limits in measuring details. Stereo object detection algorithms such as the 6D-Vison are presented to detect large static and moving objects as for example motorcycles, cars and trucks [6]. Free space detection algorithms like the Stixel World represent smaller generic objects (e.g. crash barriers etc.) and compute the free space in front of the vehicle [7].

In these works, smaller objects such as a curbstone are filtered out by minimum height constraints in order to protect the function against false positives, caused by noise or artifacts. In fact curbstones are one of the most important features to categorize the driving area in urban scenarios. Thus, it is essential to detect curbstones with a height range from 5 cm up to 30 cm to realize new functions. Other presented approaches for vehicles use Conditional Random Fields [8]. Elevation Maps [9] are used for navigation of mobile robots in natural scenarios (e.g. planetary exploration). For automotive functions it is necessary to realize a generic detection for urban road boundaries which can handle curbs and further objects. The algorithm has to detect construction sites, cars and other possible 3D-objects a vehicle can come across in urban areas or use them for orientation as well.

2 Basic Data

Stereo camera systems allow handling of generic 3D information. Their use as automotive sensors enables pedestrian protection and increases the performance of various active safety functions.

For recognition of curbs, the usage of two types of basic stereo camera data is useful: The gray or color image and dense stereo data. In a gray or color image the curb may be detected as an edge. But this information is not sufficient to definitely recognize a curb, because edges in 2D images can be the result of changes of color or brightness in road surface materials. This is the reason why it is essential to use the height information of dense stereo data as a characteristic property to detect curbs. In Fig. 1., dense stereo data transformed into a 3D point cloud is shown. The curb is clearly visible as a step in the raw data (see Fig. 1. blue line).

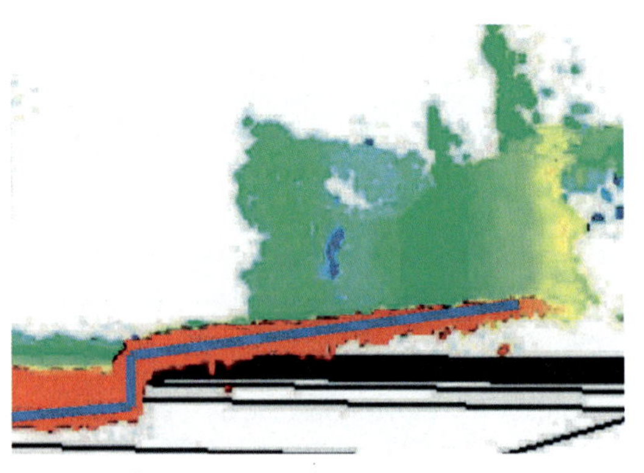

Fig. 1. Point cloud of dense stereo data

3 Recognition of Curbs

For the recognition of very small objects, the point cloud of dense stereo data is mapped into a vehicle static grid with height information (see Fig. 2.). This height map is based on the description of Elevation Maps by [9].

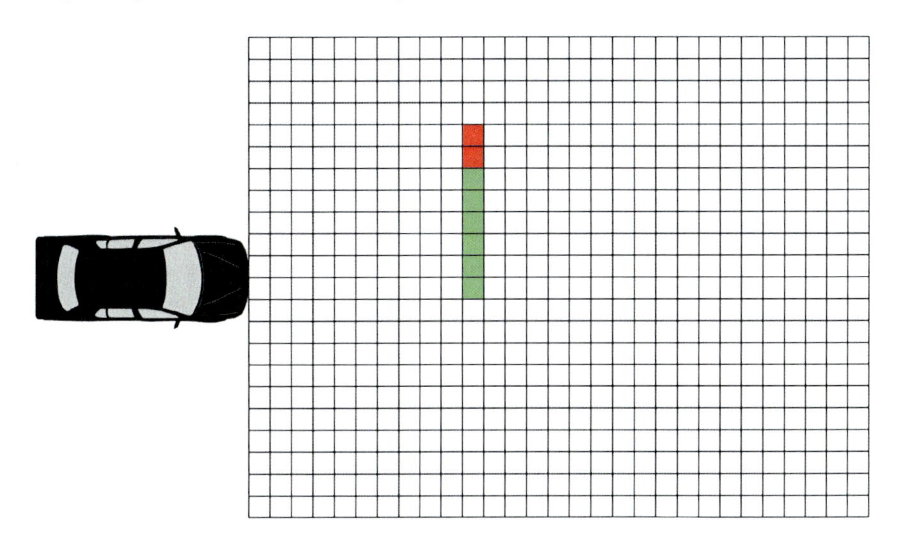

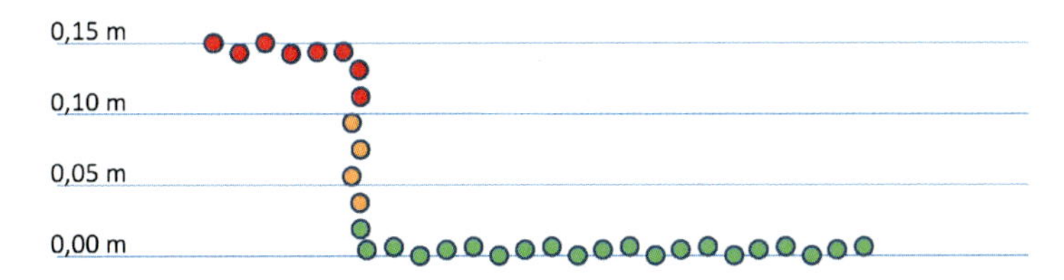

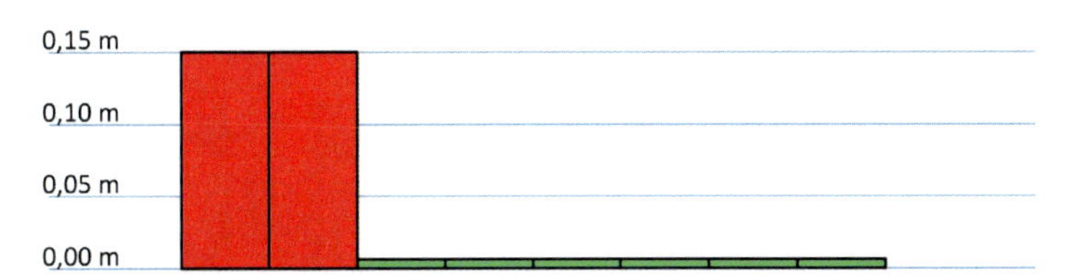

Fig. 2. Visualization of the height map (upper part), cutout: transformation 3D dense stereo information (middle part) in to a height map (lower part)

In order to recognize the curbs on the right side of the road, the height map is at first interpolated in driving direction. In the next step, the lines along the height map are scanned and possible positions of a curb are detected. To verify the curbstone position, the neighborhood of the possible position in the height map is analyzed too. The result of the curb detection is shown in Fig. 3.

Fig. 3. The left image shows an urban scenario, the right image shows the recognition of the curb (the blue points represent the cells of the height map, the detected curb is visualized in red)

Furthermore, based on the shape of curbstones a distinction is made between curbs and other objects of larger format. In Fig. 4., the detection of parking vehicles on the right side of the road is shown.

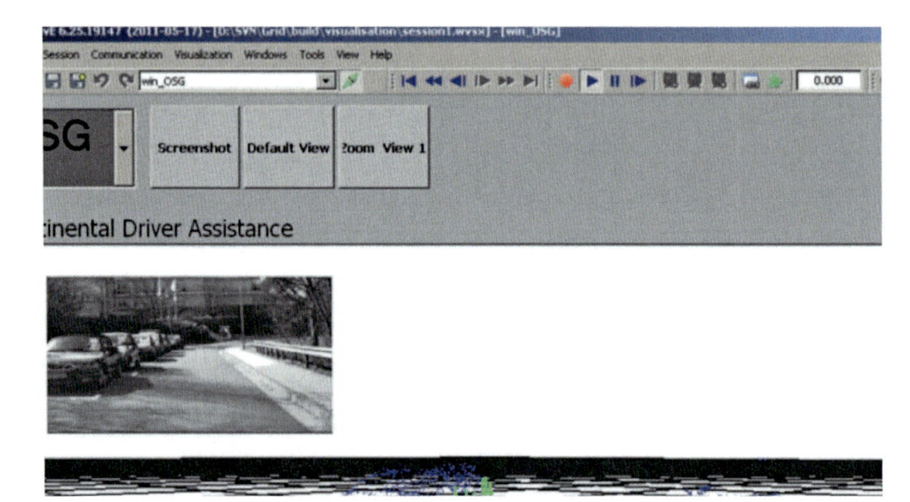

Fig. 4. Detection of parking cars on the right side of the road

4 Conclusions

It is possible to detect curbstones with a height map of up to 20 m. The next step is to increase the detection range by data fusion with an edge detection algorithm and change the algorithm to detect any possible 3D object. The computed information of the algorithm can be used for new functions to prevent tire damages (due to curbstone collision), warn the driver if the car is leaving the driving area, detect and estimate a boundary of the driving lane in urban areas without lane marking and realize lane centering functions in urban scenarios. The above mentioned applications can be implemented using data fusion with other sensors.

References

1] H. Winner, S. Hakuli, G. Wolf, „Handbuch der Fahrerassistenzsysteme", Vieweg + Teubner | GWV Fachverlage GmbH, Wiesbaden, 2009.

[2] S. Lueke, M. Komar, M. Strauss, "Reduced Stopping Distance by Radar-Vision Fusion", In: J. Valldorf, W. Gessner [Eds.], Advanced Microsystems for Automotive Applications 2007, Springer, Berlin, 2007.

[3] M. Vollrath, S. Briest, C. Schießl, J. Drewes, U. Becker, „Ableitung von Anforderungen an Fahrerassistenzsysteme aus Sicht der Verkehrssicherheit", Wirtschaftsverlag NW, Berichte der Bundesanstalt für Straßenwesen, Heft F 60, Bremerhaven, 2006.

[4] http://www.haveit-eu.org, Stand Februar 2012.

[5] S. Lueke, M. Strauss, „Exemplarische Darstellung einer hochautomatisierten Fahrfunktion im Projekt HAVEit", Beitrag zur Tagung „13. Braunschweiger Symposium AAET 2012, Braunschweig, 2012.

[6] U. Franke, C. Rabe, H. Badino, S. Gehrig, "6D-Vision: Fusion of Stereo and Motion for Robust Environment Perception", Springer, Proceedings of the 27th DAGM Symposium, 2005.

[7] H. Badino, U. Franke, D. Pfeiffer, "The Stixel World - A Compact Medium Level Representation of the 3D-World", 31st DAGM Symposium on Pattern Recognition, Frankfurt, 2009.

[8] J. Siegmund, U. Franke, W. Foerstner, „A Temporal Filter Approach for Detection and Reconstruction of Curbs and Road Surfaces based on Conditional Random Fields", 2011 IEEE Intelligent Vehicles Symposium (IV), Baden-Baden, 2011.

[9] W. Burgard, M. Hebert, "World Modeling" in Springer Handbook of Robotics, Verlag Berlin Heidelberg, Berlin, 2008.

Stefan Hegemann
A.D.C. Automotive Distance Control Systems GmbH
Peter Dornier Str. 10
88131 Lindau
Germany
stefan.hegemann@continental-corporation.com

Stefan Lueke
Continental Teves AG & Co. oHG
Guerickestr. 7
60488 Frankfurt / Main
Germany
stefan.lueke@continental-corporation.com

Claudia Nilles
Altran GmbH & Co. KG
Bernhard-Wicki-Str. 3
80636 München
Germany
claudia.nilles@altran.com

Keywords: driver assistance systems, dense stereo, curbstone recognition

Overall Probabilistic Framework for Modeling and Analysis of Intersection Situations

G. Weidl, G. Breuel, Daimler AG

Abstract

We propose a system design for preventive traffic safety in general intersection situations involving all present traffic participants (vehicles and vulnerable road users) in the context of their environment and traffic rules. It exploits the developed overall probabilistic framework for modeling and analysis of intersection situations under uncertainties in the scene, in measured data or in communicated information. It proposes OOBN modeling for the cognitive assessment of potential and real danger in intersection situations and presents schematically an algorithm for multistage cognitive situation assessment. A concept for the interaction between situation assessment and the proposed Proactive coaching Safety Assistance System (PaSAS) is outlined. The assessment of danger in a situation development serves as a filter for the output and intensity of HMI-signals for directing driver's attention to essentials.

1 Introduction

A number of research and development projects are focusing on assistance systems for preventive road safety. The goal is to alert just in time the driver in critical situations to avoid collisions completely or to reduce their severity. This is also the motivation behind the research initiative Ko-FAS (cooperative driver assistance systems) and one of its projects Ko-PER (cooperative sensor technologies and cooperative perception). The cooperative sensors are mounted both onboard of the vehicles and at critical points of the infrastructure. Their measurements are fused to provide the environment perception, the vehicle's local perception and self-localization. The exchange of information is taking place by C2X, i.e. C2C (car-to-car) and by C2I (car-to-infrastructure) communication. These data are fused by the cooperative perception to obtain a cooperatively validated set of the data, characterizing the observed situation. Finally, cognitive interpretation of an observed situation and its risk development provide an input for HMI preventive safety functions or to autonomous controllers for active interventions. Differentiation features of Ko-PER are the resolution of occlusions and the compensation of equipment heterogeneity. The basic assumption of heterogeneity incorporates, that only part of the traffic participants are able to percept their environment and to communicate (C2X, Ped2X). Thus, Ko-PER aims at socialization of perception through communication.

The Ko-PER focus is not just on one special type of maneuver for one special environment context. It aims to develop a generic framework, which covers all possible maneuvers in a certain context, dependent on the road type, e.g. intersection, highway, etc. Based on the GIDAS statistics of traffic accidents, intersections have the biggest accident record of 27% in total which requires a solution for the modeling and analysis of intersection situations. Therefore, the focus of our work is on preventive safety for intersections. Accidents on intersections are often caused by driver errors, like misinterpretation of a situation, inattention, disregard of traffic rules [1]. A main challenge here is the situation assessment. For this purpose, various methods have been studied and demonstrated for selected scenarios. The use of object-oriented causal probabilistic networks (a.k.a. object-oriented Bayesian networks OOBN) has been proposed in [2]. It has been used to deduce the probability of accident for one car in general circumstances without interaction with other road users. Bayesian Networks have been used, e.g. for lane identification and ego-vehicle accurate global positioning in intersections [3], as well as for estimation of driver intention based on topology and geometry information from the intersection map [4], [5]. Our approach for the recognition of driving maneuvers profits from the advantages of OOBNs when it comes to dealing with interrelated (or networking) road users. OOBNs offer a natural framework to handle vehicle-lane and/or vehicle-vehicle relations, which can be boosted further by exploiting the left/right symmetry of a lane-change-course [6]. Our work extends the context information with traffic rules and priority information, which is encoded in the lane attributes of the digital map as well as in the incorporated relations of type: lane-lane, lane-conflict area and road user-to-road user. The causal probabilistic treatment of situation features allows exploitation of heterogeneous sources of information and the quantitative incorporation of uncertainties in the measured signals and communicated information. This paper represents an ongoing work. The aim for the first phase has been the development of a suitable concept for situation interpretation and cognitive assessment of risk under uncertainties in intersection scenarios involving all traffic participants [7]. Moreover, a concept for multistage cognitive situation assessment to filter the HMI-output intensity for directing driver's attention to essentials will be proposed here.

2 Overall Probabilistic Framework for Situation Analysis

The situation interpretation and its risk evaluation involve the analysis of behavior patterns and recognition of maneuver intentions. This is necessary not only for the driver of the host vehicle, but also for all involved road users. Thus, an overall analysis of the intersection situation becomes a multi-objects challenge. Such overall analysis requires an overall framework for situation analysis. It consists of a number of building blocks utilizing suitable technology solutions: i) simulation of traffic scenarios for visualization and better understanding of the dynamic development of a situation; ii) generic modeling framework for risk estimation on potential hazard and real danger of a current situation; iii) derivation of context dependent information and warnings for the driver; iv) criticality evaluation of a situation as an input for an autonomous control action of the driver assistance system; v) acceptance study in a driving simulator and in real traffic. An overview concerning i) and iv) is provided in [8], [9] and i) has been extended for this work with an interface involving the integration of digital maps in open street map format (osm), and with new control algorithms to simulate a realistic physical motion of a vehicle at an intersection. The system concept is outlined in section 2.1. Some remarks on the source of localization and the use of a digital map are summarized in section 2.2. The technology solutions for ii) and iii) will be the topics of sections 3 – 4. The acceptance study in a driving simulator and in real traffic is planned to be performed with several Ko-PER networking partners to demonstrate the developed cooperative solutions.

2.1 Concept of Cooperative Situation Analysis

The situation analysis is the crucial link between sensors (environment perception) and assistance functions. It provides a fundamental basis for the development of future active safety and security systems for avoidance of accidents and/or for reduction of accident consequences. The elaboration of a concept for situation analysis requires solutions of a number of issues, which are reflected in the design of a Proactive Coaching Safety Assistance System, as shown in Fig.1. It includes several steps:

(1) Provision of input data for overall data fusion (providing evidence for the inference): a) Local vehicle perception and fusion; b) Overall vehicle fusion from both perceptions: local and from the intersection; c) Self-localization of vehicles; d) Communication of context and environment information, e.g. SPaT (phase-change of traffic lights), MAP (topology and topography of the intersection), DEN (i.e. sudden velocity change, sudden lateral displacement, pedestrian), cooperative perception messages of intersection and vehicle, posi-

tion of the objects on the intersection; e) Extended environmental model with digital map.

(2) Recognition of a situation for all percepted road users (e.g. maneuver intention, like left turn across path with potential hazard or uncritical American left-left turn, imminent collision with a pedestrian, left/right turn across bicycle path or across pedestrian walk).

(3) Concept for overall cognitive assessment including probabilistic prognosis on the development of a situation under actually observed conditions.

(4) Elaboration of an application independent evaluation of a situation (e.g. time measures for the criticality of an observed situation), which can be utilized in the cognitive assessment and by all driver assistance functions.

(5) Elaboration of an application specific evaluation, i.e. assistance concept for prioritization of control actions: HMI-information and warnings for directing the driver's attention to essentials; autonomous system's interventions.

The real challenge is in the prognosis of the intended behavior of all traffic participants; finding occlusions and plausibly informing the driver on invisible threat; as well as the proper treatment of uncertainties in measurements and/ or from communicated information. From here follows the need for developing of a probabilistic approach for cognitive assessment of hazards at intersections. And finally, provide an acceptable strategic advice on safety-awareness behavior of the driver at an intersection.

2.2 Provision of Data for Cognitive Assessment of a Situation

The localization and the use of digital map features are essential building blocks of the concept as noticed above.

Localization: Within the project Ko-PER five different methods for localization are investigated. These utilize different technologies, e.g. digital map, landmarks, Ko-TAG-transponder, tightly coupled GNSS/INS, cooperative GNSS (Global Navigation Satellite System). Finally, a fusion approach is supposed to combine the various localization methods, if necessary for better accuracy. The goal is to provide a lane precise localization of a vehicle. An accuracy of digital map and vehicle localization in the order of 0.3m has also been found as sufficient in [4].

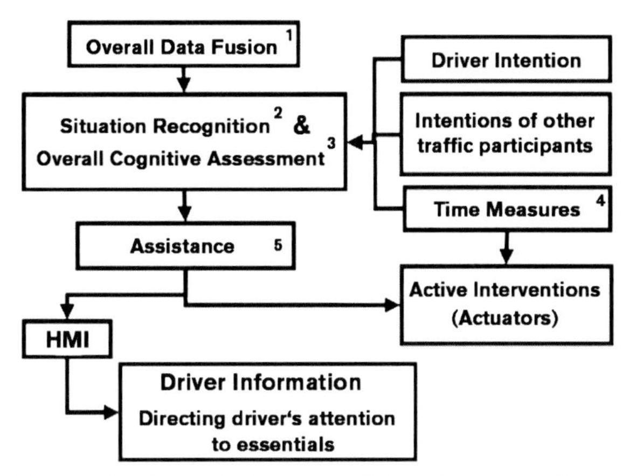

Fig. 1. Concept for Proactive Coaching Safety Assistance System (PaSAS)

Digital Map: It is installed on-board of the equipped vehicles and used twofold: for localization and as a source of context information for the situation analysis. Two intersections have been used to test the digital map in Ko-PER. The map accuracy has been ensured to be in the order of 0.1m, based on measurements of the intersection geometry, topology and reference road marks. With the development of automated geodetic measurement technology, intersection geometry measurements of similar accuracy are feasible and available in increasing coverage rate for intersections. The topography and topology are specified by the simTD-specific ways-tags [10]. An intersection maneuver-track represents a typical path through an intersection, see Fig. 2 (b). For modeling of maneuver-tracks, the following context information has been extracted from the digital map of the intersection: lane type (approach, egress); direction of lanes; vehicle lane attributes [10], extended with traffic signs and traffic rules. In addition, dependency relations have been encoded in the conflict areas of the intersection, which include relations of type motorized vehicle–vehicle, vehicle–bicycle, vehicle–pedestrian and relations between approach lane and the corresponding type of conflict areas. Moreover, "connectTo"-attributes represent relations between the approach–egress lanes, defining the possible intersection maneuver tracks starting from a certain approach lane and disappearing from the intersection at a number of possible egress lanes. A number of pair-relations' features for association of danger in a situation have been used in our approach. The relative motion of the object towards the conflict areas of the digital map is represented by the features: Occupancy time of the conflict areas of the intersection; Time-To-Enter/-Disappear into/from a conflict area; Distance-To-Enter/Disappear into/from a conflict area. The size of the cells of the conflict areas is dependent on the topology and topography of the intersection. Thus, the cells represent adaptive discretization of the conflict areas.

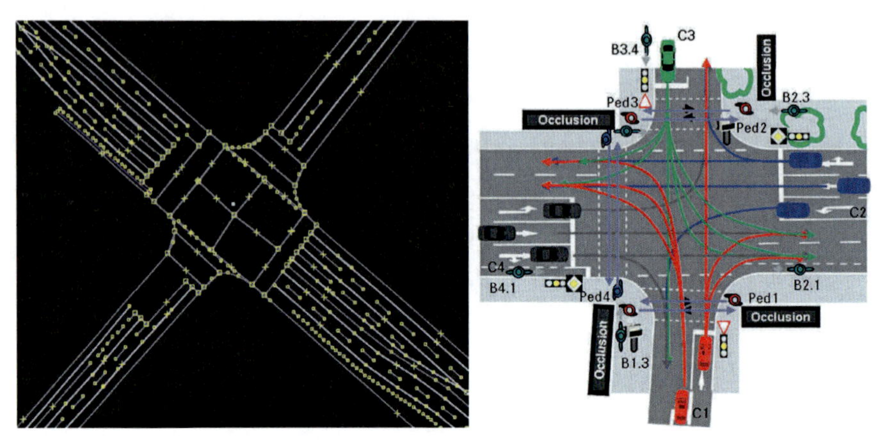

Fig. 2. (a) Digital map; (b) Generalized intersection map with possible conflict-tracks (B = bicyclist, Ped = pedestrian, C = car);

3 Situation Interpretation and Risk Evaluation

Handling of Uncertainties: The situation analysis for early recognition of traffic hazards, like other vehicle onboard applications, is challenged with the reliability of observations received by sensors, which are integrated in the vehicle. These sensors are not functioning always perfectly, dependent on the sensor measurement status, sensor visibility area, weather conditions and this is influencing the overall environment perception. Thus, there is a high level of uncertainty both in the measured data and due to complexity of the scene. This calls for probabilistic modeling, which is essential for the hazards recognition in traffic situations at an earlier stage of their development.

Hazard Assessment by Object-Oriented Bayesian Networks (OOBN): The proposed solution for recognition of objects' intentions and cognitive assessment of hazards is exploiting the features and advantages of object-oriented Bayesian networks (OOBNs) [2], [11], [12]. The advantages of Bayesian networks (BNs) involve the qualitative representation of knowledge through the BN structure of the graph, expressing the dependency relations between the probabilistic domain variables. The quantitative representation of knowledge is incorporated in the probabilities, expressing the strength of the dependency relations between variables.

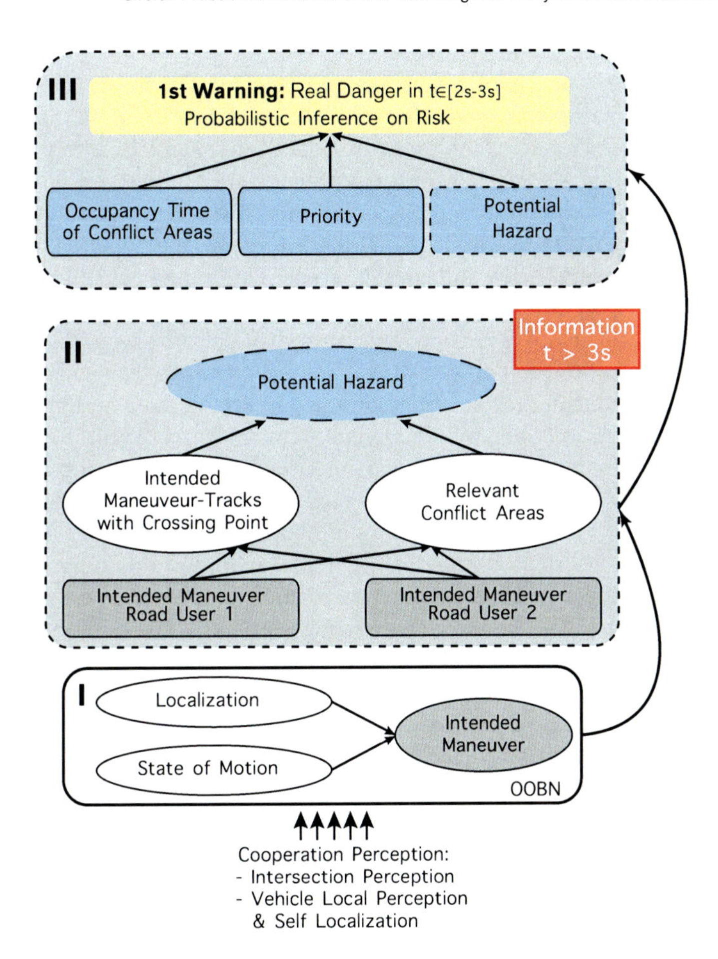

Fig.3. OOBN levels for recognition of maneuver intentions (I), potential (II) and real (III) danger between networking road users

Moreover, the OOBNs allow efficient data organization by creating model libraries for features of traffic situations. This supports the reuse of OOBN-Fragments in similar context. The advantages of BN and OOBN modeling lie in the model complexity reduction from exponential to linear growth; only relevant pairs are considered; not present objects do not contribute and thus reduce the model complexity; proper handling of uncertainties in measurements. This ensures a well motivated modeling background for situation analysis, where the static and dynamic objects provide input information for context analysis. It enables the development of an overall generic probabilistic framework for representation, modeling and assessment of hazards at intersection situations involving all traffic participants in the scene: vehicles, pedestrians and bicyclists. It utilizes OOBN, where the cooperative perception of vehicles and infrastructure is completed by the environment context information from the intersection digital maps, traffic rules and changing phase of traffic lights. Its advantage is in the possibility to combine both quantitative and qualitative relations between all involved road users (here called objects) in the scene and their environment. The qualitative representation of these objects is encoded in the hierarchical structure of the OOBN, which expresses the dependency relations between road users, lanes and conflict areas. The maneuver intention of each vehicle is derived from its lane precise localization, navigation route, traffic priority, dynamic state of motion, and control signals (brake, acceleration, etc). The intentions of unprotected road users are recognized by a separate module and are communicated by the intersection perception to all equipped vehicles. The pair wise combination of intended maneuvers serves as filter of potentially dangerous maneuvers at the corresponding conflict areas of the intersection. Thus, inference on the probability of potential conflicts in space is obtained. The features of the environment are defined by the topography and topology of the intersection, incorporated in its digital map. Conditioning on the environment allows filtering the conflict areas and the areas free for maneuver at certain time interval. The priority rules for a situation are given by the changing phase of traffic lights and the traffic signs. Combining the probability of space-time conflict together with the priority status of moving objects leads to the overall probability of hazards for an observed situation.

The developed probabilistic methodology for cognitive assessment of potential hazard and real danger in traffic situations between all road users at an intersection area consists of three levels of abstraction as shown in Fig. 3: (I) The maneuver intentions of all road users, approaching the intersection area, are cognitively assessed. Thus, intended maneuver-track-pairs with juncture point in space (i.e. potential hazard in same conflict area) are recognized. This is based on accurate localization of the road user at a certain lane or intersection area together with the corresponding state of movement. The last is deduced from movement features like position and orientation of a road user, as well as from traffic rules (light, signs) - both from the digital map and communicated, and optionally from the navigation route of a vehicle. The proposed maneuver recognition can also serve as basis for the criticality evaluation [9], providing an input for autonomous control actions. (II) At the second level of abstraction, the probability of potential hazards is inferred from a combination of maneuver intentions, i.e. probability of simultaneous juncture points of intended maneuver-tracks of the road users within the relevant conflict areas. (III) At the third abstraction level, the real risk is inferred from the cognitive assessment of the simultaneous events: potential hazard, occupation times of the conflict areas and traffic priority. Thus, it provides the inference on real space-time danger for the corresponding intersection area in the context of priority compliance. Finally, probabilistic combination of the real space-time danger with threat assessment provides the overall context-dependent inference on collision threat (Fig. 4). Both last steps prepare the system for driver information, warnings or active intervention.

4 Situation Assessment and Directing Driver's Attention

The dynamic development of a situation - from potential hazard through real danger, up to collision threat - implies four stages for directing the driver's attention: information and first warning (Fig. 3), 2nd and 3rd warning (Fig. 4).

Stage 1: provides **information** on potential hazard at time t>3s before entering (the sight line of) the intersection. This is feasible due to the fact, that on one hand, the lane related localization of a vehicle is sufficiently precise (± 0.3 m) up to 85 meters before entering the intersection. On the other hand, vehicle driving with a speed of 30 - 50 km/h passes in t > 3 sec. about 25 – 42 meters before entering the intersection. This length of turn-lane is often even longer dependent on intersection design criteria (traffic density). If just one lane is available, the probability of the three possible (left/straight/right) maneuvers, conditioned on position and orientation, is deduced from the OOBN.

Stage 2: The **first warning** on real collision danger is provided in t∈[2s – 3s] before entering the intersection. Here, the expected maneuver-pairs are colliding simultaneously in space and time with high probability. Further combination of the deduced (at stage 2) real danger with the probabilistic evaluation of kinematic motion of all traffic participants results in the threat assessment at stage 3. It evaluates the causal dependency on the probabilistic kinematic behavior of a vehicle in the presence of other road users and their kinematic behavior. The probabilistic inference is conditioned on the context of the intersection conflict areas and traffic rules.

Stage 3: The available room for maneuver is extracted from the driver profile and control values, taking into account a bunch of drivable and controllable trajectories for the 2 sec. prognostic horizon (see [13]). This algorithm is initiated with the first warning on real danger. In case, there is still room to pass an object without extreme maneuver – there are two options: the driver prefers to take self control actions or allows direct intervention of PaSAS. Thus, the

second warning on real collision threat appears in t ∈ (1s – 2s] before entering the intersection.

Stage 4: The **third warning** with autonomous intervention is executed by PaSAS in two cases: i) the driver is not taking (correct) control actions, which really reduce the threat or ii) there is room to pass an object only with an extreme maneuver (e.g. brake, steer, accelerate).

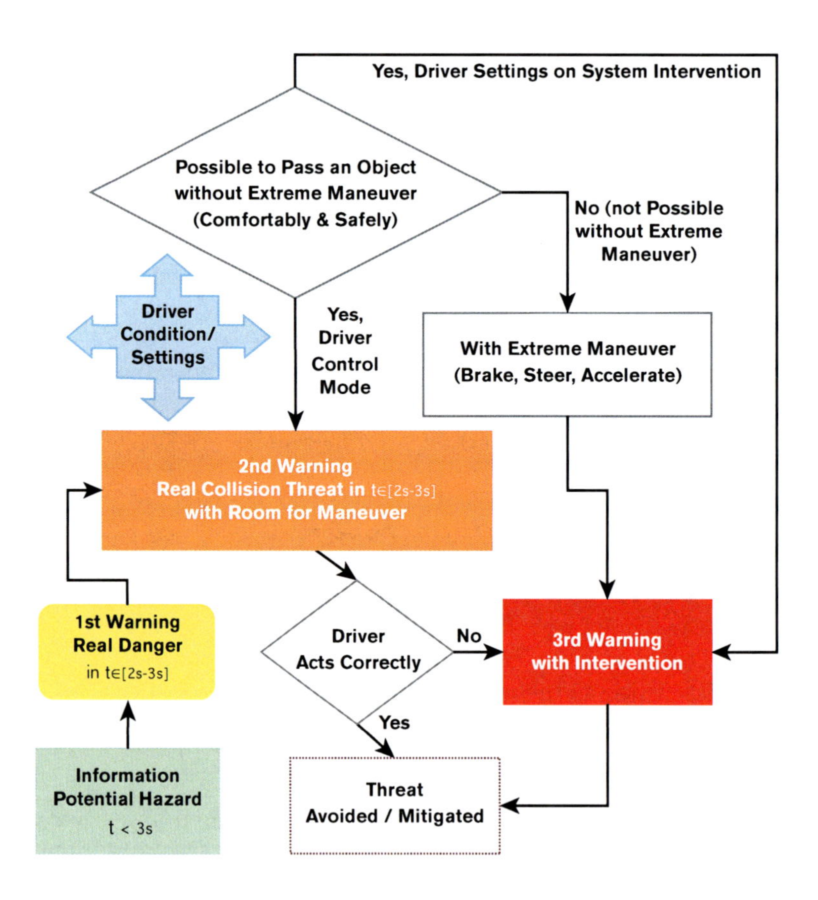

Fig. 4. Algorithm for multistage Proactive Coaching Safety Assistance System (PaSAS)

5 Conclusions

After the measurement campaigns have collected data representative of the most common intersection scenarios, the evaluation of the cooperative perception will take place. Then, the developed OOBN models will be parameterized and their performance evaluated to account on the quality of recognition of maneuver intentions, situation interpretation and on the overall assessment of risk for traffic situations at intersections. The aim is not to limit the system to solutions for independently modeled single problematic situations, but to use a generic modeling framework, which incorporates changes in all possible states of dynamic objects (traffic participants and traffic lights) in the context of traffic rules, intersection topology and topography. A generic modeling framework allows utilization of the same signal pre-processing module for the same sensor type and treatment of uncertainties in the same framework, with same accuracy. This avoids later on the need of conflict management between "island-solutions" of independently modeled situations in similar context.

The advantages of the proposed framework incorporate the proper handling of uncertainties in measurements and situation complexity, efficient combination of various (existing and under development) methods for situation analysis as input to achieve an overall hazard assessment. It is flexible to further extensions of the OOBN-approach to cover more traffic situations, both at intersections (such as obstacles on the entrance- and/or disappear-lanes) and on highway scenarios [6]. Thus, PaSAS contributes in general to the road safety, as well as to the project goal on alerting networking road users in critical hazard situations just in time.

References

[1] www.ko-fas.de, Ko-FAS Initiative, Beschreibung des Verbundvorhabens Ko-PER (Kooperative Perzeption), Fahrerassistenz und präventive Sicherheit mittels kooperativer Perzeption, 2009.

[2] D. Koller, A. Pfeffer, Object-Oriented Bayesian Networks, In Proceedings of the 13th Annual Conference on Uncertainty in Artificial Intelligence (UAI-97), pp. 302-313, 1997.

[3] V. Popescu, M. Bace, S. Nedevschi, Lane Identification and Ego-Vehicle Accurate Global Positioning in Intersections, 2011 IEEE Intelligent Vehicles Symposium (IV), pp. 870 - 875, Baden-Baden, Germany, 2011.

[4] PReVENT project - INTERSAFE subproject, "Final report (d40.75)," February 2007.

[5] S. Lefèvre, C. Laugier, J. Ibañez-Guzmán, Exploiting Map Information for Driver Intention Estimation at Road Intersections, 2011 IEEE Intelligent Vehicles Symposium (IV), pp. 583 – 588, Baden-Baden, Germany, 2011.

[6] D. Kasper, G. Weidl, T. Dang, G. Breuel, A. Tamke, W. Rosenstiel, Recognition of Driving Maneuvers by Object Oriented Bayesian Networks, 2011 IEEE Intelligent Vehicles Symposium (IV), pp. 673 – 678, Baden-Baden, Germany, 2011.

[7] Ko-FAS project – Ko-PER subproject - Midterm presentation, 2011.

[8] A. Tamke, T. Dang, G. Breuel, Integrierte Simulations- und Entwicklungsumgebung für die Ausweichassistenz zum Fußgängerschutz, In proceedings of „Automatisierungs-, Assistenzsysteme und eingebettete Systeme für Transportmittel", pp. 221- 226, 2010.

[9] A. Tamke, T. Dang, G. Breuel, A flexible method for criticality assessment in driver assistance systems, IEEE Intelligent Vehicles Symposium (IV), pp. 697-702, Baden-Baden, Germany, 2011.

[10] simTD Project (Safe Intelligent Mobility - Test Field Germany): www.simtd.de

[11] J. Pearl, Probabilistic reasoning in intelligent systems: networks of plausible inference. Morgan Kaufmann, 1988.

[12] F.V. Jensen, Bayesian Networks and Decision Graphs, Springer-Verlag, 2001.

[13] E. Käfer, G. Weidl, G. Breuel, Verfahren zur Ermittlung einer Gefahrenwahrscheinlichkeit einer Situation zwischen zwei Fahrzeugen in einem Kreuzungsbereich, Deutsches Patentschrift DE 10 2011 106 176 A1

Galia Weidl, Gabi Breuel
Daimler AG
Group Research & Advanced Engineering
Situation Analysis, Driver Assistance and Safety Systems
Dept. GR/PAA; HPC - Abt: 059 - G023 - 050
71059 Sindelfingen
Germany
galia.weidl@daimler.com
gabi.breuel@daimler.com

Keywords: situation analysis, context, risk, preventive safety, maneuver intention, threat assessment

Lane-Sensitive Positioning and Navigation for Innovative ITS Services

E. Schoitsch, E. Althammer, R. Kloibhofer, R. Spielhofer, M. Reinthaler,
P. Nitsche, Austrian Institute of Technology
S. Jung, S. Fuchs, Brimatech GmbH
H. Stratil, EFKON AG

Abstract

The goals of the project NAV-CAR are both to enable lane sensitive navigation for cars on highways and to increase robustness for high precision positioning in specific environments such as urban canyons and alpine regions where satellite based navigation systems may fail. The challenge of the project is the technical realization with a reasonable ratio of accuracy vs. costs, which is met by using sensor fusion technologies and stepwise integration of car specific data with GPS, which was implemented via an on-board unit (OBU) with CAN-bus interface. The approach was validated by test drives on urban (Vienna, A 23) as well as alpine highways (Brenner, A11/A12). Precise lane-specific trajectory reference data were derived from test drives with a special surveillance truck RoadStar. To estimate the potential impact of Galileo services as compared to existing GPS a simulation with data input from test drives in an alpine region was performed. The generation and inclusion of enhanced maps as a further option was evaluated.

1 Introduction and Overview

Based on the experiences gained in the EC FP6 project COOPERS [1] (CO-OPerative systEms for intelligent Road Safety, (www.coopers-ip.eu), co-funded by the EC), a national complementary project NAV-CAR 2, funded by the Austrian Ministry for Transport, Innovation and Technology (BMVIT) aimed at enabling lane-sensitive positioning and navigation for cars on motorways and to increase robustness for high precision positioning in specific environments, anticipating the fact that current satellite-based Car Navigation Systems often do not fulfill the high requirements for continuous, reliable and accurate satellite navigation in specific environments such as urban canyons, woodlands and mountainous regions. In order to improve Car Navigation Systems and to support new navigation-dependent applications improving safety, availability and traffic efficiency on the road that rely on a high-precision (lane-sensitive) continuous service at reasonable cost, these positioning-gaps must be filled by additional data (e.g. via co-operative infrastructure to vehicle communication, in-vehicle data and sensors, precise maps).

A robust navigational system that uses a combination of information sources and algorithms to achieve highly reliable and highly precise vehicle positioning was developed in a prototype manner. Additional sensors were integrated in an OBU (on-board unit), and vehicle-specific data (standardized and proprietary ones) were combined with the sensor data and GPS information. Requirements for the NAV-CAR 2 OBU were derived from requirements of potential services which make use of lane-specific information as well. They had been elaborated in expert workshops with highway operators (ASFiNAG, Autostrada del Brennero), a partner manufacturing devices (EFKON), drivers, emergency services and road maintenance, winter services etc., organized by Brimatech. The scenarios will be explained. The OBU sensor platform and the different levels of data to be integrated and their usefulness to contribute significantly to the overall goals will be described.

The results from test drives in two selected regions with difficult topological conditions, an urban highway in Vienna (A23, "Tangente") and an Alpine environment (A11/A12, Brenner Autobahn, Tyrol) will be presented and discussed. Additionally, a simulation of GALILEO was performed by the equipment provider pwpSystems from Germany to get an idea of the potential advantages of the different GALILEO services. The Mobility Department of AIT (ÖFPZ) was involved in providing precise calibration data with a high performance extensively equipped survey vehicle (RoadStar) and in tackling the question of generating and using precise maps in addition. Results will be assessed and used for recommendations for new in-vehicle Navigation and Positioning Systems as well as for upcoming new services to be expected with the availability of Galileo signals.

2 Services Requiring Precise Positioning & Lane-Sensitive Navigation

Initially, an assessment was done to identify which services would require precise positioning and lane-sensitive navigation. Interesting services for road operators are Road Surface Monitoring, Generating and Updating of Maps, Traffic Light Control / Regulation, Distance Measurement between two cars, Traffic Flow Management and Lane Specific Advice (lane specific speed advice, opening/closing of hard shoulders, allocation of grit or snow ploughs).

Interesting services for emergency services are exact accident localisation (transversal and longitudinal positioning accuracy is crucial, especially in case of streets that run close from each other but require different access routes). In Germany, the position of the accident is reported via TMC by the police only at the moment of their arrival at the accident site. The information has an inaccurateness of up to 15 km, even though road operators have exact maps. This inaccurateness is a result from the incompatibility of the systems of police and road operator. Also the exact localisation of the caller in case of an emergency is very important. At the moment, often the caller cannot be located. Other services interesting for Infrastructure Operators, Tolling System Providers, Emergency Services, and ITS experts have been discussed in the expert workshop and problem-centred interviews, but will not be detailed here.

3 Derived Technical Specifications

Four of the services identified in the previous chapter have been selected by the project consortium regarding technical and economical practicability and usefulness for the project:

- Generating and Updating of Maps
- Road Charging to Influence Demand (scenario: motorway interchange)
- Lane Specific Advice (such as lane specific speed advice, opening/closing of hard shoulders, etc)
- Accident Localization

These services were used as a starting point for the technical specifications.

- **Transversal position accuracy: max. +/-1 m**: The width of a lane on an Austrian motorway is 3.50 or 3.75 m (taken from [10]). If precise road maps are available, the used lane should be detected with the accuracy described above (see Fig. 1).

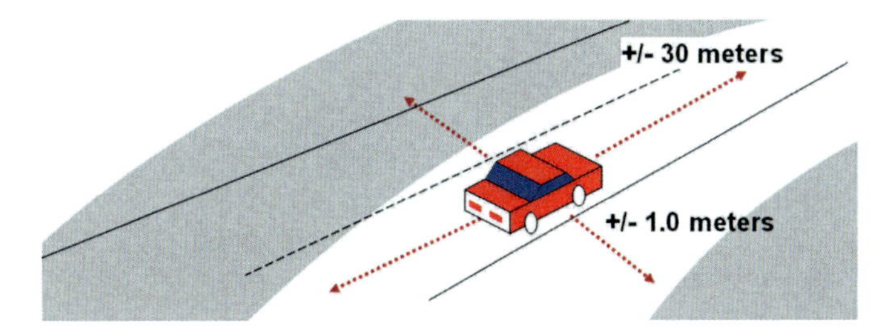

Fig. 1. Requirements for lane specific vehicle localization

- **Longitudinal position accuracy: max. +/-30 m**: This accuracy of the longitudinal position is sufficient for road charging applications and for critical sections such as on-ramp, exit or intersections (see Fig. 1).

- **Vertical position accuracy: +/-3 m**: In urban regions sometimes street sections are running in parallel or crossing in a very acute angle on different vertical positions (bridges). For a road charging application it is important to distinguish between a toll road from other non-toll roads which are situated below or above.

- **Update rate of position information: 0.8 s** - The system delivers the position coordinates in a fixed time interval. A vehicle driving a typical closely curve on an urban motorway with a speed of 80 km/h should result in a transversal positioning deviation less than 1 m in this time interval.

- **Time stamp of each position data based on the internal clock of the system:** Each coordinate data element should have a time stamp from an internal clock of the system, which is synchronised with the GPS-time. The internal clock is necessary because in the case of missing data from the GPS-receiver the internal NAV-CAR clock is used. If GPS-time is available, the internal clock will be synchronised to the GPS-time.

- **Data display:** A data display with available real time data should be implemented. The goal is to provide a direct feedback for the driver in the test scenarios while he or she is driving with the car. This feedback should only contain parameters which are directly given by the sensor or which are easy to calculate in real time and should also provide information if a sensor does not work correctly. A detailed analysis of the data is only done offline.

4 Technical OBU Implementation

The OBU consists of the following components in a low to medium price range: On-board system using AIT MECU-Board (see Fig. 2), Implementation of CAN-Interface, tested with Vector CANalyzer, Higher quality GPS system than used before in COOPERS, Inertial Sensor (6 degrees of freedom, high resolution Gyroscope fulfilling automotive robustness and temperature specifications, including temperature sensors for all axis for compensation calculations) and an Altimeter Sensor.

From the functional point of view, the implementation provides two additional features required for the test drives (data acquisition and evaluation): (a) Time synchronization with time stamp 1 ms for CAN-bus data, Inertial Sensor, Altimeter and GPS-PPS (Pulse per Second) signal, and (b) Communication with in-car PC (portable notebook), which is performing the high-level tasks (under preparation: data fusion, algorithms using and extrapolating data in case of loss of inputs, plausibility checks in case of errors of one data source etc., user interface, important results will be real-time for the driver (except Galileo simulation), validation/comparison with RoadSTAR geo-reference data).

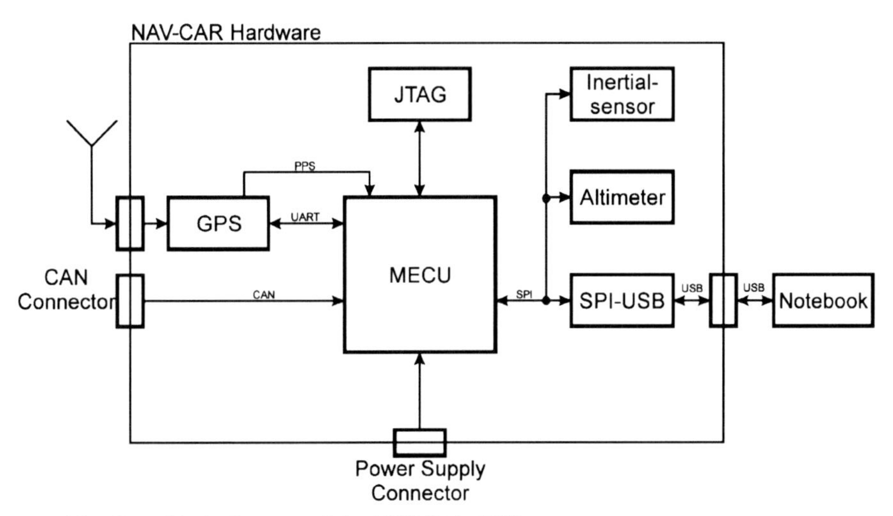

Fig. 2. Block diagram of the NAV-CAR OBU

5 Test Tracks and Drives for Urban and Alpine Scenario

For the urban motorway scenario, the A23 motorway "Südosttangente" between junction "Kaisermühlen" and junction "Inzersdorf" was chosen as test section. The A23 motorway is one of the highly frequented motorways in Europe (up to 180.000 vehicles per day) and consists of up to 4 lanes per carriageway. Short tunnels and noise barriers with heights up to 8 m provide a demanding environment with respect to satellite coverage.

A clover-leaf interchange between the motorways A23 and A3 ("Knoten Prater") is well suited for the investigation of three dimensional positioning and determination of height accuracy. Both carriageways partly run on separate bridge constructions, which lead to different absolute height of the carriageways.

The scenario "mountainous region" is situated on the A12 motorway "Inntalautobahn" and A11 "Brennerautobahn" in Tyrol. The evaluation sections form a triangle between the exits "Innsbruck Ost" and "Innsruck West" on A12 "Innsbruck Süd" on A11. The availability of a terrain model (as already used in the EC funded project "COOPERS") allows the simulation of Galileo satellite availability and the comparison of GPS and Galileo positioning accuracy. Due to the mountainous environment, GPS satellite reception is also degraded; especially the mountains on the south side of the motorway block the signals of a significant number of satellites.

6 Some Details on Selected Results

6.1 Completion of the GPS Trajectory Using CAN Data

In order to complete the trajectory obtained from GPS data which is interrupted in areas where there is insufficient reception of satellite signals (e.g. in tunnels), CAN data as well as altimeter data are used. The GPS position data (from latitude and longitude values) is obviously not correct in tunnels (see Fig. 3).

We used CAN speed data and steering angle data in order to complete the GPS position data. In order to better integrate the CAN data, GPS speed and heading data is used to obtain the position data. This is useful in order to avoid the conversion between longitude and latitude and meters and is possible because the precision of the GPS position data is sufficient enough.

6.2 Detection of Lane Change

We used the CAN data (speed and steering angle) as well as GPS data (speed and heading) to detect a lane change. In Fig. 4 the calculated trajectories based on CAN and GPS data as well as the start and destination lanes (distance 3.5 m) are depicted. Both CAN and GPS yield optimal values and are nearly identical. The z gyro value of the IMU can also be used to detect a lane change in a qualitative manner. The integrated value can be used to detect whether the car is in a parallel position before and after the lane change.

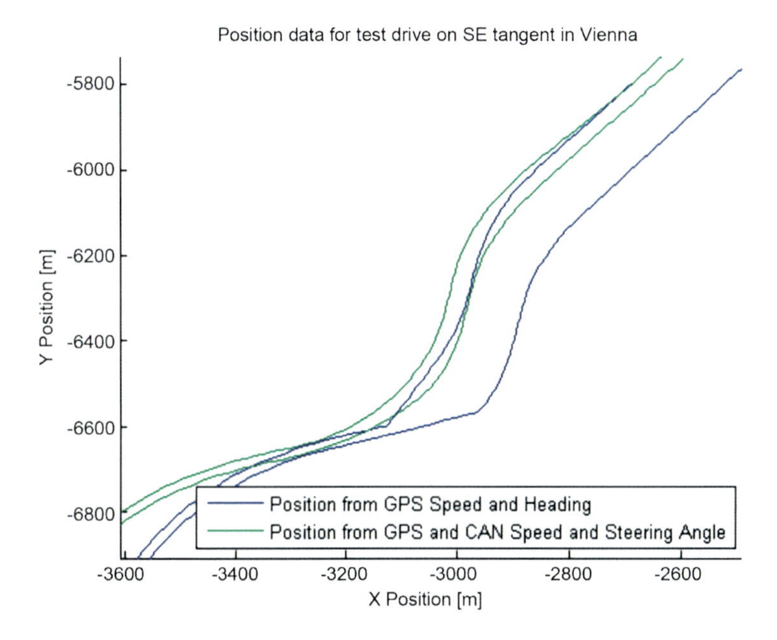

Fig. 3. Position from GPS and CAN data in tunnels and outside

Fig. 3 shows the position from GPS only (blue line) and GPS and CAN (green line) for both directions. Before tunnel 1, the blue and green lines are identical, in tunnel 1 there is a small difference (the tunnel is rather straight). For tunnel 2 the additional difference is much higher because in the tunnel there is a bend to the right. The green line (GPS plus CAN Data) is obviously correct because it corresponds to the original GPS position data.

6.3 Galileo Simulation

Galileo will offer different services for different user needs. "Open service" (OS) and "Commercial service" (CS) are two of them. Open service will be provided on two frequencies (L1B, E5a) and commercial service will be provided by three frequencies (L1B, E5b, E6b). While open service is more or less equivalent to GPS from a user point of view, commercial service will provide higher accuracy with encrypted signals, but will be liable to pay costs. There will be an availability guarantee for commercial service. These two services were investigated in NAV-CAR.

Due to the fact of lacking satellites, a real Galileo measurement is not possible today. So a simulation was chosen, based on measured data and a model of the region (mountains, satellite positions).

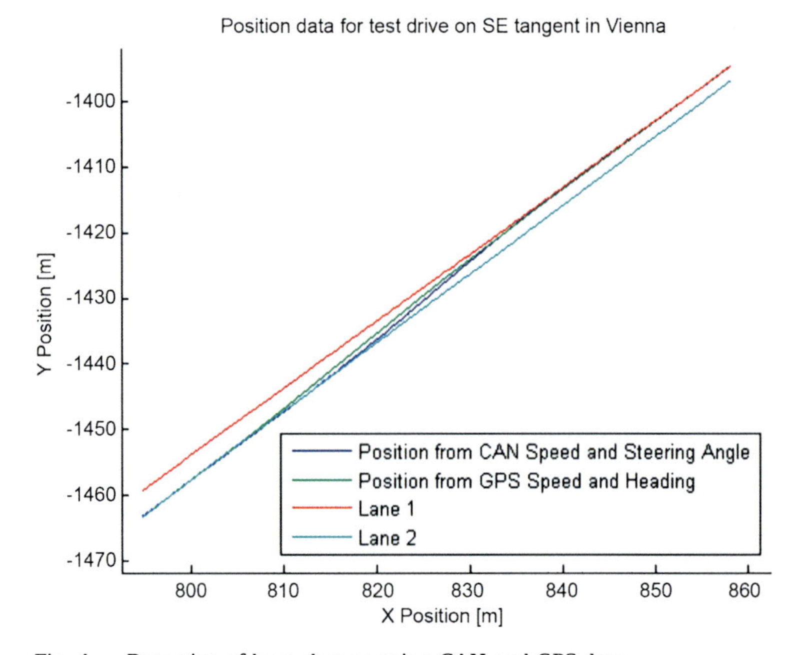

Fig. 4. Detection of lane change using CAN and GPS data

During the test trials in the Alpine region, raw observations of GPS satellites were recorded as binary streams as well as satellite ephemeris data and ionosphere correction values. Together with the reference trajectory of RoadStar and the recorded wheel speeds, the Galileo simulation was run. For the calculation using open service, an L1 receiver model was used. For the commercial service a L1 and E6 receiver model was used.

An assessment of the simulation results delivered the following:

For the lateral error, Galileo commercial service shows a distinct improvement of accuracy compared to GPS. The Galileo open service shows the same performance as GPS, which is accordance to the expectations. The primary benefit of Galileo open service is generally seen in the higher number of available satellites (27 in comparison to 24 GPS satellites) and thus better coverage in demanding environments.

Where GPS and Galileo open service have about 50 % of the points with an error better than 2 m, Galileo provides almost 75 % of the points in this error range. Regarding the requirements in NAV-CAR, the assessment shows the following results:

Lateral accuracy: Here, the requirement was defined with ± 1 m. The simulation of the open service shows that 25.4 % of all points have a smaller deviation than 1 m. Regarding the commercial service, 44.5 % perform better than the required 1 m. In contrast, GPS achieves 28.2 % of the points with a smaller deviation than 1 m.

Vertical accuracy: Here, the requirement was defined with ± 3 m. The simulation of the open service shows that 23 % of all points have a smaller deviation than 3 m. Regarding the commercial service, 39.3 % perform better than the required 1 m. In contrast, GPS achieves 56.2 % of the points with a smaller deviation than 3 m.

7 Conclusions

A short overview on the results achieved and proven by the demonstrations (test drives) reveals the following:

Results of the demonstrations at the urban highway:

- Result 1: CAN data (speed, steering wheel angle) can be used to complete GPS data (e.g. in tunnels) → Continuous trajectory is guaranteed.
- Result 2: The longitudinal accuracy of GPS can be measured using well defined points where exact GPS values are available and which can be easily detected (e.g. expansion joints) → requirements with respect to longitudinal accuracy of GPS are easily achievable
- Result 3: The position of the car on or under the bridge can be measured using GPS or altimeter → accuracy of height precise enough (3 m) for mapping on street maps.
- Result 4: The lane change can be detected both using CAN (speed, steering wheel angle) and GPS data. It can also be detected in a qualitative manner using gyro data of the IMU → Lane specific navigation possible in combination with street maps

Most important for OBU manufacturers is the fact, that using only vehicle independent data (CAN data speed, steering wheel angle, altimeter and IMU) is considerably improving OBU performance and positioning. Further vehicle dependent CAN data was examined (e.g. wheel speed) but did not result in any further improvement.

Results of the demonstrations in the Alpine environment:

- Result 1: In contrast to currently available GPS signals, the simulated Galileo commercial service positions provide promising results for the automated generation of enhanced maps with lane accuracy.
- Result 2: Regarding height information, the Galileo simulation shows varying results, which is partially due to inaccuracies of the simulation parameters.

In the final NAV-CAR 2 validation workshop (June 8th, 2011) the following issues were identified to be of crucial relevance to practice (1) Stability of data, (2) Real-time information about the degree of reliability, (3) Costs of OBU, (4) Problems related to the IMU sensor (difficult calibration for each individual IMU). Need for further research was identified with regards to IMUs (Elaboration of quality of low-price segment, (faster) self-calibration, interoperability of the software (mid-level devices in particular)).

Acknowledgments

The research leading to these results has received funding from the Austrian National Funding Authority FFG on behalf of the Austrian Federal Ministry of Transport, Innovation and Technology (BMVIT) under grant agreement 819743 of the 6th Call of the asap-Programme (NAV-CAR 2).

References

[1] COOPERS, CO-OPerative systEms for intelligent Road Safety, FP 6 project co-funded by the European Commission - http://www.coopers-ip.eu/

[2] M. Böhm, T. Schneider, Requirements on Vehicle Positioning and Map referencing for Co-operative Systems on motorways, E-Navigation, 2007.

[3] E. Althammer, R. Kloibhofer, R. Spielhofer, E. Schoitsch, Intelligent Transport Systems on the Road: Lane-Sensitive Navigation with NAV-CAR, ERCIM News, No. 84, p. 46-47, 2011.

[4] S. Fuchs, S. Jung, Spurgenaue Positionierung für innovative Anwendungen im Bereich der intelligenten Transportsysteme (ITS), The Navigation Flashlight, 01/2011, p. 3-7, Germany, 2011.

[5] E. Schoitsch, R. Kloibhofer, E. Althammer, R. Spielhofer, M. Reinthaler, S. Jung, S. Fuchs, H. Stratil, „NAV-CAR 2", Final Report, V 3.0, Dec. 2011 (public project deliverable, in English, http://www.nav-car.at/de/documents), 2011.

Erwin Schoitsch, Egbert Althammer, Reinhard Kloibhofer
AIT Austrian Institute of Technology GmbH
Safety & Security Department
Donau-City-Str. 1
1220 Vienna
Austria
erwin.schoitsch@ait.ac.at
egbert.althammer@ait.ac.at
reinhard.kloibhofer@ait.ac.at

Roland Spielhofer, Martin Reinthaler, Philippe Nitsche
AIT Austrian Institute of Technology
Österreichisches Forschungs- und Prüfzentrum GmbH
Giefinggasse 2
1210 Vienna
Austria
roland.spielhofer@ait.ac.at
martin.reinthaler@ait.ac.at
philippe.Nitsche@ait.ac.at

Sabine Jung, Susanne Fuchs
Brimatech Services GmbH
Lothringerstr. 14/3
1030 Wien
Austria
sabine.jung@brimatech.at
susanne.fuchs@brimatech.at

Hannes Stratil
EFKON AG
Dietrich-Keller Gasse 20
8074 Raaba
Austria
hannes.stratil@efkon.com

Keywords: intelligent transportation systems, navigation, positioning, on-board unit, Galileo simulation, automotive data fusion, location based services, lane-sensitive services

Vehicle Re-Identification With Several Magnetic Sensors

A.-C. Pitton, A. Vassilev, CEA-Leti
S. Charbonnier, Université Joseph Fourier - Université Stendhal

Abstract

Vehicle re-identification gives access to two essential data for traffic management: travel times and origin-destination matrices. This paper aims to evaluate the performances of a vehicle re-identification method based on Euclidean distances between vehicle "magnetic signatures" measured with several three-axis magnetic sensors. These performances are compared to the ones achieved by a single sensor and the three-dimensional (3D) Dynamic Time Warping algorithm. Moreover, the effects of a change in the vehicle orientation or in its lateral position are studied. On a data base of signatures from 25 different vehicles, the best results are obtained with the 3D Euclidean distance: 100% of the pairs of signatures are correctly re-identified without any false alarm if the vehicle keeps the same orientation and 90% if the vehicle orientation changes. The influence of the vehicle orientation to the magnetic North on the signature is therefore limited. However, the performances fall when the vehicle lateral position shifts.

1 Introduction

In 2009, the European Union made the commitment to reduce its greenhouse gas emissions by 20% by 2020 in relation to 1990 levels. In 2010, the United Nations General Assembly adopted its last resolution on improving global road safety. Meanwhile, the current increase in transportation demand leads to more and more traffic congestions, which implies more CO_2 emissions and a growing risk of road accidents. Given the limited or too expensive possibilities to extend the road network (especially in urban areas), reducing traffic congestions requires an optimization of the traffic flow. In recent years, the development of dynamic traffic management caused a need for collecting real-time traffic data and therefore modelling traffic flow on-line [1].

Vehicle re-identification consists in re-identifying a given vehicle at different locations. Re-identifying vehicles at the beginning and at the end of a route gives easy access to an accurate estimation of the Travel Time (TT), which is very useful for informing road users on traffic and for updating TT prediction models [2]. Moreover, re-identifying vehicles at an intersection allows calculating the Origin-Destination Matrix (ODM) which measures the number of vehicles coming from the street i that drove in the street j over a given period of time. The ODM is also useful for modelling traffic flow. Currently, the most common re-identification method is based on the matching of vehicles license plates that are read by cameras with automatic license plate recognition [3]. Vehicles equipped with radio-frequency identification tags (RFID tags) used for electronic toll collection can also be re-identified by their RFID tags [4]. Recently, experimentations of TT estimation calculated from the follow-up of vehicles equipped with Global Positioning Systems (GPS) and cell phones were carried out [5]. However, all these methods raise privacy issues, and the use of RFID tags requires the preliminary equipment of a sufficient percentage of vehicles for the estimation to be accurate, especially for ODM estimation, which needs a re-identification rate of almost 100%.

A rather inexpensive re-identification solution respecting anonymity is the use of magnetic sensors embedded in the road [6]. Sensors measure the vehicle's "magnetic signature" - a detailed temporal signal - when the vehicle passes over them. These signatures are generated by a disturbance of the Earth's magnetic field caused by the metallic masses composing the vehicle. They are two kinds of magnetization in a vehicle: the permanent one (m_{perm}) depending on the vehicle history, and the induced one (m_{ind}) depending on the outside magnetic field. If a large part of the vehicle magnetization comes from m_{perm}, then the magnetic signature differs from one vehicle to another. If mind prevails, the signature is linked to the vehicle orientation to the magnetic North, thus rendering the identification difficult. This eventuality will be further discussed in this paper.

The objective of this paper is to evaluate the performances of two vehicle re-identification methods using three-axis magnetic sensors. The first one is based on the Euclidean distance between three-dimensional spatial magnetic signatures of vehicles: calculating the spatial magnetic signatures from the temporal ones requires at least two sensors. The second method uses the Dynamic Time Warping algorithm to calculate the distance between two temporal magnetic signatures: this method requires only one magnetic sensor, and doesn't need speed estimation. Moreover, the effects of a change in the vehicle orientation or in its lateral position are studied.

2 Material

2.1 Data Collection

A data-base of 261 temporal magnetic signatures was collected from 25 different cars in an empty parking lot of CEA-Leti. To evaluate the effect of the vehicle orientation to the North and therefore to evaluate the part of induced magnetization in vehicles, each vehicle passed over the sensors from North to South (NS), South to North (SN), East to West (EW), and West to East (WE). In average, vehicles passed three times in each travelling directions.

Ten sensors were fixed on a 0.6m wide plastic plate lying on the road (Fig. 1). On each row, the sensors were placed 0.25m apart. Because of the plate width, the drivers could only drive in one way over the sensors: the centre of the vehicles passed almost ($\pm$ 0.1m) above the central sensor of the plate. The wireless three-axis magnetic sensors were developed by CEA-Leti from AMR sensors of Honeywell. They measured the temporal magnetic signatures at a sampling rate of 200Hz. We positioned the sensors axis this way: the X and Y axis were in the horizontal plane, with X parallel to the longitudinal axis of the vehicle and Y to the lateral one, and the Z axis was vertical.

Fig. 1. A driver ready to go above the sensors (at the top), the plate of ten sensors put on the road (down left), and one of the sensors (down right)

2.2 Data Post-Processing

The temporal magnetic signatures that are directly used in the re-identification method with a single sensor were obtained through a first data post-processing. The first step aims at normalizing the raw data recorded in Volts (V) into physical units (Tesla (T)). In order to know the gains (T/V) with precision, we calibrated the experimental readings according to the Merayo's procedure [7], which compensates also the possible non-orthogonality of the sensor axes. The second step aims at extracting the magnetic signatures from the whole magnetic signal measured. In order to only keep the magnetic field generated by the vehicle, the offset (i.e. the local magnetic field) is subtracted from the signal. Then, we applied a sliding window segmentation technique to divide the signatures [8]: if the signal in the whole window exceeds a defined threshold, a signature is detected.

Calculating the spatial magnetic signatures from the measure of temporal signatures requires at least two sensors located at different places in the direction of the vehicle path: thus, the vehicle speed v is estimated from the time delay Δt (calculated by cross-correlation) between the signatures measured by the two sensors (spaced at a known distance Δx). Then, the spatial signature is obtained from the temporal signature with a multiplication by v: while temporal signatures are warped by the vehicle speed, spatial signatures are thus independent of the vehicle speed. Eventually, the spatial signatures are resampled to one point per cm. Fig. 2 shows two examples of a temporal signature and its spatial signature, for two different vehicles: their signatures are clearly different.

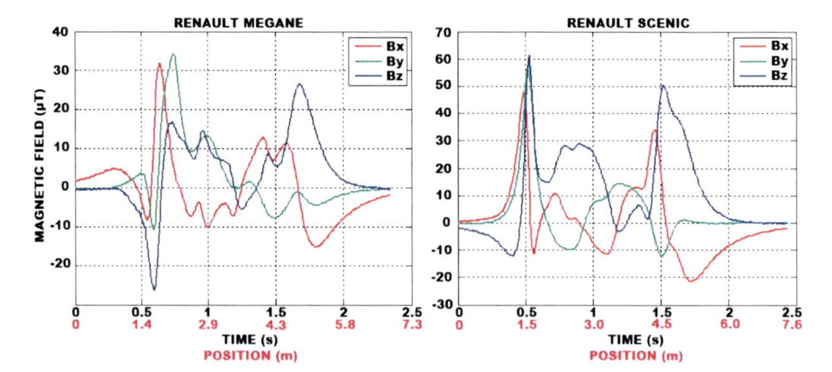

Fig. 2. 3-axis (X, Y, Z) temporal magnetic signatures recorded from a Renault Megane and a Renault Scenic, and their resulting spatial signatures: only the scale on the abscissa changes between a temporal and a spatial signature. The signatures of the Renault Megane and of the Renault Scenic are clearly different.

3 Re-Identification Methods

Our two re-identification methods act as detectors which determine if a pair of signatures comes from the same vehicle, or from different vehicles (Fig. 3). In our case, the number of pairs of signatures in the class *"same"* is much smaller than the one in the class *"different"*. For example, 66 signatures of vehicles driving North to South (NS) were measured: these signatures form 2145 ((66 * 65) / 2) pairs of signatures, with 92 in the class *"same"* and 2053 in the class *"different"*.

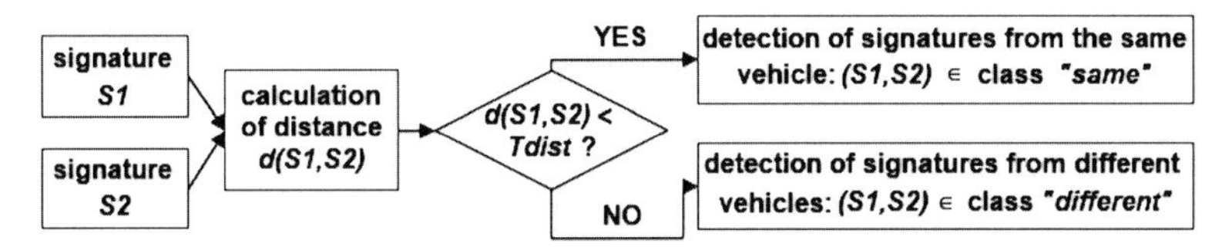

Fig. 3. The detector diagram: do these two signatures come from the same vehicle?

We used two different kinds of distances to calculate *d(S1,S2)*: Euclidean distances, and a Dynamic Time Warping distance.

3.1 Euclidean distance between spatial signatures

After the resampling to one point per cm, a pair of spatial signatures can be compared point-to-point with a standard Euclidean distance, divided by the signature length *lsign*. The signatures are first aligned to each other (with another cross-correlation): this step is necessary because if a signature is compared with a delayed version of itself, their Euclidean distance without alignment will be different from zero. Then, if the aligned signatures do not overlap entirely, they are padded with zero in order to get the same length *lsign*. Several Euclidean distances are tested: a three-dimensional (3D: axis XYZ), three bi-dimensional (2D: XY, or YZ, or XZ) and three mono-dimensional (1D: X, Y, Z).

3.2 Dynamic Time Warping distance between temporal signatures

The Dynamic Time Warping (DTW) algorithm was first introduced for speech recognition [9]. To align two temporal signals (in our case, a pair of temporal magnetic signatures measured from vehicles that may drive with different speeds), it warps their time axis and then calculates the distance between the two aligned sequences. The heart of the algorithm is based on the calculations of the distances between each point i of the first signature *S1* and each point *j* of the second signature *S2*. In a previous study, we showed that the best re-identification results were obtained with a three-dimensional distance *D(i,j)*: therefore, the results of the Euclidean distance method are only compared with those of the 3D DTW.

4 Experimental Results

4.1 Evaluation Method

The evaluation presented in this paper is based on detection curves. For each re-identification method (Euclidean or DTW distance) and for a chosen threshold *Tdist*, the percentage of True Detections (TD) is calculated: it is the percentage of pairs of signatures from the same vehicle that were correctly identified as belonging to the class *"same"*. The Percentage of Errors, or "false alarms", in the detection (PE) is also calculated for each chosen *Tdist*: it is the percentage of pairs of signatures from different vehicles that were wrongly identified as belonging to the class *"same"*. In our application of vehicle re-identification, the PE has to be very small for the estimation of traffic parameters to be accurate. The curve of TD as a function of PE is plotted thanks to the variations of the threshold *Tdist* (TD-PE curve, Fig. 4).

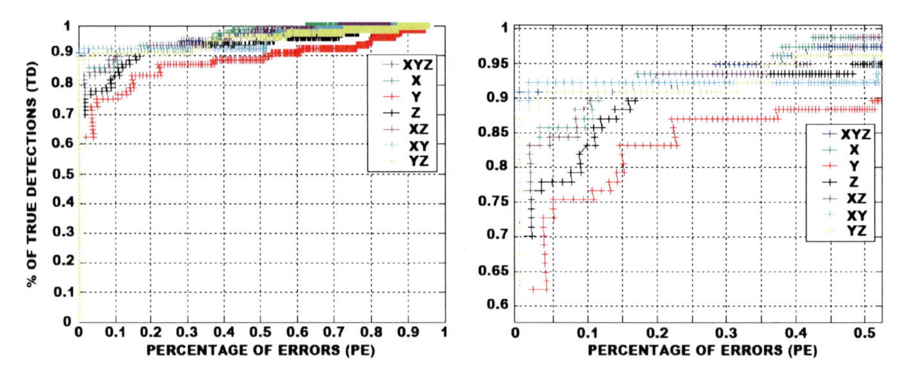

Fig. 4. TD-PE curves of the mono- (X, Y, Z), bi- (XZ, XY, YZ), and three-dimensional (XYZ) Euclidean distances, using only two sensors, and obtained with all the possible pairs of signatures recorded "East to West": the figure on the right is an enlargement of the left one. The best results are obtained for the 3D Euclidean distance.

4.2 Performances Under Different Conditions

The average performances achieved by the re-identification methods are presented in Tab. 1 under different conditions. The best results of the Euclidean distance method are obtained for the 3D Euclidean distance; therefore the results of the other Euclidean distances are not presented in this paper.

Method	Number of Sensors	Same Orientation		Different Orientations	
		TD	PE	TD	PE
3D DTW	1	84%	1%	70%	5%
3D Euclidean distance	2	90%	0%	80%	0%
3D Euclidean distance	4	100%	0%	90%	0%

Tab. 1. Percentages of True Detection (TD) and of Errors (PE) under different conditions: different re-identification methods, different numbers of sensors, and with or without vehicle orientation changes

4.3 Effect of the Number of Sensors

The Euclidean distances method was first tested with two sensors and then with four sensors (Fig. 5). With two sensors, 90% of the pairs of signatures with the same orientation (for example: pairs of two signatures recorded when a vehicle travelled from "North to South") are re-identified without any false alarm. With four sensors, the method compares this time two signals which are the sums of the signatures measured by a row of three sensors: with PE=0%, the percentage of TD is 100% of the pairs of signals. Conversely, the percentage of TD of the DTW distance method – which uses a single sensor (Fig. 5) – decreases to 84% (same orientation) with a percentage of false alarms different from zero. These worse results can be explained by the fact that the DTW algorithm compares the signal patterns of the signatures without knowing their spatial length, and therefore it may fail to differentiate a long car from a small car.

4.4 Effect of a Change in Vehicle Orientation

If we now compare pairs of signatures with different orientations (for example: a signature "North to South" versus a signature "West to East" from the same vehicle), the Euclidean distance method is able to re-identify without any false alarm 80% (two sensors) or 90 % (four sensors) of the pairs of signatures, while the DTW algorithm re-identifies 70% of them, with PE=5%. Since the re-identification rates vary only from 10 to 14% if the vehicle's orientation differs or not, (we can say that) the effect of the vehicle orientation to the North is therefore limited on the magnetic signatures.

Fig. 5. Sensor(s) used for the DTW distance (left) and for the Euclidean distance (middle: two sensors, right: four sensors)

4.5 Effect of a Change in the Lateral Position of the Vehicle on the Road

In order to simulate a lateral position shift of 0.25m between two paths of the same vehicle, we compared pairs of signatures measured by two different sensors of a row (middle sensor versus left or right sensor). For both re-identification methods, the TD rate falls by about 60%. On the supposition that the variations of a spatial magnetic signature – which is the magnetic signal measured during a longitudinal path of a vehicle – are similar to the variations of the spatial magnetic signal measured during a lateral path of a vehicle, this fall is understandable: the magnetic field of the spatial signatures shown in Fig.2 may vary a lot between two points located 0.25m apart.

5 Conclusions

In this paper, we presented two vehicle re-identification methods using magnetic sensors. We have demonstrated that the part of the vehicle magnetization coming from an induced magnetization seems small enough to allow the re-identification even after a change of direction (after a bend for example). Our re-identification method based on the Euclidean distance between a pair of spatial signatures is simpler to implement and achieves better performances than the one based on a Dynamic Time Warping distance, which admittedly requires a single sensor, but is computationally more expensive.

However, the performances of both re-identification methods turned out to be linked to the variations of the lateral position of the vehicle on the road. This matter can be settled with the installation on the road of sensors closer to each other. These results need to be confirmed and detailed with data measured in a less restrained environment. Thus, two new experiments are scheduled: the first one will be similar to the one presented in this paper but with sensors placed 0.10m apart on a narrower plate (0.44m wide), and the second one will take place in a real traffic flow.

References

[1] V. Khorani, F. Razavi, V.R Disfani., A mathematical model for urban traffic and traffic optimization using a developed ICA technique, IEEE Transactions on Intelligent Transportation Systems , vol.12, pp. 1024-1036, 2011.

[2] A. Khosravi, E. Mazloumi, S. Nahavandi, D. Creighton, J.W.C. Van Lint, Prediction intervals to account for uncertainties in travel time prediction, IEEE Transactions on Intelligent Transportation Systems, vol.12, pp. 537-547, 2011.

[3] C. Anagnostopoulos, T. Alexandropoulos, V. Loumos, E. Kayafas, Intelligent traffic management through MPEG-7 vehicle flow surveillance, IEEE John Vincent Atanasoff (JVA) International Symposium on Modern Computing, pp. 202-207, 2006.

[4] Y.A. Kathawala, B. Tueck, The use of RFID for traffic management, International Journal of Technology, Policy and Management, vol. 8, no. 2, pp. 111-125, 2008.

[5] S. Amin, et. al., Mobile century using GPS mobile phones as traffic sensors: a field experiment, 15th World Congress on Intelligent Transportation Systems, 2008.

[6] K. Kwong, R. Kavaler, R. Rajagopal, P. Varaiya, A practical scheme for arterial travel time estimation based on vehicle re-identification using wireless sensors, Transportation Research Part C, vol. 17, no. 6, pp. 586-606, 2009.

[7] J.M.G. Merayo, P. Brauer, F. Primdahl, J.R. Petersen, O.V. Nielsen, Scalar calibration of vector magnetometers, Measurement Science and Technology, vol. 11, no. 2, pp. 120-132, 2000.

[8] R. Wang, L. Zhang, R. Sun, J. Gong, L. Cui, EasiTia: a pervasive traffic information acquisition system based on wireless sensor networks, IEEE Transactions on Intelligent Transportation Systems, vol. 12, pp. 615-621, 2011.

[9] L.R. Rabiner, A.E. Rosenberg, S.E. Levinson, Considerations in Dynamic Time Warping Algorithms for Discrete Word Recognition, IEEE Transactions on Acoustics, Speech, and Signal Processing, vol. 26, no. 6, 1978.

Anne-Cécile Pitton, Andréa Vassilev
CEA-Leti
MINATEC Campus
17 rue des Martyrs
38054 Grenoble Cedex 9
France
anne-cecile.pitton@cea.fr
andrea.vassilev@cea.fr

Sylvie Charbonnier
Gipsa-lab, UMR 5216 CNRS – Grenoble INP – Université Joseph Fourier – Université Stendhal
11 rue des Mathématiques
Grenoble Campus, BP 46
38402 Saint-Martin-d'Hères
France
sylvie.charbonnier@gipsa-lab.grenoble-inp.fr

Keywords: magnetic sensor, vehicle re-identification, traffic management, induced magnetization

Components & Systems

Truck Safety Applications for Cost-Efficient Laser Scanner Sensors

M. Ahrholdt, T. Johansen, G. Grubb, Volvo Group Trucks Technology

Abstract

Cost-efficient sensor solutions are of high interest for automotive safety applications, in order to increase the use of sensor-based active safety functions and therefore improve traffic safety. In the European collaboration project Minifaros, a cost-efficient laser scanner is being developed. Particularly for heavy vehicles, cost-efficient sensors are of interest to observe objects in the close vicinity of the host vehicle. In this contribution, we will first describe the need and requirements for low-cost sensors and particularly discuss the special needs for truck applications of this sensor. Then, the target and benchmark applications for the cost-efficient laser scanner being developed in Minifaros will be described, consisting of vulnerable road user protection functions and low-speed Adaptive Cruise Control.

1 Introduction

Automotive safety systems are increasingly being used in a wider range of vehicles. One of the main objectives in Advanced Driver Assistance Systems (ADAS) is a reliable detection of the host vehicle's surrounding. While the first systems were addressing comfort functions, such as Adaptive Cruise Control (ACC), focussing on the detection of the preceding vehicle, the range of functions has been developing rapidly towards a wider range of scenarios. One main factor influencing the development of active safety systems is the availability of cost-efficient sensor solutions, which can be used two-fold: firstly, equipping a wider range of vehicles in general and secondly, addressing more specific detection tasks, which normally would not justify a dedicated sensor.

In the European collaboration project Minifaros, a laser scanner is being developed, targeting low component cost and innovative design to enable new application areas. These application areas can benefit from the high resolution obtained by a scanning laser sensor, where traditional laser scanner approaches have been too costly for specific applications. The sensor development in the Minifaros project started back in 2010, with the finalization of first prototypes expected in spring 2012. Thereafter, the sensor will be tested on a representative set of functions during the remaining project duration until end-2012. For this reason, this contribution does not contain test results yet, but rather describes the relevant truck applications for low-cost sensors.

2 Challenge and Requirements

For the development of the Minifaros laser scanner, the requirements of a cost-efficient sensor have been evaluated from the perspective of a wide range of traffic safety applications, for both passenger cars and taking specifically the needs of heavy vehicles into account [1].

For heavy vehicles, two principally different detection tasks can be distinguished, as shown in Fig. 1. One task deals with the observation of the forward area, with the sensor data to be used in a wide range of safety and comfort functions, such as Adaptive Cruise Control, Lane Departure Warning or Automatic Emergency Braking. Within this area, sensor systems have been commonly used for more than 10 years, with an increasing application range. The other detection task deals with detecting objects or persons in the close vicinity of the host vehicle. Particularly for heavy vehicles, visibility for the driver in this near area is a challenge. Sensor-based driver assistance systems are partly already available, for example for blind-spot monitoring (Volvo Lane Change Support), or being developed, such as vulnerable road user detection in the passenger-side blind spot. [2].

However, this complete detection area is today hard to address simultaneously with a single sensor solution, which is why today specific solutions are used for specific problems. Cost efficient sensors with wide opening angles can contribute to better observe this close area. For this reason, the Minifaros sensor under development has been specified to be used both for the near vicinity area, as well as – with some range limitations – for the forward detection area. The targeted specifications to support a wide range of ADAS functions are summarized in Table 1 [1].

Minifaros Sensor Target Specification	
Range	80 m
Range accuracy	0.1 m in near-field 0.3 m else
Field of view	250°
Angular Accuracy	0.25°
Update Frequency	25 Hz

Tab. 1. Minifaros Sensor Target Specifications [1]

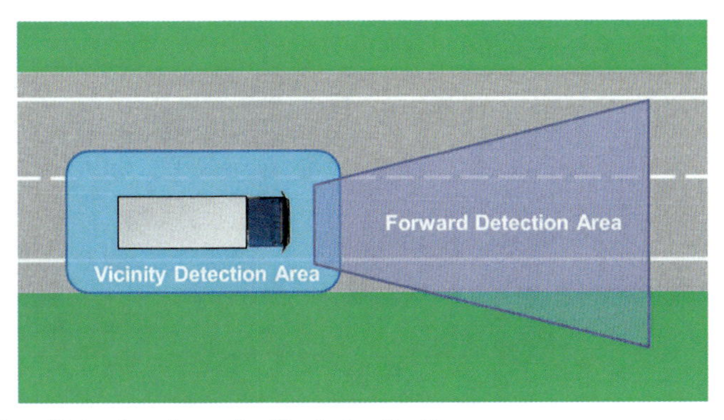

Fig. 1. Detection Areas for Truck Applications

3 Target Applications

In order to evaluate the Minifaros sensor prototype in a wide range of potential applications, target applications have been selected also based on the safety impact potential. Functions include both the detection of vehicles as well as the detection of vulnerable road users (VRU) such as pedestrians and cyclists.

For the heavy vehicle domain, three truck-specific functions have been selected:

- Stop & Go Adaptive Cruise Control
- Start inhibit
- Right-turn assistance

In the project, there will be a parallel investigation of the same sensor technology for the passenger car application portfolio consisting of:

- Pedestrian protection
- Pre-crash application
- Safe distance application

The truck test applications and their requirement on the sensor system will be described below.

Fig. 2. Volvo Truck Demonstrator

2.1 Stop & Go Adaptive Cruise Control

Adaptive cruise control (ACC) has been available for Volvo Trucks for a number of years now. It is available as a comfort function, mainly for motorway environments supporting the driver to adapt their speed according to the preceding vehicle.

The extension towards more congested traffic extends the requirements for the sensor system with the need of a higher opening angle (in order to detect cut-in situations) and good close area observation performance, particularly when the host vehicle is stationary (since distances to the target vehicle can be small in congested traffic). A first system for this purpose has been shown in the HAVEit project [3], integrating an automated queue assistance function.

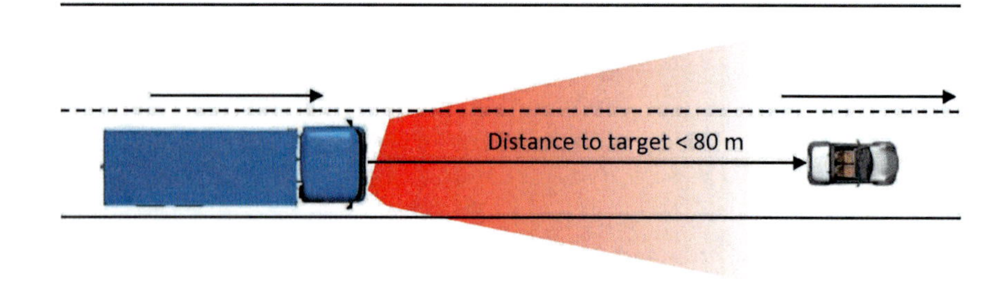

Fig. 3. Stop & Go Adaptive Cruise Control use-case

2.2 Start Inhibit

For heavy trucks, safety of vulnerable road users in the vicinity of the truck is a particular concern. One accident scenario relates to the high seat position of the truck driver (compare Fig. 2) which causes a blind spot area in front of the truck. If people are directly in front, they are visible to the driver only through mirrors. A system has been demonstrated in PReVENT APALACI [4] to inhibit the truck from starting when a VRU is in front of the vehicle. However, detection of VRU in the near vicinity of the truck is still today a challenge, which is why this situation is addressed as one of the target scenarios for the Minifaros sensor.

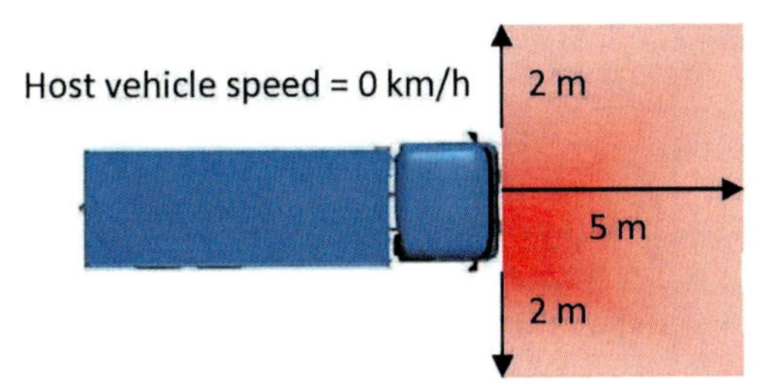

Fig. 4. Start-Inhibit use-case

2.3 Right-Turn Assistance

When turning right (We assume right-hand driving in this contribution, i.e. turning right is turning to the passenger side), safety of VRU is a particular concern for heavy vehicles. This is related to the limited visibility of the truck driver on the passenger side of the vehicle, which can only be established by different mirror combinations. Solutions for right turning assistance have been demonstrated in [2], but cost-efficient sensor technology for this wide area in proximity of the host vehicle is a challenge, which is why this use case is addressed in Minifaros.

The field of view sketched in Fig. 5 also gives an indication of the necessary mounting point. In order to observe the area for this application and the use-cases discussed previously, the Minifaros sensor will be mounted at the front right corner of the truck for evaluation.

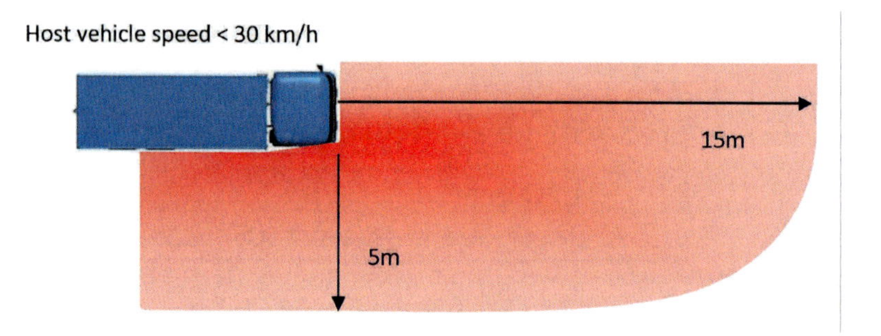

Fig. 5. Right-turn assistance use-case

3 Conclusions

This contribution has described the needs and requirements for low-cost sensors for heavy vehicle applications. Particularly for truck applications, low cost sensors have special relevance as the use of additional affordable sensors will allow better coverage of the near vicinity around the host vehicle, which is today a challenge for detection systems.

For the low-cost laser scanner sensor being developed in the Minifaros project, a set of three representative truck functions has been selected for development, test and benchmarking (and there is a set of three passenger car functions as well, not covered in detail here).

As the first Minifaros sensor prototypes are still under final development at the time of writing, the next step will be to actually test the sensor on the test vehicle under real-world conditions together with the selected applications.

Acknowledgements

The work described in this contribution has been co-funded by the European Commission, in the project Minifaros within the Seventh Framework Programme.

References

[1] Minifaros Deliverable D3.1. User needs and operational requirements for Minifaros assistance system. (2010) Available through http://www.minifaros.eu/

[2] M. Ahrholdt, G. Grubb, E. Agardt: Intersection Safety for Heavy Goods Vehicles – Safety Application Development. In: G. Meyer, J. Valldorf [Eds.], Advanced Microsystems for Automotive Applications 2010, Smart Systems for Green Cara and Safe Mobility, Springer, Berlin, 2010.

[3] G. Grubb, E. Jakobsson, A. Beutner, M. Ahrholdt, S. Bergqvist, Automatic Queue Assistance to Aid Under-loaded Drivers, ITS World Congress, Stockholm, 2009.

[4] PReVENT Deliverable D50.10b – APALACI Final Report (2007) Available at http://www.prevent-ip.org/en/public_documents/deliverables/d5010_b_-_apalaci_final_report.htm

Malte Ahrholdt, Grant Grubb, Torbjörn Johansen
Volvo Group Trucks Technology
Advanced Technology and Research
Dept 6320 M1.6
40508 Göteborg
Sweden
malte.ahrholdt@volvo.com
grant.grubb@volvo.com
torbjorn.johansen@volvo.com

Keywords: active safety, heavy goods vehicle, cost-efficient sensors, laser scanner, Minifaros

Far Infrared Imaging Sensor for Mass Production of Night Vision and Pedestrian Detection Systems

E. Bercier, P. Robert, D. Pochic, J.L. Tissot, ULIS
A. Arnaud, J.J. Yon, CEA-LETI

Abstract

In today's high class vehicles, Night Vision and Pedestrian Detection Systems take benefit of passive thermal detection based on the used Far Infrared (FIR) Imaging Sensor. Formally developed since the 1980's for night vision system used in military application, the main industrial technologies are currently producing 2D array chip formed by MEM's microbolometer integrated into vacuum package. Improvements made by ULIS over the last 10 years on microbolometers made from amorphous silicon enable today the use of FIR imaging sensors in highly demanding commercial applications such as automotive market. The high level of accumulated expertise by ULIS and CEA/LETI follows the continuous technology development roadmap, as detection material improvement, pixel pitch reduction, vacuum package technology breakthroughs, or readout integrated circuit (ROIC) on-chip innovation. The new ULIS product ¼ VGA format based on 17µm pixel-pitch, low cost vacuum package and improved ROIC enables the improvement of Night Vision and Pedestrian Detection Systems, by simplifying system design, and reducing calibration process and production time, offering high performances compatible with hundreds meters of detection range even in day or night conditions. This paper describes the technology roadmap and product improvements regarding automotive expectations.

1 Introduction

For some years the infrared sensors market have tended to follow the evolution observed in the past for visible imaging sensors. This is particularly true for cost decreasing, to address high volumes applications, while more and more performances are expected into the sensors. Indeed with smaller pixel pitch, higher resolution [3] and easier system integration, microbolometer infrared sensors intend to address new low-cost/high-volume applications such as automotive for night vision, pedestrian detections [1], [2], security / surveillance and thermography. A ¼ VGA sensor has been designed for large volume application taking into account 17μm pixel-size technology and high reliability packaging technique. We will describe first the interest of amorphous silicon for microbolometer application regarding material properties, simplified technological process and simplified sensor operation, then discuss key vacuum package and ROIC technological breakthroughs prior to presenting benefits of FIR thermal sensor improvement for automotive application, such as night vision, pedestrian detection, or HVAC control or short distance rear detecttion.

2 Key Technologies for FIR Imaging Sensors

2.1 17 μm Amorphous Silicon Microbolometer

Amorphous silicon presents attractive properties for micro bolometer applications. The amorphous silicon bolometer resistance R variation versus temperature is described by an Arrhenius law in which the activation energy E_a depends on the sensitive material.

$$R = R_0.\exp(E_a / kT) \tag{1}$$

As amorphous silicon is not an alloy, every pixel has the same activation energy (standard deviation on mean value < 0.4%) leading to a high spatial uniformity of pixels temperature behavior. Moreover, amorphous silicon activation energy E_a remains essentially constant throughout a large range of operational temperature [3]. The microbolometer resistance distribution stems essentially from resistance geometry distribution, while remaining stable regarding focal plane temperature variation. As a consequence, microbolometers' resistances follow a simple and spatially uniform Arrhenius law and hence are fully predictable, leading to easier TEC-less operation with only one gain table for non uniformity correction (NUC) to cover a broad focal plane temperature range. It's also being very predictable to simplify the algorithms required for thermography and shutter-less operation. Moreover, it leads to the use of a simplified system calibration process saving manufacturing cost for high volume system production.

Thanks to its silicon like properties, amorphous silicon is easy to integrate monolithically onto silicon substrates at temperature compatible with CMOS integrated circuit. Several key technology improvements are required to successfully scale the pixel from 25 μm down to 17 μm or less. One of which is the call for 0.5 µm lithography processing. Advanced lithography in connection with thinner films embodiment clearly gives an edge to maintain, and even to improve, thermal insulation when scaling the pixel to 17 µm while keeping a simple one-level micro bridge structure (cf. Figure 1) which leads to high operability and high manufacturing yield. This approach allows to continue to scale a single level micro bolometer architecture, taking advantage of a very simple process flow and cost saving. Amorphous silicon offers the opportunity to design a rather undemanding micro bridge structure without any extra features than the bare minimum required for the bolometer functionality. This leads to a reduced number of technological operations and results in a little number of photolithographic layers. The mere arrangement of the micro bridge leads therefore to a fast sensor, featuring pixel time constants largely below the common Figuress known elsewhere. The thermal time constant measurements of 17 µm pixels give values under 10 ms (range from 8.8 ms to 9.3 ms have been measured on the first 17 µm batches) fully compatible with 30 Hz or higher frame rate operation.

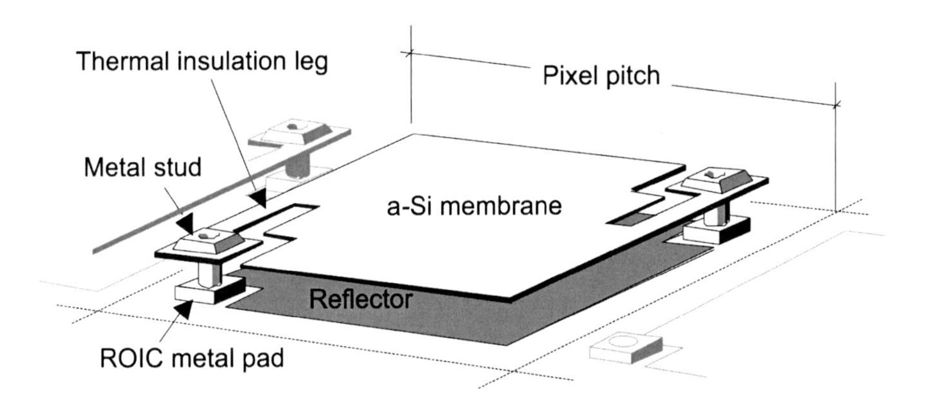

Fig. 1. Schematic of the amorphous silicon microbolometer pixel.

2.2 Packaging Techniques

In order to be sensitive to the FIR radiation input flux coming from the scene of interest, all microbolometer focal plane arrays (FPA) require thermal insulation to the convection coming from the surrounding. This insulation is done by using second order vacuum package techniques, where residual pressure is lower than 10^{-3} mbar. Moreover, microbolometer FPAs work at ambient temperature compared to other technology that work at liquid nitrogen temperature 77K typically. Thus, microbolometer FPAs do not require any expensive, power and size consuming cooling system for operation. Depending on application requirement, some microbolometer imaging sensors only use thermoelectric cooling (TEC) system integrated into metallic package for operating temperature stabilization, as shown in Figure 2 a. In 2006, ULIS introduced the first TECLESS vacuum package technology based on LCC ceramic, shown in Figure 2 b. This packaging technology is the standard used for the major part of the current IR imaging sensor production.

Fig. 2. Metallic and TECLESS ceramic packages

A new Wafer Level Packaging technique is under development to address high volume applications. Taking advantage of the silicon technology used for amorphous silicon thermometer integration on CMOS wafer and of the MEMs technology expertise, ULIS is developing an advanced technique for device vacuum integration. Figure 3 shows first WLP prototype integrated either on PCB (a) or standard JEDEC package (b). These final product are developed under AEC_Q100 Grade 3 qualification program.

Fig. 3. ¼ VGA sensor in WLP or QFP JEDEC standard configuration

2.3 Microcap Structure Packaging Technique

A description of the monolithic cap structure can be found in references [1] and [2] and is recalled in Figure 4. Microcaps are built in the continuity of the bolometer process on the ROIC wafer, in a full collective way. First, a sacrificial layer is deposited above microbolometers and a trench is realized at the periphery of each pixel. An IR window is then deposited in order to form the microcap structure. Two exhaust holes linked by an etch channel are etched through the

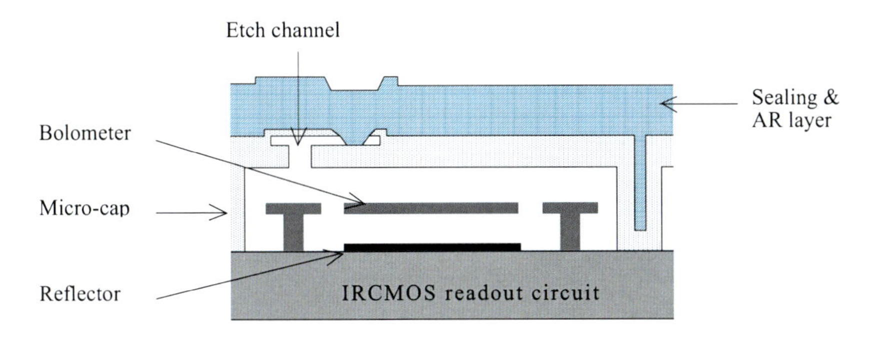

Fig. 4. Schematic cross section of the microcap

top of the microcap. The capped sacrificial layers are removed through these exhaust holes and the etch channel. Finally, the sealing and anti-reflecting layer is deposited under high vacuum to finalize the hermeticity of the micro-cap. Figure 5 shows SEM views of the microcap structure.

New advanced vacuum technology will take benefit of microcap structure: the Pixel Level Package (PLP). Basically, PLP process consists in the building of IR transparent micro-caps that cover each micro-bolometer pixel of the array. To be efficient, the microcap has to be hermetically sealed under vacuum and maintain the vacuum level in the 10^{-3} mbar range requested for nominal IRFPA operation. The main point (regarding cost reduction objectives) is that the PLP process thoroughly carried out directly on the ROIC and micro-bolometers wafer, in a full collective way. This technology will suppress the usual costly vacuum package integration of the sensor chip.

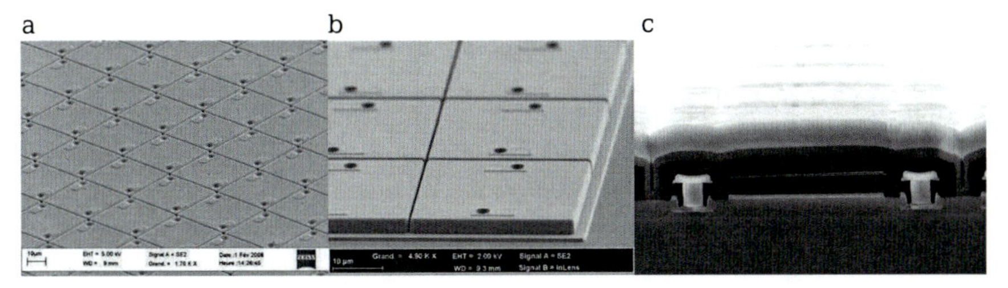

Fig. 5. SEM view of a the microcap structures: a and b show field of micro-cap, c shows a micro-bolometer pixel included in a microcap

2.4 Readout Integrated Circuit Design

The bolometer layers are processed directly onto dedicated imaging CMOS ASIC wafers. The goal of the readout integrated circuit (ROIC) associated to the sensor is to measure the thermometer resistance value of each pixel.

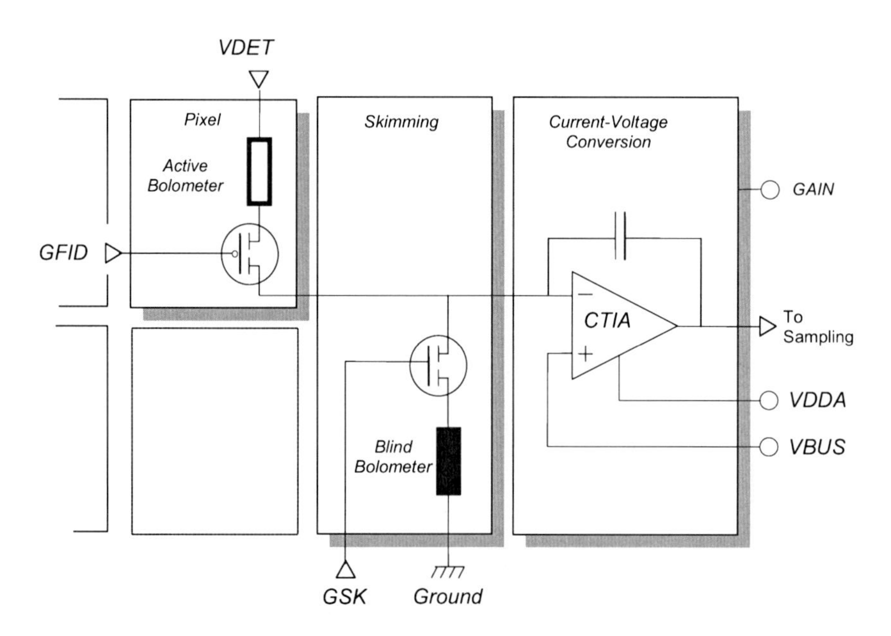

Fig. 6. Pixel readout architecture

The readout is operated in rolling shutter mode. Each imaging pixel is addressed through an injection TMOS and is coupled with a skimming blind bolometer as describe in the Figure 6 to remove a large part of the useless offset current going through the bolometer. To supply voltage samples of each pixel in video output signal, a low noise CTIA (Capacitive Transimpedance Amplifier) is used in the input stage of the readout integrated circuit (ROIC). This stage converts the current in the bolometer to a voltage value, and size the ROIC gain with the couple C_{int} and T_{int}, respectively integration capacitor and integration time. Thus the main characteristic of the bolometer sensor is the responsivity R, expressed as:

$$R = \frac{V_{pol}}{R_0} * \frac{E_a . A . R_{th} . \Delta\phi}{k . T^2} * f(T_{\text{int}}, C_{\text{int}}) \qquad (2)$$

Where E_a (activation energy), A (membrane area), R_{th} (thermal insulation of the membrane) and R_o (electrical resistance of the thermometer), are parameters that depend of the bolometer design, and particularly of the membrane area. We can observe also with this formula [2] that the only electrical parameter the responsivity depends on, is the bolometer polarization V_{pol}. As activation energy E_a is identical for every pixel in an array, it means that the focal plane temperature behavior is highly uniform and predictable leading to easier device modeling and therefore easier TEC-less and shutter-less operation.

3 FIR Sensors for Automotive Application

3.1 Thermal FIR Camera in the Automotive Field

Without any need of an illumination system, thermal FIR imaging cameras work by processing the Far Infrared radiation naturally emitted by an object, measured in long wave 8 to 14µm radio waves, and converting the differences in electrical signals that are received on a specially manufactured sensor, the microbolometer, into images. Any object that has a heat signature can be seen in total darkness or through adverse conditions such as fog, smoke, rain and snow.

Formally developed since the 1980's for night vision system used in military application, Night Vision and Pedestrian Detection Systems in today's high class vehicles take benefit of passive thermal detection based on the used of Far Infrared Imaging Sensor, thanks to affordable cost achieved over the last 10 years in the microbolometer MEMs industry.

In the last 10 years too, many projects have been conducted for automotive safety application based on the use of FIR sensor technology. In 2003, the SAVE-U project described in reference [4] was using a 160x120 35µm focal plane array to create thermal signal, signal analysed by specific algorithm to create pedestrian detection map. Studies have also been performed regarding FIR system design in the FNIR project, especially taking into account FIR sensor performances and optics influence, signal processing, pedestrian detection with cascaded classifiers, as described in reference [5]. In reference [6], FIR technology is also considered using lower resolution sensor where FIR information are fused with Near Infrared (NIR) information coming from NIR sensor.

3.2 ¼ VGA Sensor Design for Pedestrian Detection

As described in reference [4] in 2003, the SAVE-U project was using a state of the art microbolometer focal plane array 160x120, 35µm pixel pitch, achieving a Noise Equivalent Temperature Difference (NETD) equal to 56mK at F/1 in front of a 27°C scene blackbody temperature. NETD represents the lowest temperature difference that can be detected, depending on many parameters such as flux level radiated by the black body under observation, or optical aperture F/#.

Pedestrian detection and associated algorithm studied in reference [5] in 2010 are based on the use of a ¼ VGA 325x240 focal plane array with a 25µm pixel pitch. Latest developments in ULIS are especially focused on ¼ VGA microbolometer which feat the requirement of current automotive night vision and pedestrian detection system.

A new IR Focal Plane Array with 384 x 288 pixels and a pixel pitch of 17 µm has been developed under 3.3V analog power supply and 1.5V digital power supply. It reaches only 55mW of power consumption in analog video output mode. A wide electrical dynamic range (2.4V) is so maintained despite the use of an advanced CMOS node. The very low noise ROIC video output allows reaching of very high performances of the bolometer sensors. To simplify the implementation of the detector many features were integrated in the ROIC, this break is a real evolution towards the standardization of the IR detectors interfaces with digital processing systems. Therefore a 14bits ADC is integrated in the readout circuit and the digital datas are supplied on an 8bits multiplexed bus, in this mode the ROIC power consumption reaches less than 175mW. The detector configuration (integration time, windowing, gain, scanning direction...), is driven by a standard I^2C link. Like most of the visible arrays, the detector adopts the HSYNC/VSYNC free-run mode of operation driven with only one Master Clock (MC) supplied to the ROIC which feeds back pixel, line and frame synchronizations. Due to a maximum 7MHz PSYNC pixel clock the detector can work at 60Hz frame rate. On-chip OTP memory for customer operational condition storage is also available for detector characteristics. For instance the bolometer bridge biasing is achieved by internal DAC and can be stored in the factory configuration. OTP parameters stored during ULIS sensor manufacturing process support an easier integration in a camera system while uploading specific biases from sensor memory.

The specific appeal of this new product lies in the high uniformity and easy operation it provides. The reduction of the pixel-pitch turns this TEC-less ¼ VGA array into a product well adapted for high resolution and compact systems, achieving at sensor level NETD value equal to 35 mK at F/1 in front of a 27°C black body temperature, with respect of a 50 Hz frame rate compatibility and scene dynamic over 100°C typically, such as -20°C to +80°C. Corresponding results of measurement are presented in Figure 7, responsivity map in [mV/K] and b, NETD histogram in [mK].

Figure 8 presents real FIR thermal scene recorded thanks to a thermal camera based on ¼ VGA 17 µm sensor. Real scene are observed in laboratory, or at 10 meters or 5 kilometers distance.

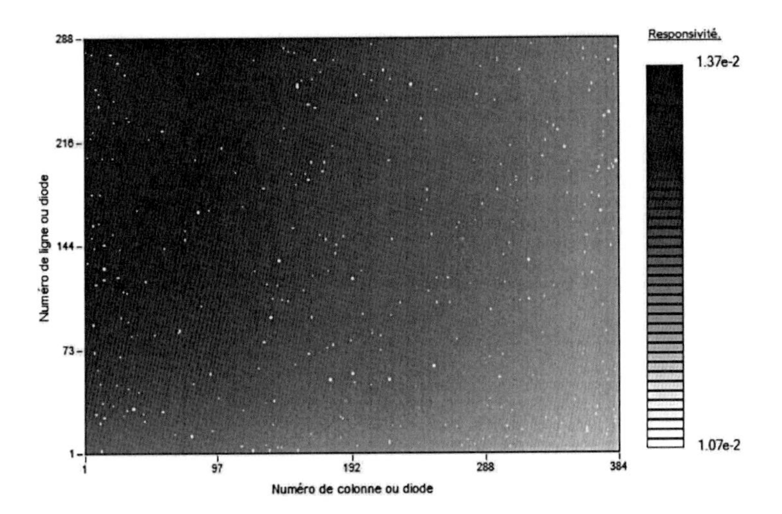

Fig. 7. ¼ VGA 17 µm performances measurement - Responsivity map [mV/K]

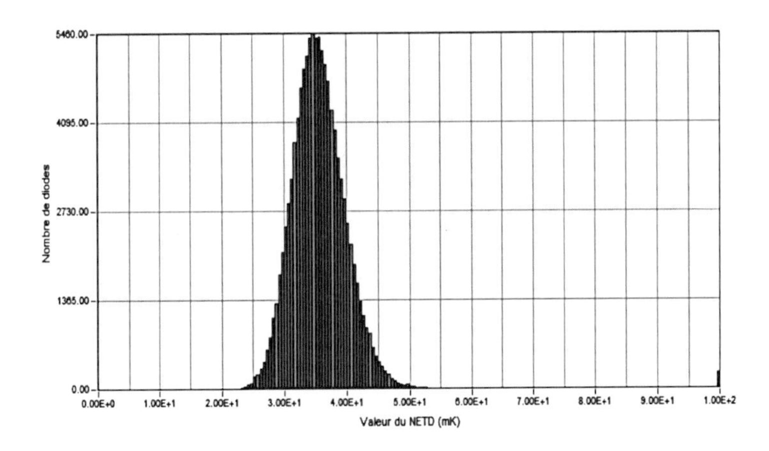

Fig. 8. ¼ VGA 17 µm performances measurement - NETD histogram [mK]

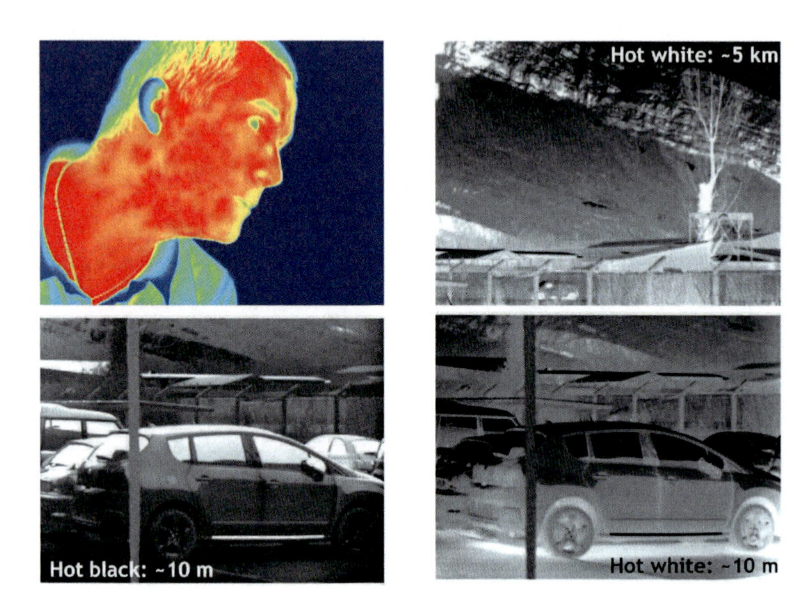

Fig. 9. ¼ VGA 17 µm performances measurement

3.3 Small Format Sensor Development for New Applications in the Automotive Field

Recent publication of the FNIR project described in reference [6] in detail additional FIR sensor cost reduction by using lowest resolution sensor. Thus, an European project named MIRTIC has been set up to develop a new line of infrared low resolution sensors adapted among others to automotive applications like imaging fusion or internal HVAC tuning or short range pedestrian detection alarm which is necessary for automotive back-viewer systems. This project will enable to accelerate this development of sensors designed to fill an unmet need in applications. Existing single-element IR sensors (pyroelectric or thermopile sensors), like those used in buildings to detect motion, lack the technological capability and pixel number to provide complex data. This includes an inability of pyroelectric single-element sensors to detect the number of people in a room, particularly if those present are immobile.

Sensor architecture developed in ULIS will take advantage of small pixel size (17 µm or less) and specifically new advanced vacuum technology: the Pixel Level Package (PLP).

4 Conclusions

Latest state of the art developments and technologies roadmap have been presented, regarding detection material, pixel pitch reduction, vacuum package or readout integrated circuit functionalities improvements. The new product ¼ VGA 384x288 focal plane array 17 µm pixel pitch take benefits of those latest technology improvements, achieving NETD performances of 35 mK at F/1, 50 Hz frame rate, compliant with the requirement of night vision, pedestrian detection system dedicated to automotive application.

Beside theses state of the art performances, amorphous silicon detection material high uniformity, behavior over a large range of operating temperatures, associated to improved ROIC functionalities enable to reduce manufacturing time and calibration process currently required by all FIR thermal system such as nigh vision system. Technology roadmap such as pixel level package will also open new application opportunities in the automotive field, based on an additional cost reduction step in the offer of FIR sensors.

References

[1] C. Trouilleau, et al., Uncooled amorphous silicon 160 x 120 IRFPA with 25-µm pixel-pitch for large volume applications, Infrared Technology and Applications XXXIII, Proceedings of SPIE Vol. 6542, 2007.

[2] C. Vieider, et. al., Low-cost far infrared bolometer camera for automotive use, Infrared Technology and Applications XXXIII, Proceedings of SPIE Vol. 6542, 2007.

[3] J.L. Tissot, et. al., High performance Uncooled amorphous silicon VGA and XGA IRFPA with 17µm pixel-pitch, Electro-Optical and Infrared Systems: Technology and Applications, proc. of SPIE Vol 7834, (2010).

[4] J.J. Yon, et. al., Infrared microbolometer and their application in automotive safety ISBN 3-540-00597-8, pp137-157

[5] S. Franz et. al., Performance evaluation of FIR sensor systems applied to pedestrian detection, Infrared Imaging Systems: Design, Analysis, Modeling, and Testing XXI, Proceeding of SPIE Vol. 7662, 766217, 2010.

[6] R. Schweiger, et. al., Sensor fusion to enable next generation low cost Night Vision systems, Optical Sensing and Detection, Proceeding of SPIE Vol. 7726, 772610, 2010.

Emmanuel Bercier, Patrick Robert, David Pochic, Jean-Luc Tissot
ULIS – BP27
38113 Veurey-Voroize
France
e.bercier@ulis-ir.com
p.robert@ulis-ir.com
d.pochic@ulis-ir.com
jl.tissot@ulis-ir.com

Agnes Arnaud, Jean Jacques Yon
CEA–LETI, MINATEC
17 rue des Martyrs
38054 Grenoble Cedex 9
France
agnes.arnaud@cea.fr
jean-jacques.yon@cea.fr

Keywords: night vision system, pedestrian detection systems, microbolometer, FIR, amorphous silicon technology

A Laser Scanner Chip Set for Accurate Perception Systems

S. Kurtti, J.-P. Jansson, J. Kostamovaara, University of Oulu

Abstract

This paper presents an integrated receiver channel and an integrated time-to-digital (TDC) converter fabricated in a 0.35 µm SiGe BiCMOS and in 0.35 µm CMOS technologies, respectively, that give the required performance for a pulsed time-of-flight (TOF) laser radar to be used in a laser scanner in automotive applications. The receiver-TDC chip set is capable to measure the positions and widths of three separate successive timing pulses with sub-ns level precision in a wide dynamic amplitude range of more than 1:10.000.

1 Introduction

The pulsed time-of-flight (TOF) laser distance measurement method is based on the measurement of the transit time of a short laser pulse to an optically visible target and back to the receiver, Figure 1. This method offers several advantages in terms of high measurement speed (<ms), high lateral and longitudal accuracy (mm-cm) and dynamic range and is thus considered to be a potential solution for perception systems in traffic and in robot vision, for example. Due to practical reasons the laser pulse length used is typically in the range of 3-5ns. This corresponds to 1-2 meters in air. On the other hand in practical applications an accuracy of a few centimeters is typically called for. Then, obviously, a stable pulse shape and the detection of a specific point of the received laser pulse are needed. This is alleviated by the fact that the amplitude of the received laser echo may vary in a range of 1:10.000 or even more.

In this work we have developed two integrated circuits which realize the functionalities of a pulsed time-of-flight laser scanner receiver. The receiver chip produces from the weak optical echo a digital timing pulse for the time-to-digital converter, whereas the TDC measures the time interval between the transmitted and received pulses. The receiver-TDC circuits can in fact measure simultaneously the time intervals to three successive pulses and also the widths of the detected pulses. The width information can be used to correct the timing walk error as is presented in more details below.

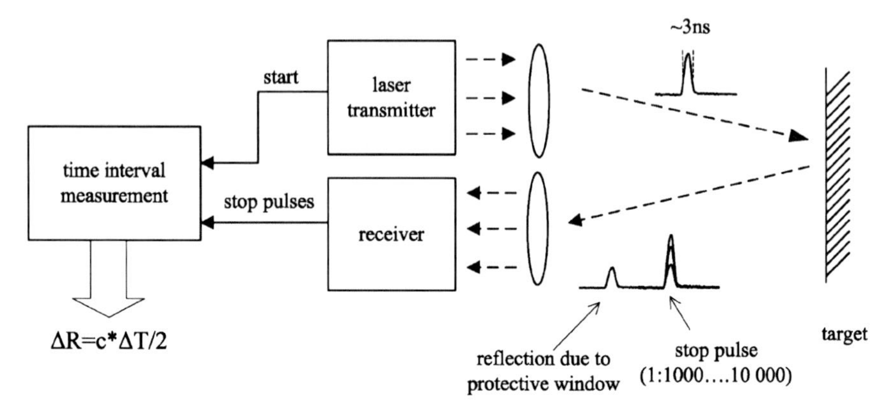

Fig. 1. Block diagram of the pulse TOF rangefinder

2 Receiver Channel

A simplified block diagram of the receiver channel is shown in Fig. 2 a). The optical echo is converted to a current pulse in an external avalanche photodiode (APD). The current pulse is converted to a voltage pulse in a trans-impedance pre-amplifier and then further amplified in a voltage-type post amplifier. A constant threshold voltage is used at the input of the timing comparator generating a logic-level timing signals for the multi-channel TDC. The TDC measures the start (generated by laser transmitter) - stop (echo or possibly multiple echoes from the target) delays and the timing pulse lengths. In addition the receiver chip includes needed bias blocks.

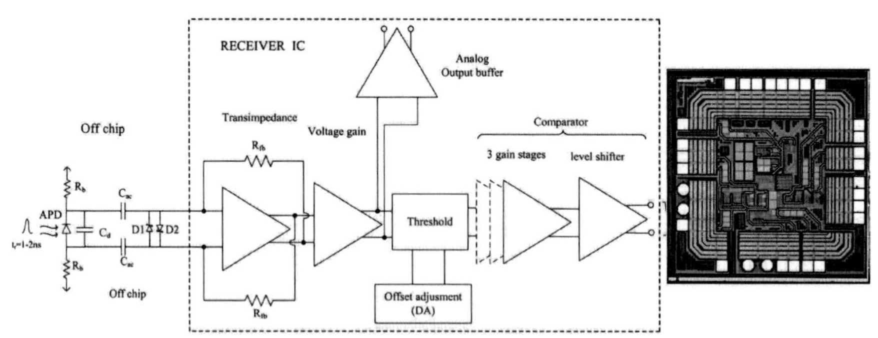

Fig. 2. Block diagram of the receiver channel (left), Photograph of the IC receiver channel (right)

A leading edge timing discrimination principle is used in the receiver channel as it allows the signal to be clipped and gives thus potentially a very wide dynamic range for the input signal amplitude. The timing comparator gives a timing signal for the TDC as the signal crosses a constant threshold voltage (V_{th}) at the input of the comparator. Unfortunately, the leading edge timing discrimination principle would produce a relatively large timing walk error in its basic configuration for the received optical echo pulses whose amplitude varies a lot due to the changes in the reflectivity, orientation and distance of objects. Timing error could be in nanosecond range (tens of centimetres) [1] which is, of course, unacceptable.

In the designed receiver-TDC configuration the timing walk error is compensated for by means of the known relation (calibration) between the generated walk error and the measured pulse length [1]. The compensation principle is shown in Fig. 3 where a constant threshold is used at the input of timing comparator. Thus the first timing mark is discriminated from the rising edge of the received signal $stop_{rising}$ and a second timing mark from the trailing edge ($stop_{falling}$). The TDC measures the time difference $start\text{-}stop_{rising}$ that is the distance information and $stop_{rising}\text{-}stop_{falling}$ (pulse length t_w for the compensation). As the walk compensation is done in the time-domain, it is operative also beyond the range where the receiver operates linearly. Thus the linear range of the receiver (typically quite low) does not limit the range over which the timing walk error can be compensated for as would be the case if the compensation would be based on amplitude measurement as in [2]. As noted above, the receiver channel and the multi-channel TDC can detect several successive pulses which may be produced as a result of a single laser shot due to the reflections from the front window or bad weather, for example.

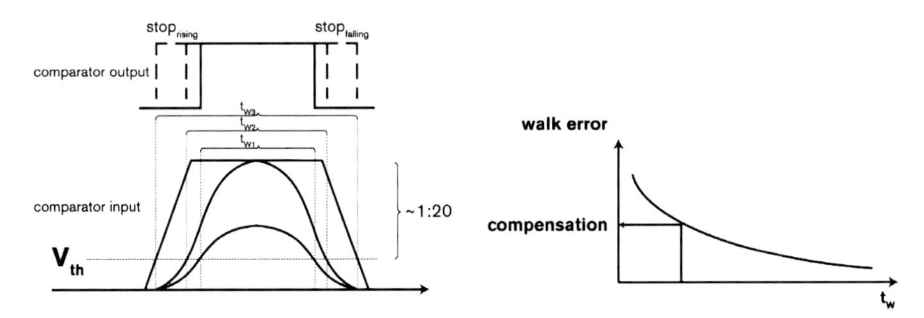

Fig. 3. Walk compensation based on pulse width measurement

3 Time-to-Digital Converter (TDC)

3.1 Operation and Architecture

The multichannel time-to-digital converter (TDC) measures the time intervals between electrical timing pulses and converts the results to digital words. The designed TDC is able to digitize the pulse positions (T_{SP1}-T_{SP3}) and widths (T_{w1}-T_{w3}) for three separate stop pulses presented in Fig. 4. As explained above, this measurement mode can be used to correct the timing walk error produced by varying timing pulse amplitude [1], [3].

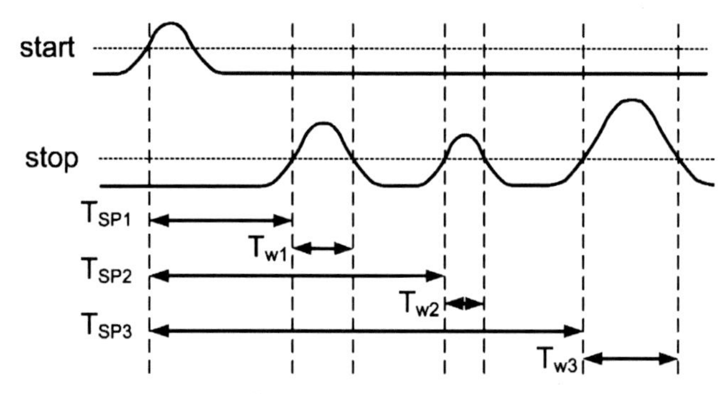

Fig. 4 Timing diagram of the TDC

The architecture of the TDC circuit is presented in Fig. 5. The low frequency (~10MHz) crystal is the only needed external component for the TDC circuit. The on-chip oscillator controls the crystal, which together form the reference signal for the measurement core. All the delays participating the measurement are locked to the cycle time of the reference clock with delay locked loop (DLL) techniques in order to prevent the resolution fluctuation with process, voltage and temperature (PVT) variations.

The time digitizing is based on a counter and a two-level stabilized delay line interpolation. A 14-bit counter counts the full reference clock cycles between the timing signals, what makes long µs-level measurement range possible. The interpolators find the locations of the timing signals within the reference clock cycles with ~10 ps resolution [4]. The TDC includes totally 7 parallel interpolators in order to detect the time positions of the start-signal and rising or falling edges of the three successive stop pulses.

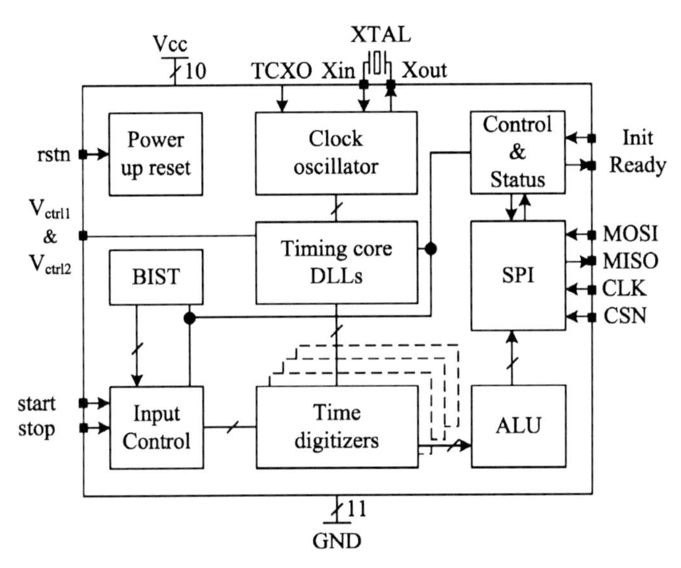

Fig. 5. Architecture of the TDC circuit

The ALU decodes the raw data and calculates the time intervals between the Start and Stop pulses. The data interface to and from the TDC is a 100MHz SPI (Serial Peripheral Interface), but Ready (measurement ready) and Init (initialize the registers) signals can be used to increase the measurement rate.

4 Measurements

The APD receiver PCB including the avalanche photo diode and both the receiver and the TDC circuits are shown in Fig. 6. The circuits are packaged in plastic QFN32 and QFN36 packages, respectively. Independent SPI interfaces allow the individual control of these devices.

Preliminary measurements have been carried out for the receiver and multi-channel TDC in the laboratory environment. Receiver channel has been measured with the on-board multi-channel TDC. Additionally, TDC performance has been verified in measurements done independently without the receiver channel. The walk error compensation was verified by measuring the compensation curve for the receiver channel. Additionally, the bandwidth, the single-shot resolution, the noise and the power consumption were also measured.

Fig. 6. APD is arranged on the bottom side of the PCB APD. All other components are placed on the top. PCB is designed by SICK. Receiver channel and TDC are developed by the University of Oulu

The receiver circuit, which is shown in Fig. 2 b), was fabricated in 0.35 μm SiGe BiCMOS technology. Fully differential structures have been used in all stages in the receiver channel. The dimensions of the receiver circuit are 1.6 mm x 1.6mm. The power consumption of the receiver channel was 132 mW. The bandwidth and the transimpendance of the receiver channel were 64 kΩ and 300 MHz, respectively. The rms voltage noise was measured from the output of the analogue output. Input referred rms noise current was about 100 nA_{rms}. The minimum acceptable input pulse current of the receiver was about 1 μA (assuming minimum acceptable signal-to-noise ratio SNR is around 10).

The walk measurements were performed by sweeping the input amplitude over the range of 1: 22.000 where the input pulse amplitude varied from 1 μA to 22 mA. For the compensation curve 2.000 measurements were averaged for each amplitudes. The measured walk error without compensation was about 2.2 ns in total which corresponds to about ± 16 cm is shown in Fig. 7 a). The walk error is compensated for by means of compensation curve measured at calibration measurement. The performance of the walk compensation was verified by measuring residual walk for random input amplitudes over the amplitude range of 1: 22.000. The measured residual walk was less than ±20 ps (< ± 4 mm) shown in Fig. 7 b).

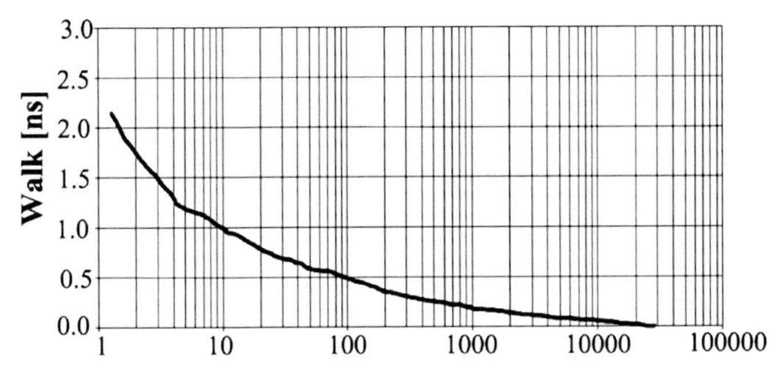

a) Input current [μA]

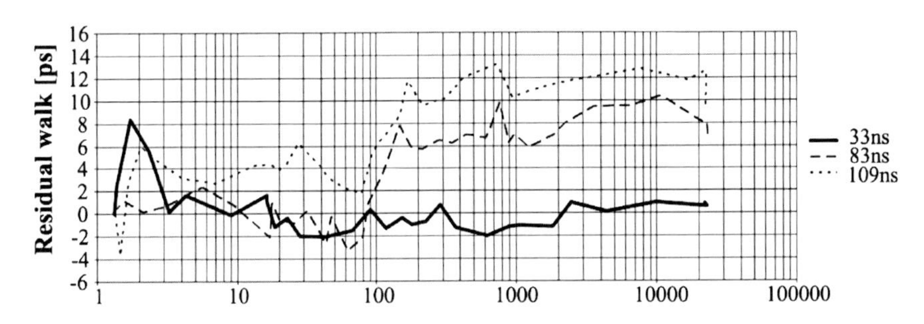

b) Input amplitude [μA]

Fig. 7. a) Measured, uncompensated timing walk error over dynamic range of 1:22 000, b) Residual walk errors for different start-stop time delays

The single-shot precision was also measured. A timing jitter was measured by measuring 2.000 single-shot measurements at a specific input amplitude and calculating the standard deviation of the timing point of the rising edge. Secondly, each of the individual measured result was compensated for by the means of a compensation curve and the single-shot precision was calculated from the obtained distribution. The worst case single-shot precision is shown in Fig. 8 including the jitter of the TDC, the jitter of from the receiver, jitter introduced by compensation and jitter caused by the measurement environment. The total single-shot precision was about 144 ps (±10 mm) in the preliminary measurements.

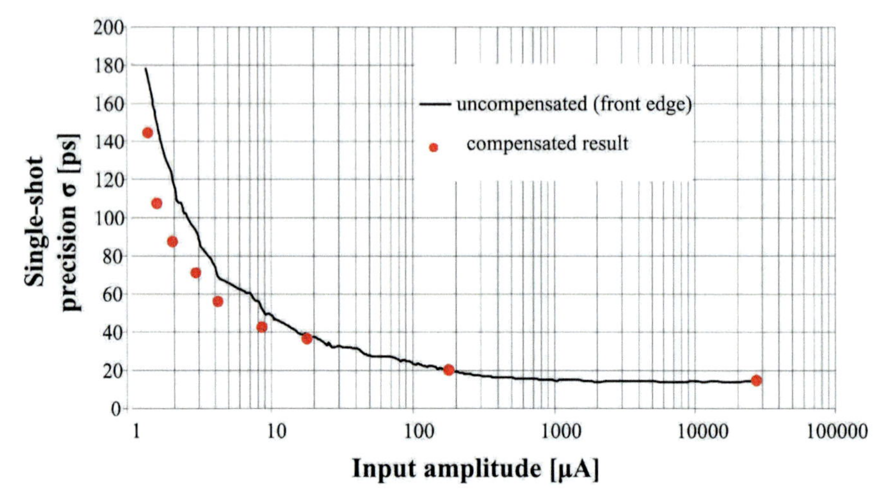

Fig. 8. Single-shot precision

The TDC circuit was fabricated in 0.35 μm CMOS technology. The dimensions of the circuit are 2.4 mm x 3.7 mm and the total area 8.89 mm^2. The power consumption at a 3.3V operating voltage, 20 MHz reference crystal and a 220 MHz internal frequency is 85 mW. The circuit was packaged in plastic QFN36 package.

Fig. 9 describes the measurement performance of the multichannel TDC. One start and three successive stop-pulses with different pulse widths and time intervals over several microseconds were generated with the Tektronix AWG 2021 signal generator. The stop-pulse cable was divided to both stop input channels of the TDC so that a longer coaxial cable was in the stop1 input path than in the stop0 input path. The measurements were repeated 100.000 times in the both measurement modes and the distributions of the results with their average values (μ) and standard deviations (σ) are presented in Fig. 9. The diagram on top of Fig. 9 shows the time intervals between the start and three successive stop-pulses and is similar in both measurement modes. The diagram above shows the stop-pulse widths. The jitter of the signal generator is present in the results.

5 Conclusions

A chip set was developed for a laser scanner application. The set includes two integrated circuits which cover the receiver channel and the time interval measurement functionalities, respectively. The chip set can detect optical echoes with a typical length of 3-4ns in a dynamic range of more than 1:10.000 with an accuracy of better than 100ps. This dynamic range exceeds substantially the typical linear range of a high-speed optical receiver and is achieved by realizing the timing walk error compensation in the time domain. Moreover, the chip set may detect simultaneously the time position of the three successive optical pulses.

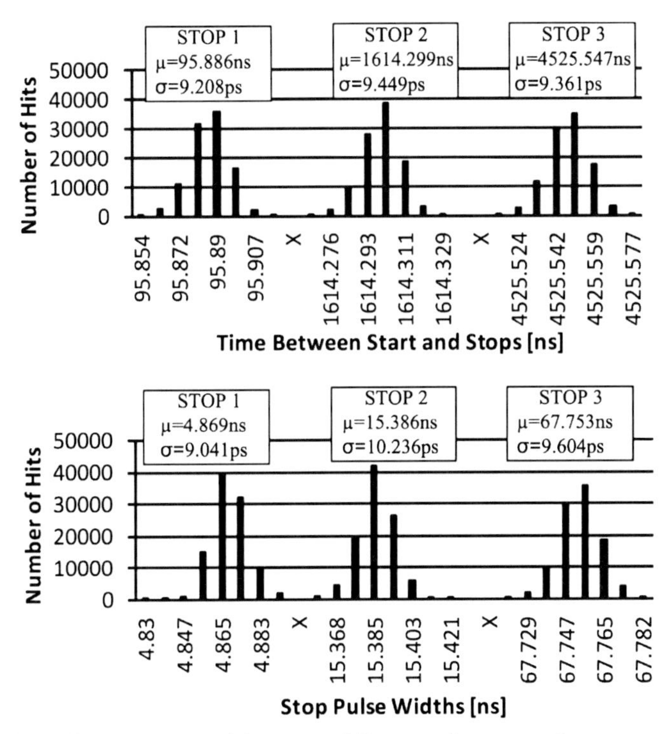

Fig. 9. Measurements with start and 3 successive stop pulses

Ackknowledgement

Minifaros is a part of the 7th Framework Programme funded by the European Commission. The partners thank the European Commission and the Academy of Finland for supporting the work of this project.

References

[1] S. Kurtti, J. Kostamovaara, An integrated laser radar receiver channel utilizing a time-domain walk error compensation scheme, IEEE Transactions on Instrumentation and Measurement, vol. 30, no. 1, pp. 146-157, 2011.

[2] P. Palojärvi, T. Ruotsalainen, J. Kostamovaara, A 250-MHz BiCMOS receiver channel with leading edge timing discrimination, IEEE Journal of Solid-State Circuits, vol. 40, no. 6, pp. 1341-1349, 2005.

[3] J. Nissinen, I Nissinen., J. Kostamovaara, Integrated receiver including both receiver channel and TDC for a pulsed time-of-flight laser rangefinder with cm-level accuracy, IEEE Solid-State Circuits, vol. 44, no. 5, pp. 1486-1497, May 2009.

[4] J.-P. Jansson, V. Koskinen, A. Mäntyniemi, J. Kostamovaara, A multi-channel high precision CMOS time-to-digital conveter for laserscanner based perception systems, IEEE Transactions on Instrumentation and Measurement, accepted for publication 2012.

Sami Kurtti, Jussi-Pekka Jansson, Juha Kostamovaara
University of Oulu
Electronics Laboratory
90014 Oulu
Finland
sami.kurtti@ee.oulu.fi
jussi.jansson@ee.oulu.fi
juha.kostamovaara@ee.oulu.fi

Keywords: LIDAR, optical distance measurement, timing discrimination, time interval measurement, Minifaros

Biaxial Tripod MEMS Mirror and Omnidirectional Lens for a Low Cost Wide Angle Laser Range Sensor

U. Hofmann, Fraunhofer ISIT
M. Aikio, VTT

Abstract

Low cost laser scanners for environment perception are a need to facilitate ADAS integration into all vehicle segments. To fulfill the need for mass-producible compact low cost laser range sensors MEMS mirrors in combination with replicable low cost plastic optics are expected to be suitable components. This paper describes concept, design, fabrication and first measurement results of a compact omnidirectional scanning system based on an omnidirectional lens and a biaxial large aperture tripod MEMS mirror. A hermetic vacuum wafer level packaging process of the resonant MEMS mirror is essential to meet automotive requirements and to achieve the required large total optical scan angles of 60 degrees in both scan axes.

1 Introduction

In order to seriously reduce the number of traffic accidents Intelligent Vehicle Safety Systems (IVSS) need to be introduced into all vehicle segments. Today IVSS are limited to a small part of the premium car segment. Future safety systems must be made affordable to penetrate all vehicle segments, since small and medium size cars are the ones dominating the road traffic and thus most of the accidents. Hence, there is a strong demand for low cost laser scanning sensors. A wide angular range and high angular resolution are key-features that laser scanning range sensors offer. But so far they have been based on expensive bulky servo motor driven scanning mirrors. It is one of the major objectives of the European funded FP7-project MINIFAROS to replace the expensive conventional scanning mirrors by low cost mass producible components [1]. MEMS mirrors batch fabricated on 8-inch silicon wafers in combination with replicable plastic optics present most promising candidates and therefore are further discussed in this paper (see Fig. 1).

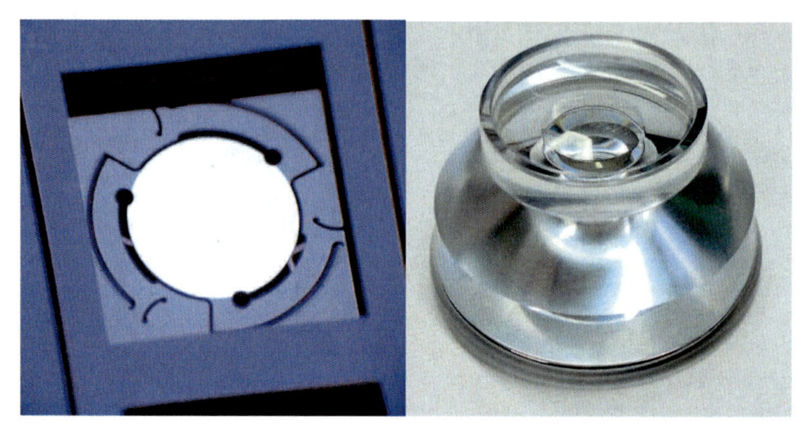

Fig. 1. Omnidirectional lens (right picture) and biaxial 7mm tripod MEMS mirror (left picture) are core components of the low cost MiniFaros laser scanner

2 Optical Sensor Concept

It is one of the goals of the MINIFAROS project to develop a laser scanner that can serve many different ADAS applications. An optical concept with an omnidirectional field of view was therefore chosen. The divergent exit beam of a fibre coupled NIR pulse laser diode is collimated and then directed on a large MEMS mirror which reflects and scans the beam on a circular trajectory along the input facet of an omnidirectional lens (see Fig. 2). After passing several internal beam forming reflections the laser beam exits the omnidirectional lens in horizontal direction. This arrangement allows to scan the beam in a horizontal plane within an angular range of 360 degrees. When the emitted pulse hits a target the laser pulse is partially reflected back and enters the omnidirectional lens. The MEMS mirror then reflects the return pulse to an avalanche photodiode. Two fundamentally different optical concepts are being investigated: A biaxial system and a coaxial system (Fig. 3). In the biaxial system the optical sender path and the receiver path are totally discoupled. While the sender path uses the front side of the MEMS mirror the back reflected laser pulse passes a second omnidirectional lens and then is reflected by the reverse side of the MEMS mirror. This configuration requires implementation of two different designs of omnidirectional lenses. The second system configuration, called the coaxial sensor, uses only one omnidirectional lens for both, sender and receiver path. Both optical paths use the same side of the MEMS mirror in a coaxial alignment. These two laser scanner concepts have their specific advantages and drawbacks. The biaxial laser scanner is by nature less sensitive to stray light than the coaxial laser scanner. Since the laser scanners

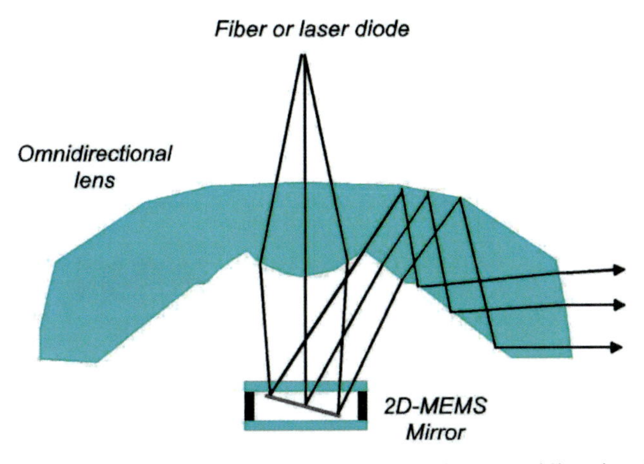

Fig. 2. Omnidirectional scanning concept based on omnidirectional lens and circle scanning MEMS mirror

operate with a highly sensitive detector, cross talk between the channels must be reduced as much as possible in order to prevent saturation of the receiver. The biaxial laser scanner however results in a slightly larger system size and is expected to be more complex with respect to alignment and the required alignment accuracy. So, finally a coaxial system approach was chosen.

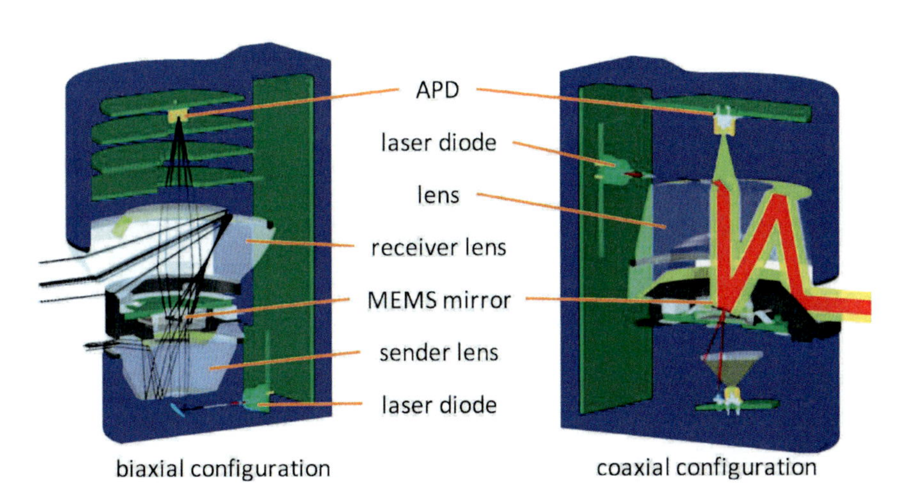

Fig. 3. Optomechanical concept of the laser scanning sensor showing the biaxial concept (left) and the coaxial concept (right)

3 Design and Fabrication of the Omnidirectional Lens

Many different lenses and system designs have been simulated and investigated with special focus on maximizing measurement range and minimization of straylight. The final design with a diameter of 60mm was realized by diamond turning in glass (Fig. 4).

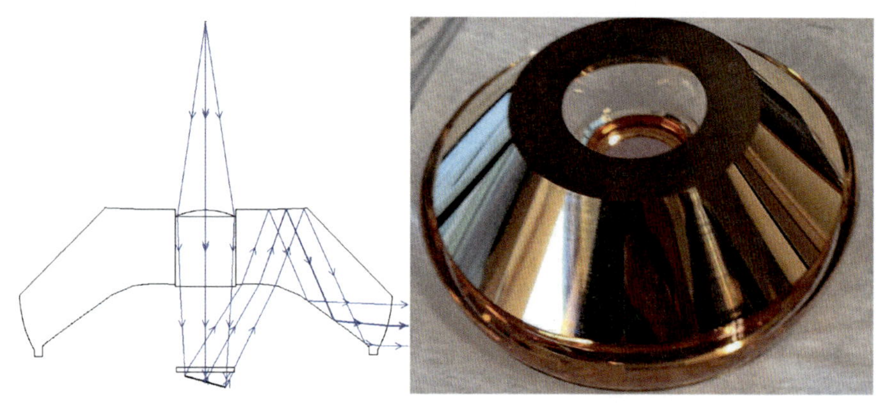

Fig. 4. Investigated optical lens design (left) and fabricated omnidirectional lens (partially gold coated)

4 MEMS Mirror Design

The MEMS mirror has to comply with the following specifications [3]:

- mirror diameter = 7mm (required in order to fulfil the measurement range of the sensor)
- circular scan trajectory (required to enter the circular aperture of the omnidirectional lens)
- mechanical tilt angle = +/-15 degrees in each axis (required to enable the compact sensor size)

For a MEMS mirror these are absolutely extreme specifications. A mirror aperture size of 7mm means a high mass moment of inertia so that only resonant actuation can be considered. But even then an electrostatic actuator would require adversely high driving voltages to achieve such large deflection angles. Since fabrication of piezoelectric drives can not yet be considered mature and since realization of an electromagnetically driven mirror with the need for mounting of large permanent magnets would lead to an unacceptably large and expensive chip level tailored solution a simple electrostatic actuation concept has finally been chosen. However, to enable such a large mechanical tilt angle of +/- 15 degrees a high Q-factor is necessary which can be achieved by vacuum encapsulation only. For the MiniFaros project that means that a wafer level vacuum packaging process has to be applied in order to minimize packaging costs.

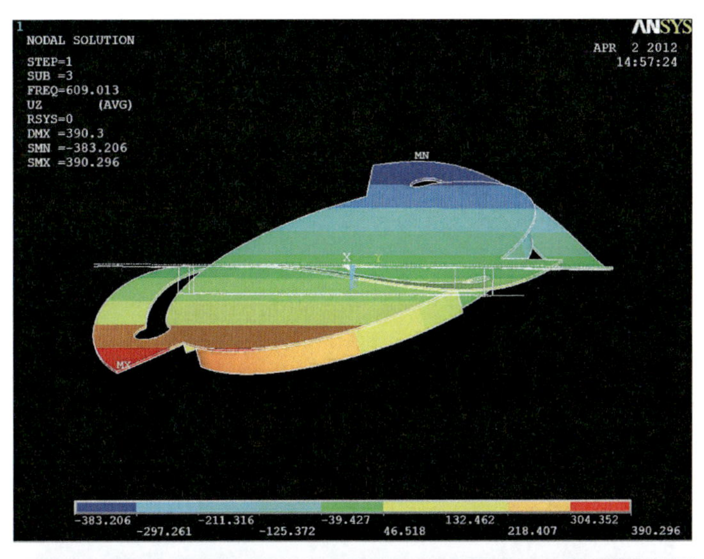

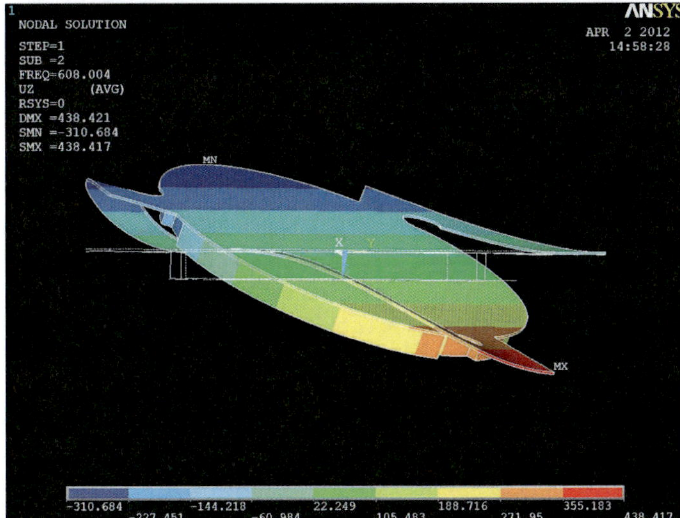

Fig. 5. FEM modal analysis of the 7mm tripod MEMS mirror. The tilting modes of the two perpendicular axes both are at 600Hz.

The next problem that arises is the large stroke if a mirror of 7mm is tilted by +/- 15 degrees. If a vacuum package is provided then the cavity needs to be deep enough to allow a stroke of roughly 1mm – provided that only the mirror is tilted. However, a biaxial mirror is required and in a standard gimbal mount configuration the actuator would easily generate a much larger stroke. Therefore, gimbal mount mirror designs were discarded and so another biaxial concept had to be found. To enable a circular trajectory a resonant MEMS mirror is required that comes with two orthogonal axes of identical resonant frequency. A solution for this has been found in a tripod MEMS mirror design consisting of a mirror plate that is suspended by three identical bending beams seperated from one another by rotation of 120 degrees.

Finite Element Analysis (FEA) has shown that to keep dynamic deformations of such a large mirror sufficiently low a 500μm thick mirror plate is necessary. The result of a modal analysis of that tripod mirror showing the natural frequency modes of the two orthogonal axes is shown in Fig. 5. Based on a thickness of the bending beams of 40μm both axes of the tripod oscillate at 600Hz enabling the MiniFaros laser scanner to scan the whole scenery of 360 degrees 600 times per second. Interlaced time of flight sampling is thus an interesting feature in comparison to conventional motorized low frequency scanning technology. Additional FEA has also been used to investigate the mechanical stress in the bending beams at the required tilt angle. The simulations show that even a deflection by +/- 20 degrees should not lead to any damage of the suspensions.

5 MEMS Mirror Fabrication

The electrostatically driven scanning micromirrors are fabricated on 8 inch silicon wafer substrates. Two 40µm thick polysilicon device layers are produced on top of a thermally oxidized silicon substrate applying epitaxial deposition. Each deposition step is followed by chemical mechanical polishing (CMP). Embedded between these two polysilicon device layers is a double silicon oxide layer that on one hand serves as a buried oxide hardmask and on the other hand is needed to electrically isolate a thin polysilicon interconnection layer that is embedded between these two layers of silicon oxide.Patterning of this interconnection layer is performed before deposition of the second oxide layer.

The buried oxide hard mask is patterned before deposition of the second 40µm polysilicon device layer applying photolithography and dry-etching. Depending on the photomask layout this etching of the oxide layer either stops at the lower polysilicon device layer or at the buried polysilicon interconnection layer.

This is an important feature for the subsequent 3D-etching of stacked vertical comb drive electrodes and furthermore, this is also important for getting a high flexibility in supplying different actuator regions with different electrical potentials. After deposition and CMP of the second thick polysilicon device layer a titanium-silver stack is sputtered on top and wet-chemically patterned to serve as high reflective mirror coating (Fig. 6a).

The front side structuring is finished by a DRIE etching process that defines the upper and lower comb electrodes, the mirror geometry as well as the bending beam suspensions. This patterning step uses a combination of a photoresist mask and the buried oxide hard mask described before. After turning the wafer, a further DRIE step etches and removes the parts of the silicon substrate underneath the MEMS actuator. The remaining thermal oxide layer generated at the process beginning is removed by HF vapor phase etching, which finally releases the MEMS devices (Fig. 6b).

This process offers large design flexibility. Besides the creation of stacked comb electrodes suspensions can be produced either with a thickness of 40µm or 80µm. The process offers the fabrication of a 80µm thick mirror plate free of stress inducing oxide layers while for other areas the two polysilicon layers remain vertically isolated by an intermediate oxide. In addition to vertical isolation an arbitrary number of laterally isolated areas such as driving or sensing electrodes can simply be produced by trench etching of the upper polysilicon layer. Each isolated area can be addressed by wires built in the buried interconnection layer.

Depending only on the photomask layout the reverse side etch can be used to provide 500 µm thick reinforcement structures underneath the mirror to enable mirror sizes of several millimeters with low dynamic deformation. The micromirror fabrication process and the mirror layout are chosen in such way that the mirror actuator is always surrounded by a closed frame of polished polysilicon without any further topography. By doing so, standard wafer bonding technologies like anodic bonding, glass-frit bonding or eutectic bonding can be applied to hermetically seal the microstructures.

a)

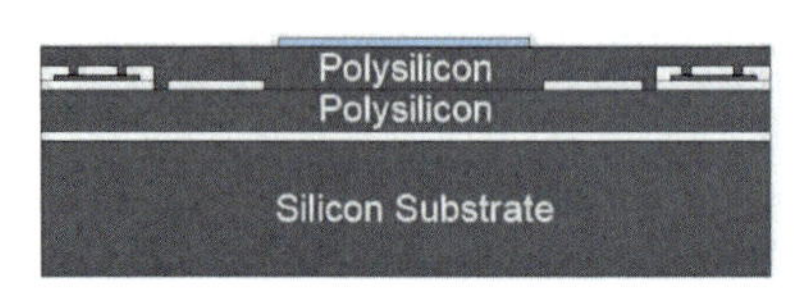

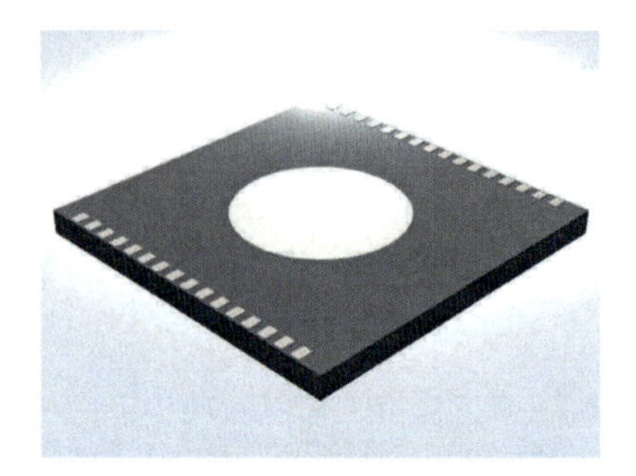

b)

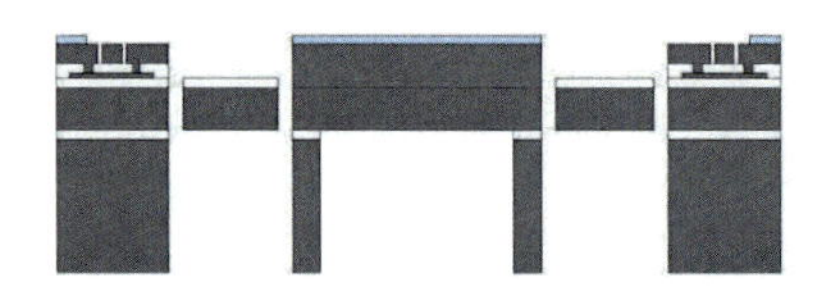

c)

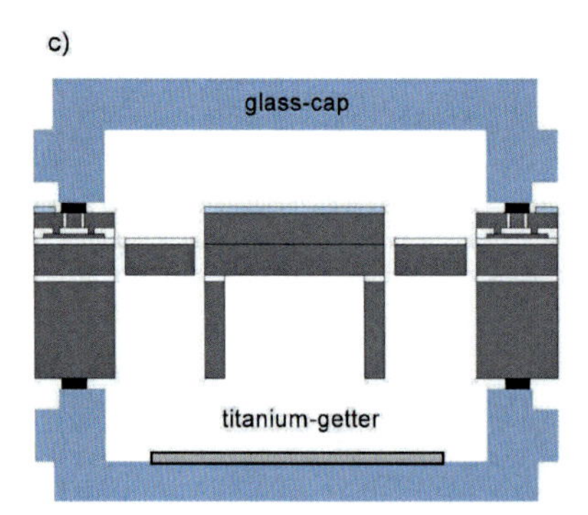

Fig. 6. Fabrication process of the vacuum packaged MiniFaros tripod MEMS mirror. The cavity depth is 1.6mm above and underneath the MEMS mirror. Glass wafers of such geometry are being fabricated by a unique glass forming process.

Fig. 7. First fabricated and wafer-level packaged tripod MEMS scanning mirrors

In the first step of the wafer level vacuum packaging process a borosilicate glass wafer having 1.6mm deep cavities is bonded to the frontside of the MEMS wafer applying glass-frit bonding. Perfect flatness and minimum roughness of the optical windows in the glass wafer can be achieved by a patented glass molding process [4]. Thereafter, a second glass wafer with 1.6mm deep cavities is bonded to the reverse side of the MEMS tripod mirror wafer. This second glass wafer is coated with a thin structured titanium getter layer to enable permanent cavity pressure levels below 1Pa after thermal getter activation (Fig. 6c). Fig. 7 shows recent results of first fabricated and packaged tripod MEMS wafers.

Acknowledgement

This work has been supported by the EC within the 7th framework programme under grant agreement no. FP7-ICT-2009-4_248123 (MiniFaros). The partners thank the European Commission for supporting the work of this project.

References

[1] K. Fuerstenberg, F. Ahlers: Development of a Low Cost Laser Scanner – the EC Project MiniFaros. In: G. Meyer, J. Valldorf [Eds.], Advanced Microsystems for Automotive Applications 2011, Smart Systems for Electric, Safe and Networked Mobility, Springer, Berlin, 2011.

[2] M. Aikio, Omnidirectional Lenses for Low Cost Laser Scanners. In: G. Meyer, J. Valldorf [Eds.], Advanced Microsystems for Automotive Applications 2011, Smart Systems for Electric, Safe and Networked Mobility, Springer, Berlin, 2011.

[3] U. Hofmann, J. Janes, MEMS Mirror for Low Cost Laser Scanners. In: G. Meyer, J. Valldorf [Eds.], Advanced Microsystems for Automotive Applications 2011, Smart Systems for Electric, Safe and Networked Mobility, Springer, Berlin, 2011.

[4] U. Hofmann, et al., Wafer-level vacuum packaged micro-scanning mirrors for compact laser projection displays, Proc. SPIE, Vol. 6887, 2008.

Ulrich Hofmann
Fraunhofer ISIT
Fraunhofer Str. 1
25524 Itzehoe
Germany
ulrich.hofmann@isit.fraunhofer.de

Mika Aikio
VTT
Kaitoväylä 1, P.O.Box 1100
90571 Oulu
Finland
mika.aikio@vtt.fi

Keywords: laser scanner, MEMS, mirror, vacuum package, electrostatic drive, tripod design, time of flight, omnidirectional lens

A Generic Approach for Performance Evaluation of Vehicle Electronic Control Systems

A. Hanzlik, E. Kristen, AIT - Austrian Institute of Technology GmbH

Abstract

The DTF Data Time Flow Simulator is developed at the AIT Austrian Institute of Technology in the course of the EU ARTEMIS Project POLLUX which is related to the design of electronic control systems for the next generation of electric vehicles. An important element of a vehicle is the communication architecture for the electronic control system. The DTF is a model based development approach, based on a modular assembly system for incremental design and analysis of electronic control systems. A current application of the DTF is related to the design and performance analysis of communication architectures for the electronic control system of an electric vehicle. This paper introduces a generic approach that shall enable companies to design and assess their control systems without the need to provide corporate know-how to third parties like tool developers.

1 Motivation and Objectives

The automotive industry operates in a very competitive market. Automotive manufacturers and suppliers are quite restrictive with the disclosure of detailed technical information about their products. One example is the communication architecture for the electronic control system of the vehicle, which consists of the physical communication network and the communication schedule. Evidence from the automotive industry shows that the design of such architectures is an arduous task that may take several months or even years for development and test. Considering the efforts that are spent in the design of such architectures, appropriate tools are a valuable support in the development and test phase. The challenge for tool developers in such a market is to provide tools for the development of electronic control systems in the absence of detailed technical information.

The DTF simulator is a simulation environment that shall support the design of electronic control systems in a very early phase of development. In a first step, a simulation model of the control system is created. Then, a set of test cases is defined that shall validate the control system against the requirements specification. The focus of these tests is on two thematic priorities:

- **Network peak load**
  The observed network peak load gives information about the stability reserves and the room for future extensions of the control system.

- **Control signal latencies**
  The observed control signal latencies show if the control system fulfills all timing requirements defined in the requirements specification.

If the tests reveal requirements violations, the simulation model is iteratively refined until all requirements are fulfilled. Possible measures are

- **Modification of the physical system structure**
  e.g. by increasing communication network bandwidth

- **Modification of the communication schedule**
  e.g. by changing the assignment of message priorities

The motivation for such an approach is to avoid costly re-design activities in the integration and test phase by anticipation of possible design weaknesses of the control system in a very early phase of development. Although the creation of the simulation model takes time, changes to the system structure can be done in the simulation environment before any hardware or software exists. Given that the model coverage is appropriate, the appearance of substantial design weaknesses in the integration and test phase of the control system is less likely than in a conventional development approach where possible design weaknesses are revealed in later development phases in which hardware and software partially exist.

The main objective of such an approach is to reduce the probability of occurrence of costly re-design activities in the integration and test phase of the control system.

2 System Model

Modern vehicle control systems are distributed systems, built from spatially separated electronic control units (ECUs) interconnected via shared communication resources [3]. ECUs are embedded systems that control one or more of the electrical or mechanical systems in a vehicle.

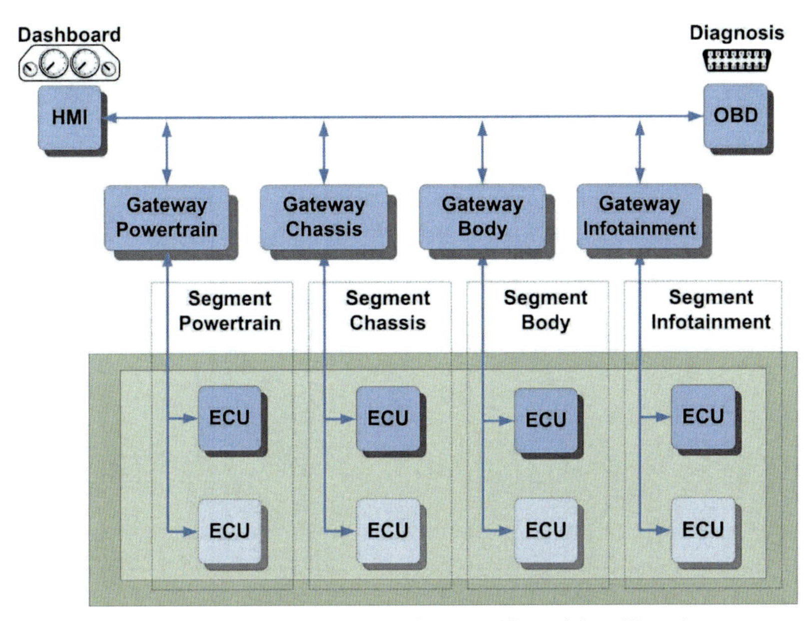

Fig. 1. Vehicle Communication Architecture (Copyright: dSpace)

For the following considerations, we assume that a communication architecture for a vehicle electronic control system can be built by repetitive use of the following components:

- **ECU.** An Electronic Control Unit (ECU) is an embedded system that controls one or more of the electrical systems or subsystems in a vehicle. An ECU receives state signals from sensors and/or other ECUs and issues control signals to actuators and/or other ECUs. Different ECUs are assigned different control tasks, like engine control, damping control or transmission control.

- **Communication medium.** A shared communication resource interconnects different ECUs. For communication, dedicated communication protocols are used, like CAN [1], FlexRay [2], LIN [5] or MOST [6].

- **Segment.** ECUs are grouped into segments, which are sets of ECUs that execute a distributed control task concurrently. The ECUs in a segment are linked via a communication medium. Typical segments in a vehicle are the powertrain, the body electronics or the infotainment segment.

- **Gateway.** A gateway links different segments and allows for communication between these segments. Further, gateways are used for gathering diagnostic information from each segment.

- **Transceiver.** Transceivers are the interfaces between the communication architecture and the vehicle electronic control system. Sensors are used to deliver signals for further computation, like the brake pedal position or the current lateral speed. Actuators are used to control mechanical or electrical components, like the brakes or the motor rotational speed.

Figure 1 shows a vehicle communication architecture providing four segments (Powertrain, Chassis, Body and Infotainment), four gateways and connections to the dashboard and the on-board diagnostics interface of the car.

3 The DTF Simulator

The DTF simulator is a discrete-event simulation [4] environment for the design and validation of electronic control systems. It is based on a modular assembly system that provides a set of primitive building blocks, the elements. Elements can be grouped together to form more complex structures, like ECUs. Each element has at least one input, one output and a propagation delay. When an element receives data at its input, the data is processed and, with the propagation delay defined for this element, the processed value is sent to the output.

The following elements are provided by the DTF simulator:

- **Sensor.** The sensor element is the input interface to the simulator. Its task is to process events from the configuration file.

- **Actuator.** The actuator element is the output interface of the simulator.

- **Processor.** The processor element processes data received from other elements, e.g. from sensors. The processor element can be provided with a user defined action routine that defines the data transformation from the input to the output.

- **Memory.** The memory element stores data received from other elements. Its task is to store data and transmit the data unchanged with a defined propagation delay.

- **Controller.** The task of the controller element is to assemble frames for transmission over a communication network, e.g. a CAN bus.

- **Extractor.** The extractor element disassembles data received from a communication network, e.g. a CAN bus.

- **Line.** The line element is responsible for simulation of the communication network protocol. Typical tasks of the line element are message transmission and arbitration of the communication network.

- **Transfer.** The task of the transfer element is to distribute received data unchanged and without delay to its outputs. It is the only element that has more than one output.

Elements are linked together to form so-called action chains. An action chain is a directed path from a source element (e.g. a sensor) to a destination element (e.g. an actuator). Figure 2 shows a simple electronic control system consisting of two sensors S1 and S2, two actuators A1 and A2, ECUs A, B C and D and a communication network BUS. The ECUs and the BUS are constituted by elements from the DTF modular assembly system. For example, the ECUs contain a Processor element Proc and a Controller element CC. The control system has four possible action chains: one from sensor S1 to actuator A1, one from sensor S1 to actuator A2, one from sensor S2 to actuator A1 and finally one from sensor S2 to actuator A2.

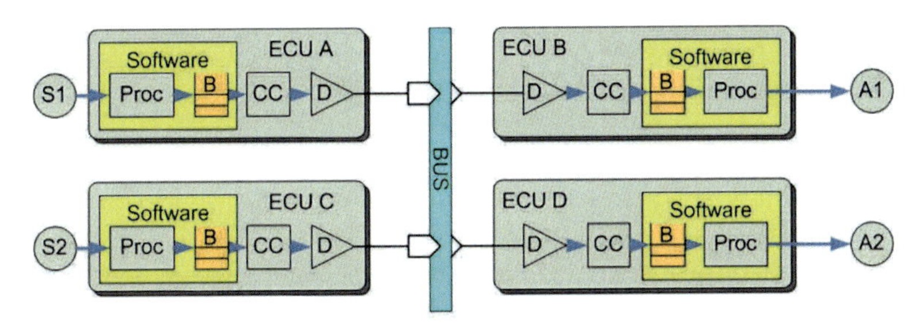

Fig. 2. Simple Electronic Control System

To assess the properties of the control system, signals are issued to the sensor elements and signal propagation is observed, both in the domains of value and time, from the sensors over the communication network to the actuators. From the time distribution of signal propagation over the communication network important information can be gained with regard to the dynamics and responsiveness of the system, especially for safety-relevant signals that usually are subjected to real-time constraints, like e.g. the brake pedal position.

The DTF simulator has been validated using a robot driving platform developed at the AIT. The robot driving platform provides a simple vehicle communication architecture, consisting of one network segment using a CAN bus for communication between the different ECUs that control the powertrain of the platform. More details about the structure of the robot driving platform and results from validation experiments using a DTF simulation model can be found in [7].

4 Vehicle Communication Architecture Simulation

To be able to assess and validate a complete vehicle communication architecture, a lot of information is required to be able to build a proper simulation model of the architecture. This information comprises the assignment of ECUs to segments, the communication medium bandwidth per segment, the requirements w.r.t. control signal latencies, possible physical constraints that may have an impact on the architecture and many others. However, this information is usually not available in detail. The automotive industry operates in a very competitive market, and the competitors like OEMs and suppliers are very reserved and cautious with the disclosure of information, which may lead to the loss of a competitive advantage. Therefore, some kind of generic approach is needed that enables companies to assess their architectures without the need to provide corporate know-how to third parties like tool developers.

Such a generic approach is introduced in the following. Based on the system model introduced in Section 2, a complete vehicle communication architecture is designed using ECUs, communication media, segments, gateways and transceivers.

Figure 3 shows a possible vehicle communication architecture for an electric car. This structure has been composed using information available from different car manufacturers, like workshop manuals. Although these sources do not reveal everything in detail, some general assumptions about the structure of such architectures can be derived from literature studies.

The proposed vehicle communication architecture (Figure 3, left hand side) consists of six network segments and two gateways. We assume an electric car with two separated power trains for the front axle and the rear axle. We therefore provide two separate network segments for the power train, SEG_PTFRONT and SEG_PTREAR, and two separate network segments for the chassis electronics, SEG_CHASSIS1 and SEG_CHASSIS2. The Power train segments are responsible for all control tasks related to power distribution and driving dynamics. The Chassis segments cover functions like damping and steering. The battery segment SEG_BATT is responsible for the provision of electric power for the whole vehicle and for power management functions like charging and power saving. The Body segment SEG_BODY covers functions like lighting, door electronics and mirror adjustment. For communication between the different segments, two gateways are provided. Gateway GW_PTCH is mainly responsible for communication between the Power train and the Chassis segments. Gateway GW_BODY connects the Body segment and the Battery segment to the system.

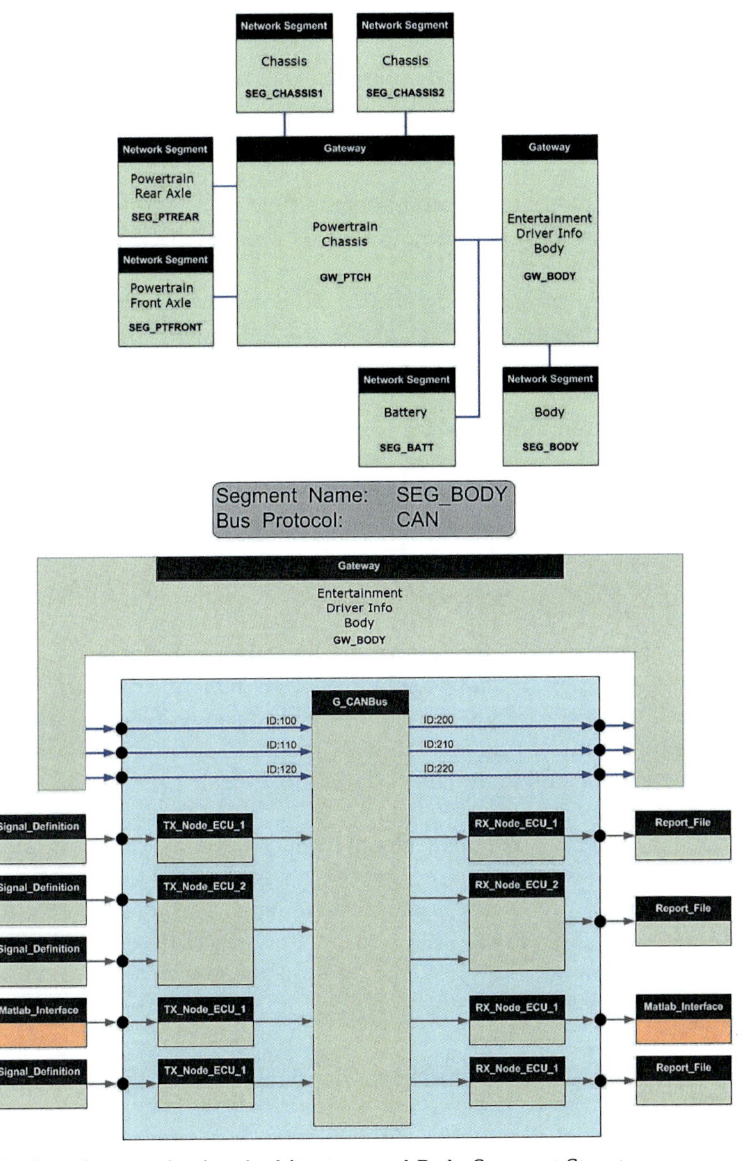

Fig. 3. Communication Architecture and Body Segment Structure

For the assessment and validation of such an architecture, we recall from Section 1 that the properties of interest are:

- **Network Peak Load**
  The observed network peak load gives information about the stability reserves and the room for future extensions of the control system.

- **Control Signal Latencies**
  The maximum observed latency for each signal within a network segment and for each signal that originates from one segment and that is delivered to another segment over a gateway.

We assume that the communication demand within the different segments is considerably higher than the communication demand between segments.

For the assessment of each segment we propose a structure like that shown in Figure 3, right hand side, which shows the Body segment. Each segment is constituted by a set of so-called TX_ECUs and RX_ECUs interconnected via a shared communication medium. In our example, we use a CAN bus. TX_ECUs receive signals from simulated sensors; these signals either originate from signal definitions in the simulation configuration file or from external simulators like MATLAB [8] (red shaded signal source in Figure 3, left hand side). These signals are packed into frames and sent to the CAN bus. The CAN bus transmits the frames sent by the TX_ECUs to the RX_ECUs and is responsible for the arbitration of the frames according to the assigned message priorities.

RX_ECUs receive the frames from the CAN bus, extract the information of interest from the frames, and calculate a signal value for the control of simulated attached actuators; these actuators are modeled either as outputs into the simulation report file or into an external simulation environment like MATLAB (red shaded signal sink in Figure 3, right hand side).

Consider for example an action chain from TX_Node_ECU_1 to RX_Node_ECU_1. Every time a signal is received at TX_Node_ECU_1 from its attached sensor, the current simulation time is recorded (t1). Similarly, every time a signal is sent from RX_NODE_ECU_1 to its attached actuator, the current simulation time is recorded again (t2). The difference between the two timestamps (t2-t1) is the control signal latency.

The determination of the latencies for control signals that originate in one segment and that are delivered to another segment is quite similar. The only difference is that the delay induced by the gateways has an impact on the overall control signal latency.

The determination of the network load of the CAN bus is quite straightforward. Every time the CAN bus is occupied by a frame sent by a TX_Node_ECU, the transmission time of the frame in transit is calculated according to the bandwidth and the message length and related to the CAN bus capacity.

This approach provides a framework for modelling a complete vehicle communication architecture. Detailed information about how many ECUs are in which segment, which protocol is used for communication and other considerations are not necessary. It's up to the user of the framework to define the properties of interest. Focusing on network peak load and control signal latencies does not require knowledge about the semantics of the control signals. Only the frame length, the message transmission rate of each TX_Node_ECU and the bandwidth of the communication medium have to be known to be able to assess the essential properties of the vehicle communication architecture: the network peak load and the control signal latencies.

5 Conclusions

The main purpose of the DTF simulator in the context of the POLLUX project is the design and assessment of vehicle control architectures for the next generation of electric cars. It shall be possible to anticipate and to avoid possible design weaknesses in early development stages, ideally right away from the requirements specification. The aim is to be able to take all relevant design decisions before implementation to avoid costly re-design activities in later development stages, when hardware and software are already available.

Although details of the final architecture design are not available in very early development stages, an assessment of the overall communication architecture with regard to network load and control signal latencies can be done by experimental evaluation of different architectures. These simulated architectures are based on requirements that define upper bounds for the network load and upper bounds for each control signal. Such a simulation based design process shall lead to considerable development time and cost savings due to minimization of re-design efforts.

References

[1] K. Etschberger, (Hrsg.), CAN Controller Area Network – Grundlagen, Protokolle, Bausteine, Anwendungen. Hanser Fachbuchverlag, Deutschland, 2001.

[2] FlexRay Communications System Protocol Specification Version 3.0.1. Available from http://www.flexray.com.
[3] W. Zimmermann, R. Schmidgall, Bussysteme in der Fahrzeugtechnik – Protokolle, Standards und Softwarearchitektur. Vieweg+Teubner, 4. Auflage, 2010.
[4] J. Banks, J. Carson, B. Nelson, Discrete-Event System Simulation. Prentice Hall, 2000.
[5] A. Grzemba, H. Von der Wense, LIN-Bus. Franzis, Deutschland, 2005.
[6] A. Grzemba, MOST. The Automotive Multimedia Network. From MOST25 to MOST150. Franzis, 2011.
[7] A. Hanzlik, E. Kristen, DTF - A Simulation Environment for Communication Network Architecture Design for the Next Generation of Electric Cars. In: G. Meyer, J. Valldorf [Eds.], Advanced Microsystems for Automotive Applications 2011, Smart Systems for Electric, Safe and Networked Mobility, Springer, Berlin, 2011.
[8] A. Downey, Physical Modeling in MATLAB. Green Tea Press, 2008.

Alexander Hanzlik, Erwin Kristen
AIT Austrian Institute of Technology GmbH
Donau-City-Str. 1
1220 Vienna
Austria
alexander.hanzlik.fl@ait.ac.at
erwin.kristen@ait.ac.at

Keywords: simulation, automotive control systems, data and time flow, communication network performance analysis, electric vehicles

Outlook for Safety and Powertrain Sensors

R. Dixon, IHS iSuppli

Abstract

In the past 15-20 years, sensors have grown organically and reached relatively stable penetration rates with very low growth, at least in the mature automotive markets. Any new stimulus therefore comes from the adoption of safety and basic powertrain sensors and regulations and mandates that force adoption to full penetration in different regions. The current paper provides a 5-year snapshot of the market opportunity for safety and powertrain applications for the silicon sensors based on Micro Electro Mechanical System (MEMS) technology and also magnetic sensors used in position, rotation and angle measurements, and looks at the underlying drivers in detail.

1 Market Overview

IHS iSuppli estimates that its silicon sensor coverage—accelerometers, gyroscopes, pressure, flow, linear position, rotation, angle, etc.—accounts for about 75% to 80% of the key automotive sensor insertion points. In the discussion that follows, IHS iSuppli excludes body and chassis sensors, which includes a large number of applications for the former, but is relatively small in revenue terms. Nor does IHS iSuppli include oxygen sensors, knock or oil quality (typically piezoelectric), temperature and humidity, and cylinder combustion sensing (e.g. ionic). Advanced Driver Assistance Systems (ADAS) including radar, cameras, ultrasonic parking sensors, etc. are covered by a different group within IHS iSuppli. Note also that all revenues for sensors are captured for first level packaged components.

With these caveats in place, we can still comfortably say that the automotive sensor market is entering a stage of fast growth: overall the market now grows at a compound annual growth rate (CAGR) of 11% from 2010 to 2015, exceeding $4.2 billion in 2015. These expansion rates are above average, as prior to the early 2009 recession, sensor sales had been running at about 7- 8% on average.

What drives this growth is the legislation requiring improvements in safety and emissions, which, as this is best achieved with the help of sensors, has have become the motor of the automotive MEMS sensor economy. In 2011, about 55% of the above-described MEMS sensor revenues alone were accorded to safety domain applications, compared to about 30% for sensors used in powertrain. The balance is accorded to body and chassis applications.

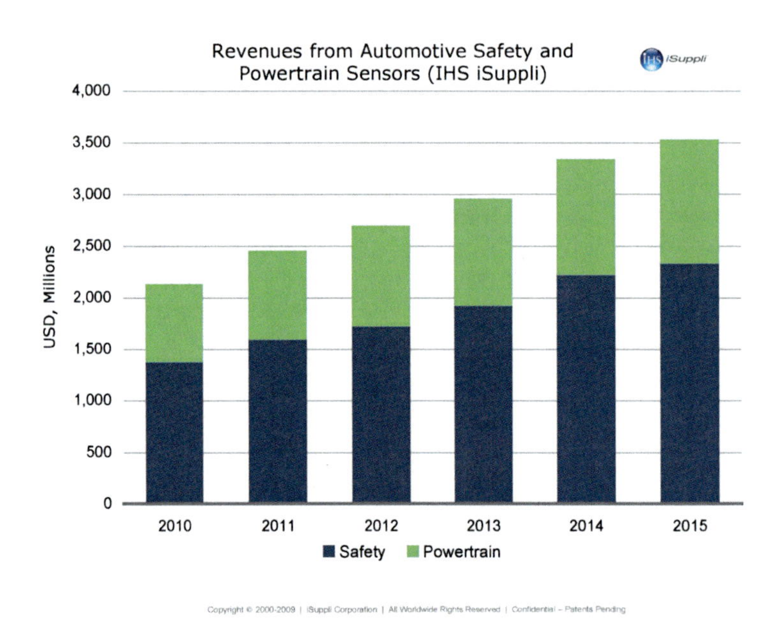

Fig. 1. Sales for safety and powertrain sensors (IHS iSuppli)

Overall the market for both the safety and powertrain domains was $2.1 billion in 2010 and will reach $3.5 billion in 2015. This is a CAGR of 10.6%.

2 Safety Sensors

2.1 Safety Mandates as Market Engine

Safety mandates are acting as the glaciers that sculpt the market and supplier landscape, propelling the market as regulations demanding full fitment rates spread worldwide for safety systems like Electronic Stability Control (ESC) and Tire Pressure Monitoring Systems (TPMS).

2.2 Electronic Stability Control

The ESC system is an extended ABS system that has been around since 1995. For the ESC system, a number of sensors are needed:

- Accelerometers (2x 1-axis or 1x 2-axis), gyroscopes (1-axis) for the inertial cluster (IMU)
- Pressure sensor(s) (1x or more of approx. 200 bar) for the brake modulator
- Steering angle sensor (1x Hall or AMR, optical...)
- Wheel speed sensors (4x, Hall or AMR)

The prices of these components—especially gyroscopes and low g lateral X, Y accelerometers—are still relatively high. Thus the market for ESC related sensors is the largest, growing from $577 million in 2010 to $997 million in 2015, a CAGR of 11.5%.

For safety mandates like ESC, the U.S. has taken the lead, but has been recently joined by the EU zone, the Republic of Korea, Australia, Canada and most recently, Japan. By late 2012, all new vehicles selling domestically must therefore be fitted with ESC systems. Other recent regulations include ejection mitigation, a U.S. mandate that aims to prevent passengers exiting the car in an accident, and facilitating the accelerated adoption of roll sensors to prepare seat restraints and activate the required airbags.

2.3 Price Demands Accelerate Innovation

Safety mandates lead to accelerated market growth. This in turn leads to a more dynamic sensor supply chain, which is good for sensor suppliers. And while some established players like Systron Donner have left the market, others with more aggressive price structures like Panasonic (which can leverage binning via very high volume manufacturing for cameras) have had considerable success with gyroscopes, for instance. The major ESC sensor suppliers comprise Bosch, VTI, Denso, Panasonic ADI, MEMSIC, Goodrich (Silicon Sensing Systems) for inertial sensors, and Sensata, Bosch and Denso for pressure sensors used in the brake modulator. Steering angle sensors is provided by companies like Allegro, Micronas, and Infineon—the latter is a major supplier of Hall Effect wheel speed sensors, along with NXP. Systems are provided by Bosch, Continental, TRW, ADVICS, Autoliv, Beijing West, Mando and others.

Innovation has also been stimulated as a result of high price pressure caused by safety mandates: an example is the adoption of multiple sensors in single packages (eliminating packages and ASICs) and co-location with other sensors, such as accelerometers and gyroscopes with high g accelerometers inside airbag ECUs. These so-called "combo sensors" are now beginning to enter the market from VTI and Bosch and will assume significant volumes in 5 years. The chances for sensors integrated on the same piece of silicon—the ultimate low cost device—are unlikely to materialize at any point in the future due to technical challenges.

2.4 Tire Pressure Monitoring

Tire pressure monitoring systems or TPMS have also been in the market for some years. Two systems are found; an indirect system that relies on algorithms to calculate wheel radius (hence if the tire changes) from wheel speed measurements in on-board ABS systems, and a direct measurement using MEMS pressure sensors (the bulk of the market today). Having started as fairly simple systems with a pressure die, companies like Infineon and GE now supply pressure sensors, accelerometer wake up switches, control ASIC and additional RF electronics in a first level package.

The U.S. (via some strong lobbying from the insurance companies and National Highway Traffic Safety Administration) has been out front in leading the regulated adoption, having passed a law (FMVSS No. 138) requiring the full fitment of such systems into law since 2007. Europe, Korea and significantly, China, have followed. China is applying a rapid phase-in program that will see full adoption from mid-2015, starting in 2012 with a fraction of engine sizes over 1.6 l. China's impact will accelerate the fitment rate to 73% of vehicles worldwide by 2015.

In European circles, discussions currently center around which of these technologies will dominate in the European market when TPMS is required on new models in 2012 and all new vehicles in 2014. Indirect systems provide by nature relative values and require the system to be reset when the tires are inflated, although spectrum analysis allows all four tires to be monitored.

Key suppliers of the direct systems include Schrader and Continental, while Nira Dynamics is a proponent of the indirect system, so far adopted by Audi and BMW.

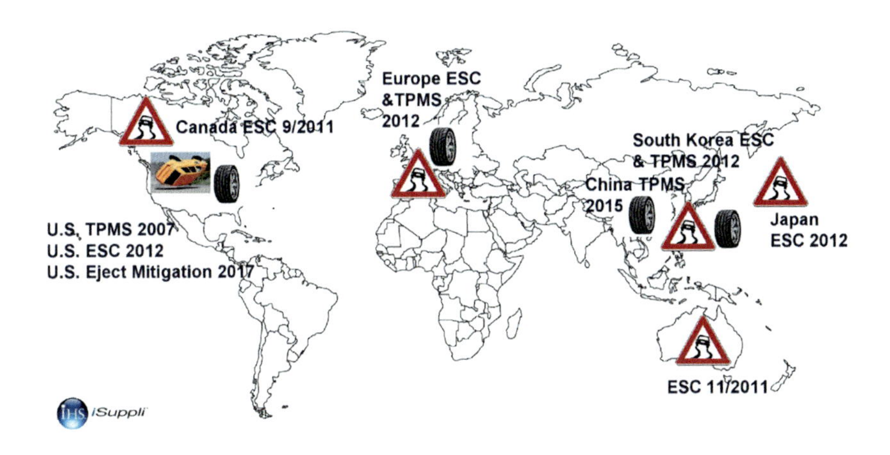

Fig. 2. Location and timeline of safety mandates affecting safety sensors (IHS iSuppli)

3 Powertrain Sensors

The market for powertrain sensors tracked by IHS iSuppli was equal to $760 million in 2010 and will grow to over $1,200 million in 2015, a CAGR of 9.6%.

At least 35 applications based on MEMS or magnetic sensors are identified for powertrain. Key sensor insertions include manifold air pressure sensors (MAP) and barometric air pressure (BAP) fitted in the main ECU for engine management applications, mass air flow sensors for diesels, Exhaust Gas Recirculation (EGR) position and pressure, high pressure metal membrane devices for common fuel rail pressure, transmission sensing and diesel particle filters. Stop start systems are an opportunity for crank angle sensors. Many other kinds of measurements are made, including the level of various fluids or position of throttle valves and acceleration pedals, many kinds of motors and battery monitoring sensors.

In the powertrain domain, emission regulations continue to keep manufacturers on their toes. The advent of the standard Euro VI in 2015 with more stringent particle number limits will almost certainly impact gasoline direct injection engines, which produce high levels of particles compared to diesel direct injection engines. Although fundamental engine based alternatives are also investigated, these regulations (in force by September 2015 for all models) will mean GDI engines may have to be fitted with gasoline particulate filters in the same way as diesel powered vehicles.

While companies concentrate on the newest European and U.S. standards, emerging countries fill in the space behind. China's regulation is now advocating Euro V emission standards (Chinese standards are based on European standards) to reduce pollution, starting with Beijing at the end of 2012. China adopted China 4 (Euro IV) for spark ignition engines nationwide in July 2011, and diesels will follow suit in 2013.

China is particularly important for suppliers of pressure sensors for manifold air measurements. Bosch has made a good business in recent years and is a major supplier of this kind of sensors to the region. BAP sensors and common fuel rail pressure sensors capable of measuring 2000 bar (higher in the newest systems) are also needed for diesel engines, although China has not yet embraced this type of powertrain in the same manner that Europe has.

Generally, sensors are placed where measurement improves control and performance. As every gram of CO_2 or 1/10 litre of fuel counts, system developers look more and more to sensors to trim these key metrics. Therefore there is increasing interest in more precise control of transmission (especially now manual and multi-gear double clutch transmissions). Bosch has recently introduced a high pressure MEMS device (up to 70 bar) for this kind of application, while Hall, Reed switches and permanent-magnetic linear Contactless Displacement (PLCD) technology is used to provide fork position information (e.g. for stop-start systems).

There is also a trend to electrification of the many pulley-driven motors in a vehicle (main and auxiliary water pump, oil pump, etc.), replacing brushed motors with brushless DC motors where possible. These electronically controlled motors operate on a need basis and require commutation switches (Hall Effect or AMR) for their control; they also may use current sensors, e.g. the steering column motor. Stop-start systems require current (shunt resistor or Hall) and temperature measurement in order to ascertain if the battery has sufficient charge for the restart. Clearly, hybrid electric vehicles will use a number of such devices (up to 6).

The other major trend, engine downsizing via turbo-charging, requires MAP-like sensors operating at up to 3 bars and also position sensing devices, especially in the case of variable position turbochargers.

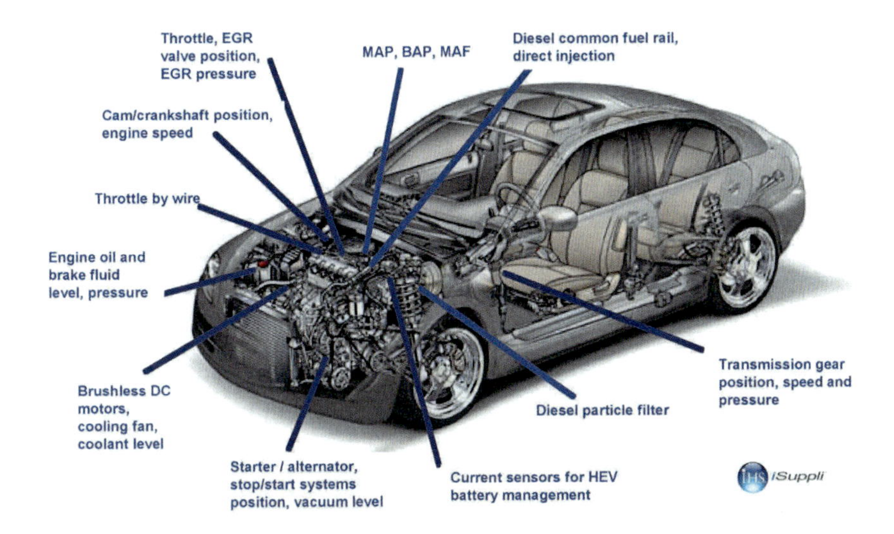

Fig. 3. Major applications for powertrain sensors (IHS iSuppli)

The major suppliers in this domain are the same as for safety applications but include Denso, Bosch, Sensata, Melexis, Fuji Electric and Kavlico for pressure sensors, and Hall Effect IC suppliers like Allegro, Micronas, AustriaMicroSystems, Infineon, NXP, Melexis and AKM.

4 Supply Chain

4.1 One Year after Japan Earthquake: Darwinism in Action

The Japan earthquake and tsunami in March 2011 and floods in Thailand later that year represented a challenging time for the automotive industry. IHS Automotive forecasts [1] indicated that depending on how long the production delays would continue, as many as 2 million fewer vehicles would be produced in Japan as a result of diverse parts shortages. A concrete example comprised an innocuous device: the hot wire flow sensor. During the crisis this engine management device (in this case from Hitachi) held up production as far away as Opel's plant in Germany, which makes the Corsa.

As for automotive MEMS sensors, a significant portion of the world's market, an estimated 24% in 2011, is tied up in Japanese companies. Major suppliers comprise Denso (airbag accelerometers, pressure sensors for engine management) and Panasonic (gyroscopes for vehicle stability, GPS navigation); smaller players include Murata (gyroscopes for GPS navigation systems), Kyocera (tire pressure monitors), Mitsubishi Electric (e.g. airbag accelerometers), Fuji Electric (manifold pressure sensors) and Nicera (thermopiles to detect temperature in the interior).

There were several reasons why the sensor market may therefore have stalled, but did not. Mainly, all parts of the chain adapted very quickly. First, Toyota, Honda, Nissan et. al. did a great job finding new sources to mitigate the destruction caused by the earthquake and associated electricity blackouts. Unfortunately, these OEMs were also hit by the Thailand floods in November 2011, mitigating much of the re-sourcing work up to that point (IHS Automotive). Tier 1s also maintained the supply without accumulating too much additional sensor inventory by internally qualifying new sensor suppliers (in some cases in record time) to ensure a continued supply to their customers. This factor has diversified the supply chain as Tier 1 companies now consider additional sources to mitigate unforeseen circumstances. The sensor suppliers were able to meet the challenge, with some interesting micro-features:

- Companies like Bosch with a strong reputation for quality (combined with a reputation for not being the cheapest) recovered market share as their customers learned to appreciate a supply base with the manufacturing scale to absorb market disruptions, unforeseen or otherwise

- Some companies are also observing greater acceptance, e.g. Silicon Sensing Systems, currently phasing out an older ESC gyroscope, reports increased interest in new products, e.g. among Japanese Tier 1s. In the past, new product designs would have met higher resistance entering what is a highly conservative market.

4.2 Importance of the New Market Economies

China remains the region with the highest potential for vehicle sales over the next few years. China's meteoric growth saved the automotive market from even worse disaster and bolstered an already rapidly expanding 2010—Chinese consumers bought 31% more cars than 2009, itself a banner year.

However, 2011 saw several major Government-backed subsidy schemes expire, which dampened sales, but this lull appears now to be ended; the region will contribute the highest vehicle sales growth of 8% from 2010 to 2015. The sensor penetration rates are also currently much lower than in developed markets (electronic contents of 12% compared to 18% for U.S.) but are fast rising. Chinese vehicles currently feature about half of the number of sensors of a high end European or American equivalent, for instance, although this situation is rapidly changing.

China is particularly important for pressure sensors used in manifold air measurements, TPMS, accelerometers for airbags, although not yet for ESC applications, which would benefit gyroscopes, accelerometers and pressure sensors. The Chinese buying public wants vehicles with western standards, which will increase electronics adoption rates going forward.

Other regions will also grow in the coming years, especially South America. The Indian market will likely remain low-tech from a sensor perspective, while the fate of Russia's automotive market is currently unclear.

5. Conclusions

The current paper has described how, after a period of organic growth lasting around 15-20 years, the new market drivers for sensors in the car are ultimately driven by safety mandates and emissions regulations on the one hand, and emerging economies requiring initially basic sensors enabled functionalities on the other. In the 5-year snapshot provided, these are the major forces propelling the market, but with a wider horizon one sees the new wave of safety sensors will be founded on driver assist (ADAS) systems, a trend that is already in motion, i.e. using cameras, radar, ultrasound sensors and so forth to improve current safety systems and allow the car to assist and react faster to the situation than the driver. This trend will not lead to replacement but rather an augmentation of today's sensors.

The electric vehicle poses the biggest threat to powertrain sensors; it will eliminate some sensors but add new devices that manage the electric motors. Hybrid variants with combinations of combustion engine and electric powertrain will exist requiring more or less sensors in the internal combustion engine, underlining the fact that sensors will remain a key electronic component of today's and tomorrow's vehicles.

References

[1] IHS Automotive releases Japanese light vehicle production update, 31/3/2010 http://www.ihs.com/products/Global-Insight/industry-economic-report.aspx?ID=1065929302

Richard Dixon
IHS iSuppli
Spiegelstr. 2
81241 Munich
Germany
richard.dixon@ihs.com

Keywords: automotive MEMS and sensors, markets, safety sensors, powertrain sensors, ESC, TPMS, pressure sensors

List of Contributors

List Of Contributors

List of Keywords

List of Keywords